LANDSCAPE DESIGN
Theory and Application

Join us on the Web at
agriculture.delmar.com

LANDSCAPE DESIGN
Theory and Application

Ann Marie VanDerZanden, PhD
Iowa State University

Steven N. Rodie, ASLA
University of Nebraska-Lincoln

THOMSON
DELMAR LEARNING

Australia • Brazil • Canada • Mexico • Singapore • Spain • United Kingdom • United States

Landscape Design: Theory and Application
Ann Marie VanDerZanden, PhD, and Steven N. Rodie, ASLA

Vice President, Career Education Strategic Business Unit:
Dawn Gerrain

Director of Learning Solutions:
John Fedor

Acquisitions Editor:
David Rosenbaum

Managing Editor:
Robert L. Serenka, Jr.

Product Manager:
Christina Gifford

Editorial Assistant:
Scott Royael

Director of Production:
Wendy A. Troeger

Production Manager:
Mark Bernard

Senior Content Project Manager:
Kathryn B. Kucharek

Technology Project Manager:
Sandy Charette

Director of Marketing:
Wendy Mapstone

Channel Manager:
Gerard McAvey

Marketing Coordinator:
Jonathan Sheehan

Art Director:
Joy Kocsis

Cover Images:
Getty Images Inc.

Cover Design:
TDB Publishing Services

© 2008 Thomson Delmar Learning, a part of the Thomson Corporation. Thomson, the Star logo, and Delmar Learning are trademarks used herein under license.

Printed in Canada
1 2 3 4 5 XXX 11 10 09 08 07

For more information contact Delmar Learning,
5 Maxwell Drive, PO Box 8007, Clifton Park, NY 12065-2919.

Or you can visit our Internet site at
http://www.delmarlearning.com.

ALL RIGHTS RESERVED. No part of this work covered by the copyright hereon may be reproduced or used in any form or by any means—graphic, electronic, or mechanical, including photocopying, recording, taping, Web distribution or information storage and retrieval systems—without written permission of the publisher.

For permission to use material from this text or product, submit a request online at http://www.thomsonrights.com
Any additional questions about permissions can be submitted by e-mail to thomsonrights@thomson.com

Library of Congress Cataloging-in-Publication Data

VanDerZanden, Ann Marie, 1966–
  Landscape design : theory and application /
  Ann Marie VanDerZanden, Steven N. Rodie.
       p. cm.
  Includes bibliographical references and index.
  ISBN 978-1-4180-1286-1
  1. Landscape design.   I. Rodie, Steven N. II. Title.
  SB472.45V364   2007
  712—dc22
                                          2007021523

NOTICE TO THE READER

Publisher does not warrant or guarantee any of the products described herein or perform any independent analysis in connection with any of the product information contained herein. Publisher does not assume, and expressly disclaims, any obligation to obtain and include information other than that provided to it by the manufacturer.

The reader is expressly warned to consider and adopt all safety precautions that might be indicated by the activities herein and to avoid all potential hazards. By following the instructions contained herein, the reader willingly assumes all risks in connection with such instructions.

The publisher makes no representation or warranties of any kind, including but not limited to, the warranties of fitness for particular purpose or merchantability, nor are any such representations implied with respect to the material set forth herein, and the publisher takes no responsibility with respect to such material. The publisher shall not be liable for any special, consequential, or exemplary damages resulting, in whole or part, from the readers' use of, or reliance upon, this material.

To my husband, Joe, and children, Nina and Peter.
A. M. VanDerZanden

To my wife, Amy, and children, Erin and Kyle.
S. N. Rodie

CONTENTS

PREFACE xi
ABOUT THE AUTHORS xiii
ACKNOWLEDGMENTS xv

CHAPTER 1
LANDSCAPE DESIGN: TOOLS OF THE TRADE 1

Getting Started: Drawing Surface, T-Square, Paper Types, and Tape 2
- Drawing Surface and T-Square 2
- Drawing Papers 2
- Drafting Tape 3

Drawing and Erasing: Pencils, Pens and Markers, Erasers, and Dusting Brush 4
- Drawing Leads 5
- Millimeter or Mechanical Pencils 6
- Colored Pencils 7
- Pens and Markers 8
- Erasers 10
- Dusting Brush 11

Measuring: Scales and Rolling Ruler 12
- Architect's Scale 13
- Engineer's Scale 15
- Rolling Ruler 16

Angles and Curves: Triangles, Protractor, Compass, and Curves 17
- Triangles 17
- Protractor 17
- Compass 18
- Irregular Curves 19

Time Savers: Lettering Guide and Templates 20
- Lettering Guide 20
- Templates 20

Digital and Computer-Aided Design and Drafting 22
- Types of Computer-Aided Design Tools 23
- Making the Purchase 27

Summary 29
Knowledge Application 29
References 29
Online Resources 29

CHAPTER 2
LANDSCAPE GRAPHICS 30

Importance of Graphics 31
Choosing Between Drafted and Freehand Graphics 33
Graphics—Black and White Mechanics 36
- Line and Dot Quality 38
- Line Weight Hierarchy and Contrast 40
- Value Contrast 43
- Texture 46

Graphic Symbols and Definitions 52
- Plan View Graphics 52
- Additional Graphic Methods 62
- Analysis and Concept Graphics 73

Color Basics 73
- Common Media 77
- Principles for Using Color Effectively 78
- Applications 81

Freehand Drawing and Sketching 82
- The Right Frame of Mind for Accurate Sketching 83

Drafting 85
Lettering 85
- Criteria for All Lettering Styles 86

Sheet Mechanics and Layout 89
- General Layout 89
- Title Block 89
- North Arrow and Drawing Scale 89
- Additional Information and Formatting 93

Graphics Applications for Specific Drawing Types 93
- Base Map 94
- Site Inventory and Analysis 94
- Functional Diagram 94
- Preliminary Design 94
- Final Design 97

Computer and CAD Graphics 100
- Benefits of Computer-Aided Design 100
- Computer Graphics Trade-Offs 101
- CAD Graphics Summary 102

Other Graphics Applications and Considerations 103
Summary 104
Knowledge Application 105
Literature Cited 105
References 105

CHAPTER 3
DEFINING LANDSCAPE DESIGN AND THE FOUNDATIONS THAT SHAPE THE DESIGN PROCESS 106

A Landscape Design Definition 107
  Aesthetic and Functional Components of Design 107
  Additional Components That Define Landscape Design 107
Common Misconceptions about Landscape Design 108
  Misconception: Landscape Design Is a Product
  Reality: Landscape Design Is Both Product and Service 109
  Misconception: Good Designers Do Not Have to Think about Design
  Reality: Good Design Requires Time and Energy 109
  Misconception: Design Is a Rare Talent
  Reality: Design Is a Learned Skill 112
  Misconception: Good Designers Have an Obvious Signature Style
  Reality: Personal Style Is Universal 112
  Misconception: Good Designers Work Best Alone
  Reality: Good Designers Appreciate the Benefits of Additional Viewpoints 112
  Misconception: Quality Design Is Unique
  Reality: Quality Design Is Grounded in Creative Adaption 113
  Misconception: Good Designers Do Not Need to Follow the Basic Aesthetic Design Principles
  Reality: Good Designers Know the Basic Design Principles Well and Know How to Stretch Them 113
Design as a Process 113
  Accept Situation 116
  Analyze 116
  Define Desired Outcomes 118
  Generate Ideas 118
  Select 118
  Implement 118
  Evaluate 118
  CASE STUDY: Applying the Design Process 119
Relationships of People and the Outdoor Environment 121
  Landscape Preference 121
  Restorative Value of Landscapes 127
Summary 129
Knowledge Application 129
Literature Cited 130
Reference 130

CHAPTER 4
THE PRINCIPLES OF DESIGN: DESCRIPTION AND APPLICATION 131

How We Learn to Design 132
  Awareness 132
  Perception 132
  Imagination 133
  Expression 133
Elements of Art 133
  Line 133
  Form 133
  Color 135
  Texture 139
  Space 140
Landscape Design Principles 140
Overarching 141
  Simplicity 142
  Blending Form and Function 144
  Reflecting Local Elements of Architecture, History, Nature, and Sense of Place 144
Aesthetic Landscape Design Principles 146
  Order or Design Framework 146
  Repetition 150
  Rhythm 151
  Unity 154
  Balance 154
  Proportion and Scale 156
  Emphasis 159
Aesthetic Principles Applied 159
  Visually Connect Structures to the Landscape 160
  Define Space: Mass and Void 160
  Direct People Physically and Visually 161
  Designing Outdoor Living Spaces to Meet the Physical, Cognitive, and Restorative Needs of People 162
Functional Design Principles 163
  Working with the Existing Topography 165
  Creating a Useful Outdoor Living Space 166
  Accounting for Maintenance 167
  Addressing Irrigation Needs 167
  Incorporating Sustainability 169
  Providing Wildlife Habitat 170
Summary 170
Knowledge Application 170
Literature Cited 171
References 171

CHAPTER 5
LANDSCAPE DESIGN PROCESS 172

The Landscape Design Process 173
Design Process Case Study 173
Client–Designer Communications and the
Sales Process 173
    Making Initial Contact with the Client 174
    Design Planning Questionnaire 174
    CASE STUDY: Introduction 175
    Meeting with the Client for the First Time 176
    The Budget Question 180
    CASE STUDY: Client Feedback 182
Site Documentation 182
    Photos and Videos 182
    Developing a Base Map 183
    CASE STUDY: Base Map Information 196
Site Inventory and Analysis 198
    Capitalizing on the Homeowner as a Resource 199
    Two Distinct Steps: Inventory and Analysis 199
    Documenting the Information 201
    Combining Site Inventory and Analysis with
    Sustainable Design 214
    Summary: Site Inventory and Analysis 215
    CASE STUDY: Site Inventory and Analysis 215
Developing a Design Concept and a Design Program 216
    Design Concept 216
    Design Program 217
Functional Diagrams 217
    CASE STUDY: Design Concept and Design
    Program 218
    Functional Diagram Elements and Information 219
    Developing a Functional Diagram 220
    Summary: Functional Diagram 223
Preliminary Design 224
    Purpose and Characteristics 225
    CASE STUDY: Functional Diagrams 227
    Graphic Style and Content 241
    Summary and Design Process Flexibility 241
Final Design Package 241
    CASE STUDY: Selecting a Design Theme 242
    CASE STUDY: Preliminary Design 244
    Final Design Checklist 245
    The Final Design or Master Plan 246
    CASE STUDY: Final Design 247
Summary 249
How Four Design Professionals Would
Approach a New Landscape and a Landscape
Renovation 249
    Scenario #1—New Homeowners in a New Subdivision 249
    Scenario #2—Renovating an Existing Landscape 252

Knowledge Application 255
Literature Cited 255
Online Resources 256

CHAPTER 6
PLANT SELECTION: BLENDING FORM AND FUNCTION 259

Site Analysis and Plant Selection 260
Environmental Considerations 260
    Plant Hardiness Zone 261
    Heat Zone 262
    Frost Occurrence 263
    Seasonal Rainfall 263
    Soil Characteristics 264
    Sun Exposure 266
    Microclimate 267
Aesthetic Qualities 267
    Plant Types 268
    Plant Lifecycles 269
    Four-Season Plants 271
    Form 271
    Texture 276
    Color 280
    Flowering and Fruiting Habit 284
    Mature Size 285
Maintenance Requirements 287
    Woody Plants 288
    Herbaceous Plants 288
Functional Uses in the Landscape 290
    Defining Spaces 291
    Views 293
    Utility Purposes 295
    Aesthetic Purposes 297
Conceptual Plant Communities 298
    Trying to Replicate Nature 300
The Urban Growing Environment 300
    Soil 300
    Light Patterns 301
    Humans 301
Native Plants vs. Non-native Plants 302
Summary 304
Knowledge Application 304
Literature Cited 305
References 305

CHAPTER 7
HARDSCAPES: SELECTION AND USE 306

Hardscapes in the Ground Plane 307
    Patios 307
    Decks 307

Walkways 309
　　　Driveways 310
　　　Steps 312
　Hardscapes in the Vertical Plane 332
　　　Fences 333
　　　Walls 334
　Hardscapes in the Overhead Plane 343
　　　Pergolas and Arbors 344
　　　Gazebos 346
　Landscape Lighting 347
　　　What Lighting Adds to a Landscape 347
　Garden Features and Embellishments 350
　　　Water Features 352
　　　Fire Pits and Outdoor Fireplaces 356
　　　Garden Embellishments 356
Summary 362
Knowledge Application 362
References 363

CHAPTER 8
PRICING THE PROPOSED PROJECT 364

Charging for Your Expertise 365
Proceeding after a Successful First Meeting 366
Pricing the Design Only 366
　　Developing a Contract for Design Services 366
A Final Meeting with the Client 370
Landscape Contracting Terminology 371
　　Differentiating Between Cost and Price 371
　　Differentiating Between Estimate, Bid, Proposal, and Contract 371
The Design/Build Business 375
　　Designers Can Benefit from Collaborating with Installation Professionals 375
Pricing for Design and Installation Services 375
　　Design Services 375
　　Materials Costs 376
　　Labor Costs 376
　　Overhead 378
　　Contingency 378
　　Profit 379

Developing a Bid 380
　　Managing Change 380
The Country Landscapes Business Model 381
　　The Process 382
　　What the Client Sees 382
　　Comparing Estimated to Actual Costs 382
　　Using Experience to Develop Estimates 383
Computer Software for Estimating and Bidding 384
　　EasyEst Estimating Software 385
Manual Measuring and Calculating for Estimating 386
　　Accuracy Is Essential 386
Summary 388
Knowledge Application 388
Literature Cited 388
Reference 388

APPENDIX
CASE STUDIES: LANDSCAPE DESIGN SCENARIOS 389

Case Study #1 Old Neighborhood Residence 389
　　Base Map and Site Photos 389
　　Owner Wish-List and Feedback 389
　　Site Inventory and Analysis 393
Case Study #2 Acreage Residence 394
　　Base Map and Site Photos 394
　　Owner Wish-List and Feedback 394
　　Site Inventory and Analysis 399
Case Study #3 Suburban Neighborhood Residence 400
　　Base Map and Site Photos 400
　　Owner Wish-List and Feedback 400
　　Site Inventory and Analysis 404
Case Study #4 Large Lot Residence 406
　　Base Map and Site Photos 406
　　Owner Wish-List and Feedback 406
　　Site Inventory and Analysis 409

GLOSSARY 411
INDEX 417

PREFACE

Throughout our combined 23 years of teaching introductory landscape design, we have developed and refined curriculum to meet student needs and course objectives and to fit within the constraints of a college semester. With this experience behind us, we have chosen to write a textbook that reflects what we have learned to be important in such a course. The eight chapters of this text are focused on a select group of topics that provide the foundation for good landscape design. We realize that for some students this may be your only exposure to landscape design, while for others it may provide the foundation for additional coursework. Regardless, our aim in developing this textbook was to provide a sound framework for all landscape design students.

The chapters are organized to provide sequential coverage of the topic of landscape design. Chapter 1 discusses design tools used by landscape design professionals, and Chapter 2 describes how to use these tools to graphically represent a landscape design concept. Chapter 3 introduces the topic of design as a process and how human preferences affect landscape design components. Chapter 4 describes the basic elements of art and their application to aesthetic landscape design principles. In addition to describing the aesthetic principles of design, this chapter also describes how functional design principles need to be considered in concert with aesthetic principles and how landscape preference influences the application of the aesthetic principles. In Chapter 5, the process of landscape design is discussed in detail, including strategies for interacting with the client, selling a landscape concept, and creating a design from start (concept development) to finish (final plan). With the foundation for design principles and design process in place, Chapters 6 and 7 provide detailed descriptions of plant material and hardscape material selections, respectively. And finally, Chapter 8 discusses strategies for pricing the landscape and examples of landscape business models. In addition, it highlights a successful landscape design/build company to provide concrete examples of how and why it has been successful.

In addition to the eight chapters in this book, we have also included four case studies in the appendix. Each case study will provide the opportunity to apply knowledge learned and skills gained to actual landscape design dilemmas, putting into practice what you have learned.

As we wrote the text, we both reflected on our first landscape design courses and how at times the concepts and terminology seemed so foreign. With that in mind, we tried to tailor this text to describe and illustrate the underpinnings of good landscape design, provide guidelines and applications of design principles, include rules of thumb to illustrate points, and incorporate our individual design experiences to help personalize the process. We have also included a wide variety of figures and graphics throughout the book to help students visualize critical content as well as proper techniques. Many of the figures are borrowed from past student work in our introductory design courses. Design knowledge and skills are truly attainable for students who are engaged in their initial design training, and our hope in using student work is to show what is possible.

An overarching goal is for you as a beginning designer to learn that there often is not a single, right way to solve a design dilemma or to simply create a stunning outdoor living space for a client. Each situation is unique. Each provides opportunities to apply your own artistic ideas and understanding of basic horticultural principles to address the constraints of the site and to meet the needs of the client.

ABOUT THE AUTHORS

Dr. Ann Marie VanDerZanden is an Associate Professor of Horticulture and a member of the Iowa State University Horticulture Department. She has taught landscape horticulture courses, including herbaceous plant identification, landscape design, landscape construction, and landscape contracting and estimating, for 12 years. In addition to her teaching responsibilities, she is an extension specialist for the nursery and landscape industry, and she previously served as an extension specialist in consumer horticulture. Her research interests include undergraduate pedagogy and using new technology to enhance the learning experiences of students and nursery and landscape professionals.

She is a seasoned writer and has published numerous manuscripts on teaching and extension outreach projects in peer-reviewed journals as well as general interest articles in garden magazines, including *American Nurseryman, Iowa Horticulture,* and *Northwest Woman.* She recently wrote a chapter on landscape design for the Iowa Nursery and Landscape Association's Certified Nursery Professional program training manual and another chapter on landscape design for the Iowa State University Master Gardener training manual. In addition to writing, she has a number of speaking engagements each year, reaching members of the nursery and landscape industry and home gardeners.

Professor VanDerZanden completed her academic training in horticulture science and earned her BS and PhD from Washington State University and her MS from Cornell University. Prior to joining the faculty at Iowa State University, she was a faculty member at Oregon State University and Illinois State University. She has worked in the green industry in retail nursery sales and production, landscape design, landscape sales, and landscape estimating.

She lives in Ames with her husband, two young children, and big yellow lab. She enjoys exploring and photographing new gardens, but finds the greatest joy spending time in her garden with her family.

Steven N. Rodie is an Associate Professor and Extension Landscape Horticulture Specialist in the Department of Agronomy/Horticulture at the University of Nebraska–Lincoln (UNL). He teaches courses in the Pre-Horticulture Program at the University of Nebraska at Omaha (UNO), including Landscape Plants I and II, Landscape and Environmental Appreciation, and Introduction to Landscape Design. In addition, he coordinates coursework and advises students in the transfer programs offered at UNO through the UNL College of Agricultural Sciences and Natural Resources. He maintains a courtesy appointment as Associate Professor in the Department of Biology at UNO.

His primary extension program emphases are sustainable landscape design and the use of native and adapted plants in Midwest landscapes. He develops and delivers presentations, workshops, and extension publications tailored to a wide range of audiences, including turf and golf course professionals, arborists, grounds and parks managers, landscape designers, garden center employees, Master Gardeners, and the general public.

Professor Rodie brings a variety of practical experience to his teaching and extension programming, including 9 years of landscape architecture practice and 6 years of nursery and landscape contracting experience. He currently serves as an elected member of the Board of Trustees of the American Society of Landscape Architects and as an appointed professional member of the Nebraska State Board of Landscape Architects. He is a registered Landscape Architect in Nebraska, Kansas, and California; and he has degrees from Colorado State University (BS in Forest Management) and Kansas State University (MLA).

His life in Omaha is enriched by an understanding wife, two tolerant grown kids, two demanding Airedale terriers, an addiction to model railroading, and an abundance of beautiful parks for sharing plant walks with energetic students.

ACKNOWLEDGMENTS

In writing this textbook, we realized just how many people have shaped our teaching, research, and outreach efforts in the area of landscape design. Countless green industry professionals—including landscape architects and designers, landscape contractors, nursery men and women, and businesspeople—have contributed to our professional development. Past instructors and former students have significantly influenced our pedagogical growth and improvement. This unique mix of individuals has directly or indirectly had a hand in creating this text.

We are also grateful to our friends and colleagues who generously gave their time to this project. Their contributions came in the form of discussing concepts for the book, answering specific questions or providing suggestions, generously sharing their extensive collections of images and other teaching tools for use in the book, and in general providing moral support and good humor when needed. In particular, we would like to thank Iowa State University Horticulture Department faculty Cynthia Haynes and Jeffery Iles; Tom Cook at Oregon State University; Tigon Woline, Iowa State University student; Jim Mason and Deb Carlson at Country Landscapes; Lynn Kuhn, Linda Grieve, and Dan Canova at Perennial Gardens; Ross and Beth Brockshus at Del's Garden Center; Shawn Davis; Sherry Levine; and students in HORT 380 and 381. Special acknowledgments to those on the Nebraska side of the Missouri River include Richard Sutton and Anne Streich at the University of Nebraska; Bryan Kinghorn and Anne Houser at Kinghorn Gardens; Chad Friesen at Mulhall's Nursery; and UNL horticulture students Karen Richards and Alyssa Stapp.

Most importantly, we thank our families. Editing and proofreading were only the beginning. Joe, Nina, Peter, Amy, Kyle, and Erin—all supported us and understood missed dinners, late nights and weekends in front of the computer, shortened vacations, and missed family outings; they made this book possible. We hope you all know how invaluable you were to this project.

Landscape Design: Tools of the Trade

CHAPTER 1

Just as masons, carpenters, and artists have a unique set of tools used in their work, so does the landscape designer. Historically, the tools used by landscape designers were similar to those used by engineers and architects: a drawing board, T-squares, pencils and erasers, tools for making angles and curves, templates, and scaled rulers. Today, the contents of the designer's toolbox have expanded to include laptop computers, digital cameras, scanners, and graphic manipulation and design software. This chapter will focus on the most common tools used by designers.

OBJECTIVES

Upon completion of this chapter you should be able to

1. differentiate between drawing tools used by a landscape designer.
2. explain appropriate uses for these drawing tools.
3. demonstrate how to combine different tools to achieve the desired design effect.
4. read dimensions on a scale (architect's and engineer's).
5. explain the basic advantages and limitations of CAD.

KEY TERMS

T-square
parallel ruling straightedge
trace
vellum
architect's scale
engineer's scale
French curve
template
CAD

Each landscape designer assembles his or her own set of tools, which he or she selects because of the type of work he or she does. It is not uncommon for designers to develop a preference for a particular product or brand. As a beginning designer, it is important to experiment with different tools and decide which best meet your needs. Designers use drawing tools to transform a mental image of the landscape design into an illustrated version that will sell the design concept to a client (requiring artistic styling) and convey instructions precisely to installers (requiring accuracy).

GETTING STARTED: DRAWING SURFACE, T-SQUARE, PAPER TYPES, AND TAPE

Some of the basics a designer needs to get started include an appropriate drawing surface, paper, something to hold the paper in place during the drawing process, and basic tools that allow a designer to draw horizontal and perpendicular lines.

Drawing Surface and T-Square

Landscape designers need a large, flat, smooth work surface (Figure 1-1). The surface needs to be large enough to accommodate an 18″ × 24″ sheet (the minimum size used for most residential designs). The next larger sheet size (24″ × 36″) may be required for larger projects. A flat surface to accommodate this size provides more flexibility. The surface should also have a straightedge that a **T-square** can run along (Figure 1-2a) so a series of horizontal parallel lines can be drawn.

An alternative to a T-square is a **parallel ruling straightedge,** or parallel bar (Figure 1-2b). This tool works similarly to a T-square, except that it is fixed to the drawing surface and moves over the area via small cables or guides. The bar remains locked at a consistent angle across the drawing as it is moved up and down on the drafting table. Small nylon or steel rollers under the bar keep it elevated to minimize smudging. Some designers prefer the flexibility of using a T-square while others feel the sliding straightedge is easier to manage and provides more control. In either case, when a triangle is set along the straightedge, the designer can draw perpendicular lines.

Drawing Papers

Landscape designers use a variety of paper types (Figure 1-3). During the initial brainstorming process, a designer uses thin, translucent paper called tracing paper to sketch out ideas. Sometimes, this paper is also referred to as **trace**, onionskin, or bumwad. Trace is sold in rolls up to 50 yards long and in widths from 12 to 42 inches. A roll of tracing paper allows a designer to tear off paper in various sizes. A designer often draws the base map, which is a scale drawing of the house and property, on a sheet of **vellum** and then lays trace over that to experiment with design concepts.

Vellum is a high-quality paper made from 100 percent, pure white, rag stock. Rag stock refers to paper with cotton fiber content between 25 and 100 percent. In the case of vellum, it is made completely of cotton fibers. It is sturdy and can withstand a lot of handling and erasing without leaving marks. Gridded vellum contains a grid of light blue lines that will not reproduce at regular photocopy or diazo (blueprinting) process contrast settings and is available for different scales (8 lines per inch for architect's scale drawings and 10 lines per inch for engineer's

FIGURE 1-1

This large, smooth drawing surface provides adequate room for the paper and keeps drafting tools close by. *(Kory Beidler)*

scale drawings). It is especially useful, since it sets the scale of the sheet, allows quick calculations of line lengths and areas, and provides guidelines for straight lines and lettering. Vellum is best used with lead drawings because ink is typically difficult to erase from vellum. Mylar is a trademark name for a thin, strong, polyester film, and it is suitable for use with ink. Errors made in ink on Mylar can be removed using an ink eraser. Mylar is heavier and more expensive than vellum; both reproduce well when photocopied or blueprinted.

How Paper Type Impacts Reproduction of the Drawing.
Vellum and Mylar are both semitransparent, which allows the light used in the blueprint copy process (or diazo process) to pass through the sheet and burn the drawing image onto the blueprint paper. Since the diazo process will recognize any opaque mark or change in drawing thickness, corrections to vellum or Mylar sheets using regular photocopy methods such as Scotch tape, White-out, etc., will create shadows or black-out areas. Design offices and drawing duplication companies are gradually phasing out the diazo process, with the advent of more affordable large-format photocopy machines and direct CAD plotting or printing on bond paper. It is still available, however, and produces less expensive copies.

Photocopying done on large-format machines does not require semitransparent paper like blueprint duplication does, and these machines can copy from an opaque sheet. Large-format photocopies allow more freedom in editing and correcting. As in any photocopy process, however, if the drawn image is light and the machine contrast has to be darkened, then irregularities, smudges, and correction can be easily seen.

Drafting Tape

Drafting tape is a low-adhesion tape used to fasten the paper to the drawing surface. The lower adhesion holds the paper to the board while the designer is working, but it is also easily removed without tearing the paper. Drafting tape is available in a continuous roll or as precut circles called drafting dots (Figure 1-4). When you

FIGURE 1-2

A traditional T-square (a) and a sliding straightedge (b) both allow a designer to drawn parallel horizontal lines. *(Ann Marie VanDerZanden)*

(a)

(b)

are using a continuous roll, it is best to cut the tape, leaving a clean edge that is less likely to stick to the underside of drawing equipment.

DRAWING AND ERASING: PENCILS, PENS AND MARKERS, ERASERS, AND DUSTING BRUSH

After the paper has been selected and secured to a suitable surface, you can begin drawing. Because landscape designs use a standard set of line weights (thick, medium, and thin) designers use different leads and pens to achieve them. Some

FIGURE 1-3

Paper types commonly used in landscape design. From top, trace; vellum; and gridded vellum. *(Ann Marie VanDerZanden)*

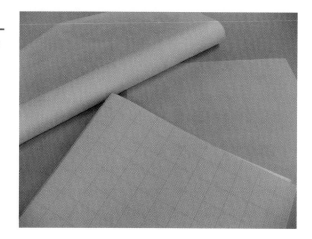

FIGURE 1-4

Drafting tape and drafting dots have low adhesion but securely affix the paper. *(Ann Marie VanDerZanden)*

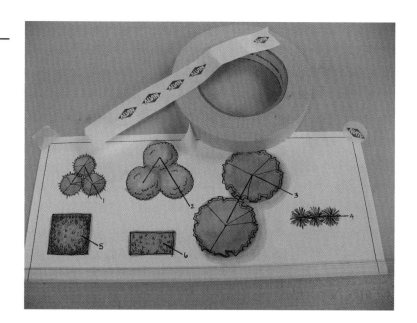

designers prefer using graphite (pencil) for their designs while others prefer to work with ink. Pencil is easier to erase and modify than ink, but it is also less permanent than ink, and it may not reproduce as clearly on drawing copies. Regardless of whether pencil or pen is used, when the inevitable mistake happens, the designer can use different erasers to fix it.

Drawing Leads

Drawing leads come in varying degrees of hardness. They run from 8B, which is very soft, to 9H, which is very hard (Figure 1-5). Softer lead (e.g., 4B) results in a thick, dark line that smudges easily because of the relatively large amount of graphite that it applies. Softer lead also needs to be sharpened more often to maintain a fine point at the tip. Harder lead (e.g., 3H) is used to draw light, thin

FIGURE 1-5

Lead weights range from very hard to very soft (a). These pencil tips show differences in lead size and color. From top to bottom: 6H, 2H, H, 6B, 7B, and 8B (b). *(Ann Marie VanDerZanden. Used with approval from STAEDTLER, Inc.)*

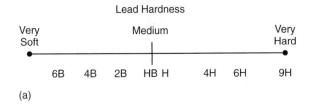

(a)

(b)

lines. Harder lead tends to smudge less but also tends to break more often as additional pressure is applied. For drawing guidelines and drawing features that should be barely visible, the light lines from hard leads work well.

Lead is available as a regular pencil or as individual pieces designed for use in a lead holder (which looks like a mechanical pencil) (Figure 1-6). One drawback to this system is that the lead must be sharpened (or pointed) frequently. Lead is pointed in a pencil pointer by placing the tip of the lead in the pointer and rotating the top. After the pencil is pointed, the tip is stuck into the cleaning pad on top of the pointer to remove excess graphite shavings. Most designers will use a couple of different lead weights (e.g., H and 3H) on a regular basis. Changing lead in a pencil is both time consuming and messy, so it is best to have a separate lead holder for each lead type. It is faster and easier to change holders than to change leads.

Millimeter or Mechanical Pencils

Instead of the lead holder system described above, some designers prefer millimeter pencils. These mechanical pencils are available with refillable leads in a variety of widths and hardness. Typical widths are 0.3, 0.5, 0.7, and 0.9 millimeters; the hardness typically ranges from HB to 2H (Figure 1-7). A nice feature of these pencils is that they do not require frequent sharpening like the other leads. They provide a consistent and uniform line width for drafting as long as they are held vertically. Having an assortment of these pencils in varying widths and preferred lead hardness makes it easy to change line widths during the drawing process.

FIGURE 1-6

Individual leads are used in a lead holder. A designer switches out leads when different line weights are required in the design. *(Ann Marie VanDerZanden. Used with approval from STAEDTLER, Inc.)*

FIGURE 1-7

Millimeter pencils are available in different lead thicknesses. *(Ann Marie VanDerZanden. Used with approval from STAEDTLER, Inc.)*

Colored Pencils

Colored pencils can make a design come alive. Appropriately applied color conveys depth and texture that is not evident in black-and-white designs (Figure 1-8). Colored pencils are relatively easy to use, and with practice they can be blended to create a wide range of color hues and values. Quality pencils are relatively inexpensive compared to high-quality graphic markers; and their soft, wide leads are easier to use than less expensive hard-leaded, colored pencils. A basic palette of pencil colors should include a range of colors focusing on natural tones such as browns, yellows, blues, and greens, as well as a blender pencil. White and black are also valuable for shading and highlighting. A good drawing reference, such as those listed at the end of this chapter, will provide lists of specific colors.

FIGURE 1-8

Colored pencils add a dramatic effect to these plant symbols. *(Courtesy of Anna Vold. Ann Marie VanDerZanden)*

Pens and Markers

Some designers prefer to use ink instead of pencil in their designs. Ink creates a sharper line definition compared to graphite, but mistakes are more difficult to fix due to its permanence. Successful ink removal will vary, depending on ink density, pen brand, and paper type. It is important to use an ink eraser and to erase mistakes quickly when using vellum or tracing paper. Once the ink dries, it is difficult, if not impossible, to remove completely. To minimize ink errors, designers use three strategies: draw the initial design using a hard lead (4H) and once satisfied with the design, redraw it using ink and erase any obvious remaining guidelines; use nonphoto blue lead to draw the design and then trace over it with ink (the nonphoto blue lines will not reproduce in the photocopying process); and draw the design on a draft sheet of tracing paper, then retrace freehand with ink on a overlay of vellum or tracing paper.

Like mechanical pencils, ink pens are available in a range of widths and styles. Technical pens are refillable, have specific widths and hard, wear-resistant tips for width consistency (jewel-tips are required for Mylar since its surface quickly wears down metal tips), and can be used for all types of drawing. They are relatively expensive and can require high maintenance to keep the small tips free of dried ink. These pens were the primary ink drafting tool prior to computer plotting. A variety of nonrefillable ink pens are available that provide a wide selection of tip sizes and types, including metal, fiber, and felt (Figure 1-9). Drying time, ink density, and tip longevity vary with each brand, but they are relatively inexpensive. As a designer, it is worth experimenting with the different types to find the best tool.

One caution with ink pens is that they can smear when not used carefully. The combination of paper type and ink type can also influence smearing. For example, vellum paper is available with a slightly toothy (refers to surface roughness) finish or a smooth plate finish. The smoother finish is more likely to smear. Paper that is toothier (rougher), such as one with high rag content, is less likely to smear, but the ink will bleed into the fibers, resulting in a fuzzy line edge. Using blotters or

FIGURE 1-9

Technical pens have nibs in different sizes, which results in lines of different thicknesses. *(Ann Marie VanDerZanden. Used with approval from STAEDTLER, Inc.)*

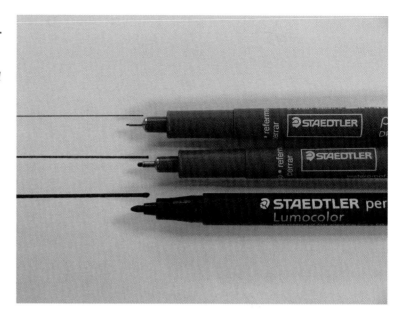

FIGURE 1-10

A blotter is placed on this drawing to prevent smearing the ink as the designer continues to work. *(Courtesy of Benjamin McIntosh. Ann Marie VanDerZanden)*

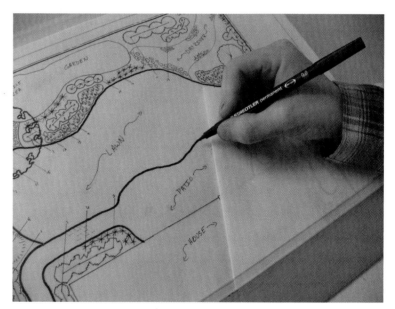

elevated straightedges made specifically for ink use (look for those listed as inking templates or with inking edges) minimizes ink smearing by keeping the edges slightly above the paper so capillary action does not draw ink under the equipment as the tip is pulled along the edge (Figure 1-10).

Like colored pencils, markers add color to a design. Graphic markers are available in large and fine nibs (Figure 1-11). Some brands have shaped nibs that can be used to draw multiple line widths, and some include different nib widths on each end of the marker. The large nib is usually a chisel tip while the fine nib is rounded. The large nib applies a generous amount of ink. A blender marker has no color of its own and is used to blend colors together, which softens the contrast.

FIGURE 1-11

Graphic markers generally come with large chisel tip nibs and smaller rounded nibs. *(Ann Marie VanDerZanden)*

FIGURE 1-12

Eraser types commonly found in a designer's tool box. Clockwise from top: dry cleaning bag, ink, kneadable, rubber, and plastic. *(Ann Marie VanDerZanden)*

Markers lend a colorful and artistic effect to a design, but it requires practice to use them skillfully. They are also a significant investment and need to be replaced on a regular basis as they dry out. It is normally best to purchase markers individually, focusing on the colors most often used in landscape drawings, such as a range of light to dark greens, browns, and grays. Marker sets are less expensive, but they tend to include many colors that have limited use in landscape drawings.

Erasers

Designers commonly use different erasers, each with a special purpose (Figure 1-12). It is important to have the right type of eraser, based on the type of drawing tool and drawing surface. Erasers vary in color, hardness, and material. While rubber

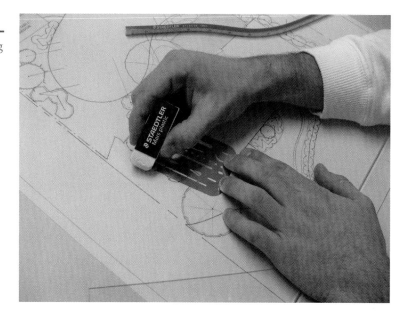

FIGURE 1-13

Eraser shields can be very useful when erasing lines that are close to each other. *(Ann Marie VanDerZanden)*

erasers are still used, white plastic erasers work effectively with less smudging. In addition to these essentials, some designers prefer a kneadable eraser that softens pencil line weight or completely removes pencil lines. Designers who work with ink often use a plastic ink eraser, which has an ink-erasing compound embedded in the eraser. This eraser works best on Mylar and can be used with limited success on vellum or tracing paper.

Erasing should be done with caution. Many hours of tedious drawing can be undone in the blink of an eye when an improperly used eraser accidentally grabs the paper and shreds it beyond repair. Stretching the section of paper to be erased between two fingers and holding it tightly reduces the likelihood that the paper will wrinkle or tear. Applying too much pressure can also damage the paper or even cause the eraser to heat up and leave a smudge. It is important to keep the eraser clean by frequently rubbing it vigorously on scrap paper to remove any graphite on its surface.

The designer can use an eraser shield when erasing a line that is near other lines. Select the opening that best fits the area to be erased, and place it over the line to be erased (Figure 1-13). Hold the shield firmly, and erase.

Dusting Brush

Inevitably, designers change their mind or make a mistake during the design process. Because of this, a dusting brush is a good investment for any designer, particularly the beginner (Figure 1-14). Designers use this specialized brush to remove eraser debris from the drawing surface without smudging the drawing. Sweeping away eraser crumbs with your hand often results in unsightly smudges.

KEEPING THE DRAWING CLEAN. Designers draw many landscape plans in pencil. Unfortunately, the graphite in the pencil lead is prone to smearing and smudging. The most elegant and beautiful designs can be ruined if the final drawing is covered with smudges. For the end product to look professional, it must

FIGURE 1-14

A dusting brush is a better alternative than the side of your hand when removing eraser bits. *(Ann Marie VanDerZanden. Used with approval from STAEDTLER, Inc.)*

have crisp, dark lines and be free of smudges. The crisp lines will reproduce well when copies are made. Unfortunately, smudges on the original will also reproduce. There are a number of methods that can limit smudging, including the following:

- Keep all tools clean. The T-square, scale, triangles, and templates pick up graphite as they rub over the pencil lines. Wiping these tools frequently with a dry cotton cloth will remove the graphite build-up.
- Place a clean piece of paper over areas that have already been drawn to prevent smearing. When possible, fill in the drawing details starting in the top left corner and working toward the bottom right corner. (Left-handed designers should start in the top right and move to the bottom left.) This should limit smearing from your arm and hand.
- To avoid smearing, use a dusting brush when removing eraser crumbs from a drawing.
- Some designers choose a dry cleaning pad or drafting powder to limit smudges. A dry cleaning pad is a loosely woven sack filled with eraser crumbs. The pad is rubbed lightly over the surface of the drawing to remove graphite particles. If too much pressure is applied, then the lines are lightened. Drafting powder is lightly sprinkled over the drawing while the designer is still working on the design. It prevents drafting tools from making direct contact with the graphite already on the paper, and it absorbs skin oil.

 ## MEASURING: SCALES AND ROLLING RULER

Landscape designers create scaled drawings of the landscape they design. They use a scale to represent the actual dimensions of the property in a reduced size. The scale used depends on the desired size of the final drawing and sometimes the size of the paper the design will be drawn on. Two scales that designers commonly use are the **architect's scale** and the **engineer's scale** (Figure 1-15; Figure 1-16). Although a scale may seem difficult to use at first, with a little practice it becomes

FIGURE 1-15

An architect's scale (top) and an engineer's scale (bottom) are both commonly used in landscape design. *(Ann Marie VanDerZanden)*

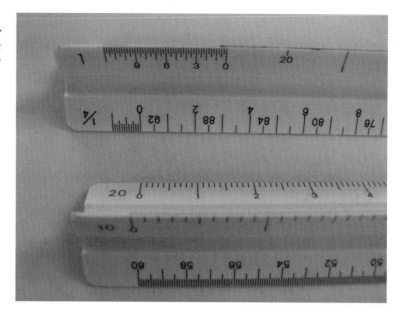

FIGURE 1-16

These four views show what a line 40 feet long looks like when drawn using two different engineer's scale and two different architect's scales. *(Thomson Delmar Learning)*

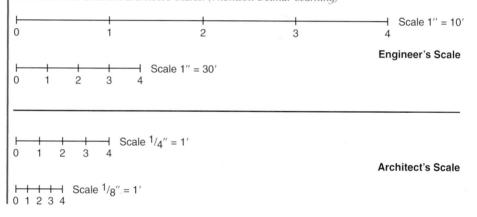

easier. Designers should be familiar with both scales, understand the differences between them, and be able to combine information from varying scales into a selected scale for use on a project base map.

Architect's Scale

An architect's scale contains 11 different scales, all of which represent 1 foot. For example, the $\frac{1}{4}$ scale means that $\frac{1}{4}$ inch represents 1 foot on the drawing. The scales on an architect's scale are: $\frac{1}{16}$, $\frac{3}{32}$, $\frac{1}{8}$, $\frac{3}{16}$, $\frac{1}{4}$, $\frac{3}{8}$, $\frac{1}{2}$, $\frac{3}{4}$, 1, $1\frac{1}{2}$, and 3. In order to accommodate all of the choices, most scales overlap on the same edge; one scale reads from left to right, and the other reads from right to left.

The architect's scale is divided into feet along the entire length; only the foot division at the end is divided into inches or fractions of an inch. For example, on the $\frac{1}{2}''$ scale, the section at the end is divided into 24 parts. Each part represents $\frac{1}{2}$ inch. Table 1-1 shows the divisions found at the end of each scale.

Scale Is Easier Than It Looks

Using and understanding scale in landscape drawings are two of the most difficult concepts for beginning designers to learn, but with a little practice scales can be mastered.

It is critical to mentally retain the sense of scale being used in a drawing. One way is to compare elements in the drawing to familiar examples. For example: the drafting table is 3 feet by 5 feet; the floor tiles in the studio are 1 foot square; a favorite deck area at a local restaurant is 10 feet by 25 feet. Another method is to draw a scaled legend of common landscape elements to keep a visual scale cue close at hand. The legend might include: several tree and shrub circles from small to large, a typical-sized picnic table, and a square of known area such as 100 square feet. Occasionally glancing at these rules-of-thumb as the design develops may save many headaches later on that would result from using the wrong scale.

TABLE 1-1

Scale equivalents for marks to the left or right of zero on an architect's scale.

SCALE	EACH MARK REPRESENTS
$3'' = 1'$	$\frac{1}{8}''$
$1\frac{1}{2}'' = 1'$	$\frac{1}{4}''$
$1'' = 1'$	$\frac{1}{4}''$
$\frac{3}{4}'' = 1'$	$\frac{1}{2}''$
$\frac{1}{2}'' = 1'$	$\frac{1}{2}''$
$\frac{3}{8}'' = 1'$	$1''$
$\frac{1}{4}'' = 1'$	$1''$
$\frac{3}{16}'' = 1'$	$1''$
$\frac{1}{8}'' = 1'$	$2''$
$\frac{3}{32}$	$2''$

To measure a length, both the foot increments and the inches or fractions of an inch are used. For example, to measure 3′5″ on a $\frac{1}{4}'' = 1'$ scale, start with the 0 mark and count 3 feet to the left of the 0 and 5 inches to the right (Figure 1-17). In some cases, you may need to estimate the number of inches if the scale used is not divided into the correct units. In many cases, this type of estimation, at the inches level, is acceptable in landscape design. For example, the width of a planting bed or turf area can vary in size by a few inches. However, in the cases

FIGURE 1-17

This line represents 3′5″ on a ¼″ scale. *(Ann Marie VanDerZanden)*

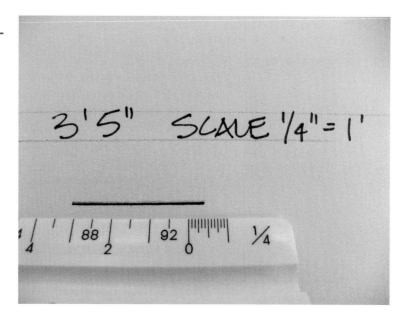

when patios or decks are measured, it is critical that the measurements be accurate to ensure that appropriate hardscape materials will be purchased and that the hardscapes will be constructed correctly.

Engineer's Scale

The engineer's scale is divided into decimal parts, or units of 10. The divisions are 10, 20, 30, 40, 50, and 60 parts to the inch. The scale marked 10 means that the inch is divided into 10 parts. In landscape design, it means that 1″ = 10′. Using the 20 scale, 1 inch on the drawing would be equal to 20 feet. The scales can also be extended for extremely large sites (1″ = 100′, where each increment on the 10 scale represents 10 feet). Another example is for small sites (each increment on the 50 scale represents 0.5 feet, or 1″ = 5′), where an enlarged scale such as for a detailed flower garden is necessary. An engineer's scale is efficient for measuring large land areas, and it is the standard for the civil engineering and land surveying professions.

DECIDING WHICH SCALE TO USE. Selecting the type of scale to use on a drawing is based on several considerations.

- It is most efficient to use the same scale the property information was drawn in. This may have been done using an architect's scale (if it was drawn by an architect or contractor) or in engineer's scale (if it was developed by a surveyor or engineer).

- Engineer's scale produces a smaller drawing, which may be useful in fitting a project to a specific sheet size. For example, engineer's 10 scale is 80 percent as large as architect's ⅛″ scale.

- Architect's scale, because it is slightly larger, may be a better choice for showing additional landscape detail if sheet size is not an issue. An architect's scale is more user friendly for most homeowners as well, since it can be translated to actual dimensions using a tape measure and common 12-inch ruler.

So . . . What Scale Is It?

One of the most frustrating things that can happen to a designer is to be handed a detailed base map with no scale on it, or just a written scale that is obviously incorrect. Several options exist to make the information useable. If any legible measurements are noted, they can be used to approximate the scale. If measurements are not available and access to the site is available, the designer can measure specific map elements (the side of a house, for example) in the field and transfer them to the map to identify a scale. If all else fails, the designer can use standard site features (4-foot sidewalks, 8-foot single-car garage doors) to approximate a proper scale for the drawing. With the common availability of photocopy enlargement and reduction, the designer should exercise caution when picking up any plan copy, and he or she should verify the scale. Any incorrect scaling will ultimately result in embarrassing and costly changes to the design.

FIGURE 1-18

A rolling ruler can be used to draw parallel lines, curves, and large diameter circles. *(Ann Marie VanDerZanden)*

- Architect's and engineer's scales are never used interchangeably on the same drawing. Occasionally, section/elevation drawings or enlargements of plan view areas will need to be enlarged to show detail, but the same type of scale should be used and should be clearly labeled to identify the change.

Rolling Ruler

A rolling ruler is a relatively new addition to the designer's tool box (Figure 1-18). This multiuse tool is similar to a straightedge, and because of the rolling feature it allows a designer to make a series of parallel lines at multiple angles without the need for a T-square. A designer can also use a rolling ruler to draw large circles, but it is not as effective at drawing smaller diameter circles. Although this can be a time-saving tool, it does take some practice to become adept at using it.

FIGURE 1-19

Triangles are used to draw specific angles as well as perpendicular lines. *(Ann Marie VanDerZanden)*

ANGLES AND CURVES: TRIANGLES, PROTRACTOR, COMPASS, AND CURVES

Crisp angles and gentle curves make up the bedlines and hardscapes in most residential landscapes. Sometimes the lines are freeform and irregular, and other times the curves need to be repeatable. Designers rely on a set of tools described below to help them draw these different lines. As previously noted, if the designer will be using ink for drawing, inking edges are preferred on all drawing equipment to minimize smearing.

Triangles

Two commonly used triangles in landscape design are the 45-degree and the 30–60-degree triangles (Figure 1-19). They are available in a variety of sizes, and a triangle that is at least 12 inches long makes it easier to draw long vertical lines. A triangle's size is categorized by the length of the longest side of the right angle. By combining the 45- and 30–60-degree triangles, additional angles such as 75, 105, and 135 degrees can be drawn (Figure 1-20). Adjustable angle triangles are also available. They are more expensive, but they can be used for all angles and provide flexibility in matching uncommon angles.

Protractor

A protractor provides an additional method to draw angles beyond the typical sizes and combinations of 30, 60, and 45 degrees (Figure 1-21). A protractor is useful in drawing very precise angles and is accurate to a single degree. A designer can also use a protractor to determine existing angles on a drawing, which is important when transferring information onto a base map and when ensuring that information is accurate for installation.

FIGURE 1-20

Combining different triangles results in more angle options. In this case a 30-60-90 triangle (a) is combined with a 45-degree triangle (b) to create a 75-degree angle. *(Ann Marie VanDerZanden)*

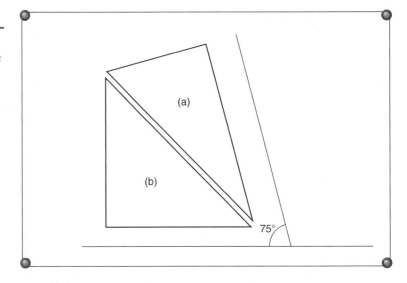

FIGURE 1-21

A protractor allows a designer to draw precise angles, including those not found on typical triangles. *(Ann Marie VanDerZanden. Used with approval from STAEDTLER, Inc.)*

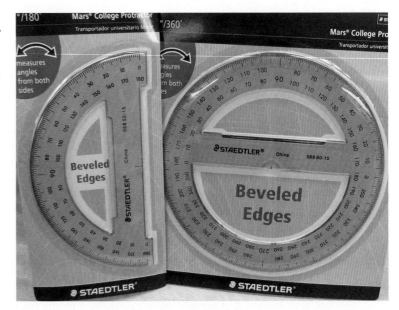

Compass

A designer uses a compass to draw circles and arcs. The top is twisted to adjust the distance between the legs. One leg is a pin used to anchor the instrument; the other leg holds the lead or pen (Figure 1-22). A compass is used when you do not have a circle template or when the diameter of a plant symbol is greater than the largest circle on a template. This tool is also useful when triangular spacing of plants is desired.

TRIANGULAR SPACING. Some designs require exact placement of plants in order to achieve a formal and precise look. One way to achieve this accurate placement is by using triangular spacing. Using the design's scale, place two marks at the desired plant spacing. These marks represent the center of plants 1 and 2

FIGURE 1-22

A compass is useful when drawing relatively large-diameter circles and is essential to illustrating triangular spacing. *(Ann Marie VanDerZanden)*

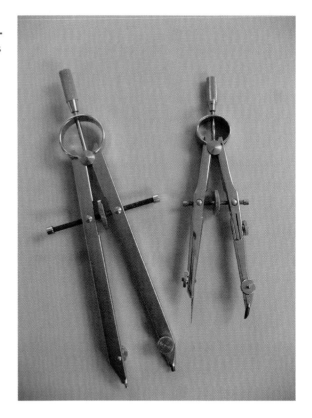

FIGURE 1-23

These three plants were drawn using the triangular spacing technique. The center of each plant is noted (1, 2, 3). The intersecting arcs made by anchoring the compass at point 1 and then at point 2 are shown just below point 3. The triangular spacing is emphasized by the dashed line. *(Ann Marie VanDerZanden)*

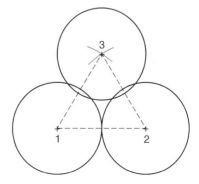

(Figure 1-23). Label the marks as 1 and 2. Anchor the compass pin on mark 1, and make a light arc above. Move the compass pin to mark 2, and repeat overlapping the arcs. Where the two arcs intersect is the center of the third plant (3). The center of each of the three plants is equally spaced from each other. This process can be repeated throughout the area to be planted. When it comes time for installation, this same process can be repeated in the planting bed by using a stake, an appropriate length of string, and landscape paint. This is a quick and easy way to guarantee correct plant spacing during installation.

Irregular Curves

Most landscape designers have two types of curves in their toolbox: a rigid curve and a flexible curve. They use these tools to draw noncircular curves, which

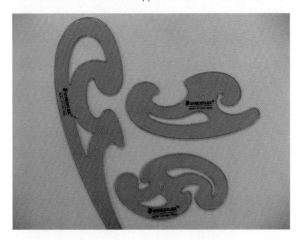

FIGURE 1-24

French curves, available in different sizes and configurations, are used to draw noncircular curves. *(Ann Marie VanDerZanden. Used with approval from STAEDTLER, Inc.)*

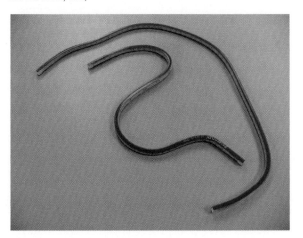

FIGURE 1-25

Flexible curves can be used to create freeform-shaped curves. *(Ann Marie VanDerZanden. Used with approval from STAEDTLER, Inc.)*

have arcs with varying radii and cannot be drawn with a compass or circle template. The rigid curve templates are made from a hard plastic and come in different sizes with the curves already cut into the plastic. These rigid curves are often referred to as **French curves** (Figure 1-24).

The flexible curve is useful for broad, sweeping curves, but it does not work well for curves on a tight radius. It can be used freeform or to connect three or more points to make a line (Figure 1-25). It can also be laid along a bedline that has already been drawn and used to copy that particular effect. Rather than relying on a flexible curve to draw curvilinear bedlines, some designers prefer to draw them freehand. The flexible curve can be cumbersome to use.

TIME SAVERS: LETTERING GUIDE AND TEMPLATES

Designers use a number of time-saving tools, such as lettering guides and templates, that allow them to do repetitive tasks quickly and accurately. As a designer, it is important to be efficient with your time in order to maximize your output for a client.

Lettering Guide

The Ames® lettering guide is a standard in many designers' tool boxes (Figure 1-26). It is used in conjunction with a T-square or parallel bar to draw parallel lettering guidelines at fixed widths. The distance between lines is easily adjusted to allow for large letters or smaller letters, depending on the designer's preference. In addition to the upper and lower guidelines, a midline can also be drawn, which is helpful for beginners practicing hand lettering.

Templates

A **template** is a thin, flat, plastic tool with openings of different shapes cut into it. The pencil or pen is placed into the opening and moved along the outline.

FIGURE 1-26

The Ames® lettering guide allows a designer to quickly draw lettering guidelines at different widths. *(Ann Marie VanDerZanden. Used with approval from STAEDTLER, Inc.)*

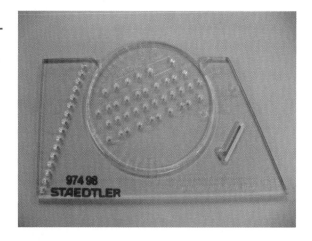

FIGURE 1-27

Templates are available in a number of different configurations. *(Ann Marie VanDerZanden)*

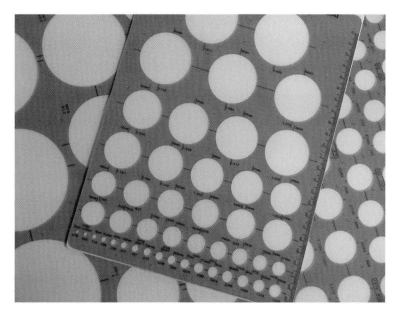

There are many different kinds of templates, including some designed specifically for landscapes (Figure 1-27). Inking templates, which should be used when drawing with pen, have small bumps the on the bottom side to hold the template above the drawing surface and prevent the ink from smearing. (Taping pennies to the bottom side of a regular template can also make it usable with pen.) Circle templates are ideal for landscape design because they allow a designer to draw circles to represent different plant sizes and then to customize each circle to represent a specific plant type (Figure 1-28). Landscape templates provide limited size and symbol options.

Other available design templates include useful landscape symbols, such as lawn and deck furniture, a basketball hoop, lighting, hedges, paving stones, timbers, walls, edging border, and fences. There are countless template options available. Templates are available in different scales, such as $\frac{1}{4}''=1'$ and $\frac{1}{8}''=1'$. If most of the design work is done in a particular scale, then using a template in that scale makes it quicker and easier to find the desired circle size.

DRAWING A PLANT SYMBOL TO SCALE. So just how do you draw a circle the correct size to represent a particular plant? It is as simple as using your scale

FIGURE 1-28

Simple circles drawn using a circle template can be transformed into artistic plant symbols. *(Courtesy of David Madsen. Ann Marie VanDerZanden)*

FIGURE 1-29

Using a $\frac{1}{8}'' = 1'$ scale, this circle represents a plant with a 10′ spread. *(Ann Marie VanDerZanden)*

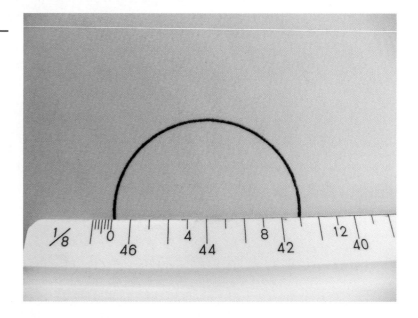

and circle template. If the design is in the scale of $\frac{1}{8}''=1'$, and the tree you are drawing has a mature spread of 10′, place your scale over the circle until you find the one that measures 10′ across on the $\frac{1}{8}''$ scale (Figure 1-29). The circle itself will measure $1\frac{1}{4}''$ across.

DIGITAL AND COMPUTER-AIDED DESIGN AND DRAFTING

In addition to the hand drawing tools described in this chapter, designers also use a number of digital and electronic tools, such as digital cameras, scanners, and a variety of computer software programs. Advances in technology have made these design tools relatively easy to use and more affordable. The particular tool(s) you choose will depend on the type of design work you plan to do, as well as your interest and adeptness in using these technologies.

Purchasing these digital technologies can be a significant investment, particularly for small firms. However, finding a software program or suite of programs that allows you to design in both two and three dimensions, generate a cost estimate, track materials, and provide some level of scheduling and project management can ultimately be a significant cost savings. When evaluating these tools, it is important to consider programs that can be expanded to accommodate how your business evolves over time.

Although computer-aided design (**CAD**) can be quick, relying heavily on a computer before a designer has a mastery of basic design principles and the sequential and iterative design process often leads to poorly conceived and unrefined design concepts. For beginning designers, drawing on paper requires focus and thoughtful reflection as the design develops, which is easily missed in the rapidity of computer-based design. However, developing hand drafting skills at the complete exclusion of developing digital manipulation skills will likely limit a designer's employability. A blending of hand and digital drawing skills makes a designer more marketable since he or she is able to work across multiple mediums and efficiently manage time spent on a project.

Another consideration when using computer-aided design is what the end products will look like. Clients often have an expectation of how the design will be graphically represented at both the draft and final stages. Although computer software and other digital design tools can allow a designer to be more efficient, the graphic quality of the end product may not be acceptable to the homeowner. New programs or add-ons to design program standards like AutoCAD mimic the look of hand-drawn graphics and may be more desirable to homeowner clients.

Types of Computer-Aided Design Tools

The variety, type, and function of computer-aided design tools are similar to tools used in hand drafting. Each tool performs a unique part of the design process, and it can be combined with other digital or traditional hand tools to create a design. For discussion in this text, we have categorized tools as digital capturing tools, design software, linked design and estimating software, and graphic-enhancing software.

DIGITAL CAPTURING TOOLS. Digital cameras and video cameras are quick and reliable tools to document site information during the site survey. Digital images taken in a sequence around the site provide efficient documentation of existing site elements, and they can be an invaluable reference during the design process, particularly when you are drawing the base map.

Further, designers can then use these images in design and graphic-manipulation software programs to provide additional views of the design for the client, and to help him or her visualize what the project might look like (Figure 1-30) before detailed drawings are completed.

DESIGN SOFTWARE. Some design programs are stand alone, while others are AutoCAD based. AutoCAD (meaning automated computer-aided design) is the

FIGURE 1-30

A digital photograph of this site was imported into a design software package. Plant images from the software's database were placed on the image to give the client a rough design idea. *(Ann Marie VanDerZanden)*

FIGURE 1-31

Sketchup software was used to create this three-dimensional structure. Within Sketchup the image can be manipulated to show multiple views. *(Grant Thompson)*

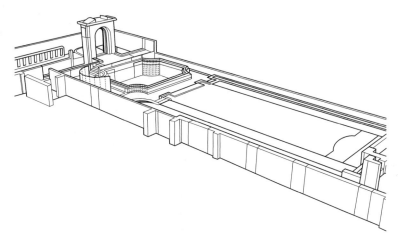

design industry standard software with numerous add-ons available for enhanced functionality. Originally developed in the early 1980s, AutoCAD was one of the first CAD programs for personal computers. AutoCAD and third-party add-ons, or stand-alone software such as Microstation and LANDCADD, are used for most landscape architecture and civil engineering design. They provide a useful and complex range of tools for site assessment, irrigation design, grading and drainage calculations, and planting design.

Most residential design/build firms do not need a complete version of AutoCAD but rather can benefit from other CAD programs, such as AutoSketch, QuickCAD, Sketchup, and PlantARE. All of these programs have interfaces that are relatively easy to use. AutoSketch has a complete set of CAD tools, including extensive content libraries for deck and landscaping icons and hatch and fill patterns. The software can be used to create a range of drawings, from conceptual sketches to technical illustrations. QuickCAD allows a designer to customize the interface and to add more features and functions as he or she becomes more adept at using the program. Sketchup allows a designer to create 3-D models of the design (Figure 1-31).

FIGURE 1-32

The plan view of a design was drawn using PRO Landscape design software. *(Tigon Woline)*

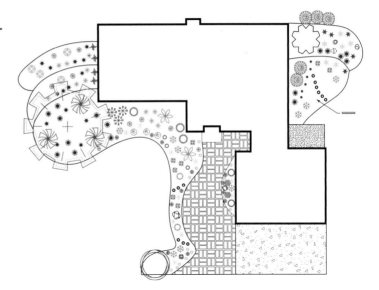

The PlantARE program is based on multiple modules, which perform different functions (i.e., a design module, a planting plan module, and a plant database). This program requires AutoCAD (or another Autodesk product) in order to run, and it is compatible with Photoshop. Keep in mind that as a designer you may outgrow the capabilities of design programs with limited features, and you may want to consider a more robust program for your initial purchase.

Some general features to look for in design software are: a plant symbol library that is searchable by hardiness zone, flower color, size, etc.; the ability to do macro programming to automate repetitive steps in design; the ability to create customized patterned hardscapes; and the choice of output in a hand-drawn appearance or a computer-drawn appearance. The ability to look at the design in plan and elevation view simultaneously is also a useful feature that helps a client visualize what the design will look like once it is installed. In addition to the actual planting composition, some design software programs offer the option to design an irrigation system and an outdoor lighting system. The lighting system is illustrated at night so the full effect of the lighting can be seen. Other programs will generate a corresponding maintenance plan based on the plants specified in the design.

LINKED DESIGN AND ESTIMATING SOFTWARE. Design/build firms can benefit from using software that can generate planting plans, site plans, construction drawings, plant schedules, cost estimates, and even plant maintenance plans. Two examples of these multifunctional programs are PRO Landscape and DynaSCAPE. The PRO Landscape program consists of three integrated modules: Planner, Image Editor, and Proposal. Using the Planner module, a designer can create a 2-D site plan including plantings, hardscapes, and irrigation (Figure 1-32). The extensive plant database is searchable, and a custom database of plants suitable to your hardiness zone can be created. Within the Image Editor module, a designer can create a photorealistic view of the project by using a digital image of the

FIGURE 1-33

Using the Image Editor feature of PRO Landscape, the design created in Figure 1-32 is transformed into a photorealistic view. This view can help a client visualize the project better than the plan view. *(Tigon Woline)*

site (Figure 1-33). The Proposal module automatically generates bids directly from the drawing created in the Planner module.

The DynaSCAPE program provides a slightly different group of features, depending on the version purchased. The Professional Edition and AutoCAD versions are the most powerful; they provide a complete set of 2-D drawing tools, including an extensive library of predrawn plant and hardscape (paving, water feature, irrigation, lighting) symbols and the ability to automatically create a quote, material estimate, and plant-care information. The plant-care information is created when the plant materials list is linked with the program's online plant encyclopedia. Designers particularly interested in providing extensive plant information to clients can take advantage of DynaSCAPE's ability to include color images of plants and detailed plant descriptions.

Programs that allow you to easily generate a materials estimate for the landscape design and create a bid can save a significant amount of time for design/build firms. Look for programs that include an easy-to-use area calculator and dimension tools and a customizable database that allows for input for local material prices.

GRAPHICS ENHANCING SOFTWARE. Although AutoCAD now includes a number of features to enhance the graphic quality of the design, paint and image processing programs like Adobe Photoshop and Illustrator and Microsoft Picture It are commonly used to enhance the graphics of a design. An AutoCAD drawing, or other file format of the design, can be imported into these programs; and color, line weight, and other features can be added (Figure 1-34). A designer can also use these programs to create before and after images. For example, importing an image of a client's house into Photoshop and erasing the existing landscape, or a portion of it, can be an important sales tool because it shows a client the possibilities created by a blank canvas (Figure 1-35).

FIGURE 1-34

This simple CAD drawing with minimal color (a) was imported into Adobe Photoshop, and additional color was added (b). *(Tigon Woline. Joseph VanDerZanden)*

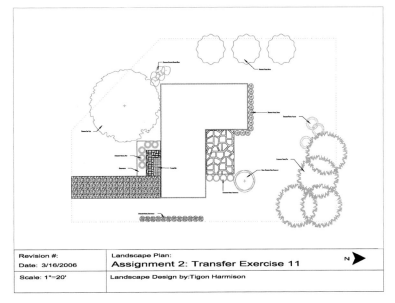

(a)

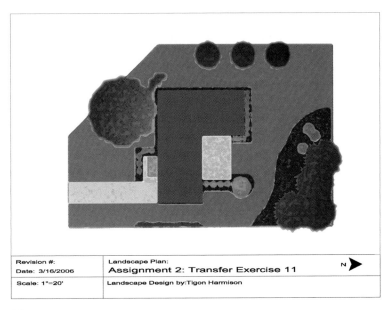

(b)

Making the Purchase

Each of the various computer-aided design tools has advantages and disadvantages, based on features available, ease of use, cost, and ability to update. Many manufacturers provide free trial copies of design software to potential customers, who can use the software for 30 days to explore the different features. Other manufacturers provide tutorials on the Internet that highlight the software's components.

FIGURE 1-35

By importing a digital image into Photoshop and using a few basic features of the program, "before" (a) and "after" (b) images of a landscape can be created. *(Joseph VanDerZanden)*

(a)

(b)

One of the most important features to look for in any software purchase is a manufacturer that provides good customer support and one that is continually working on product improvements and software upgrades to take advantage of new technologies and computer capabilities.

SUMMARY

Landscape designers use a variety of tools to illustrate a design concept and produce a design drawing. This tool box has expanded dramatically with the use of computers, digital cameras, and other forms of technology. Although some tools have specific uses, many tools used by a designer are based on personal preference. Experimenting with different tools is important for beginning designers in order to establish personal preference for the tools that work best for each type of work.

KNOWLEDGE APPLICATION

1. Describe examples and appropriate uses of the following:
   - paper types
   - pencils
   - pens and markers
   - erasers
   - irregular curves
   - templates

2. Differentiate between an architect's scale and an engineer's scale.

3. Draw lines of the following lengths, using an architect's scale.
   - Using a scale of $\frac{1}{8}''=1'0''$, draw lines of: 10′8″; 20′6″; 45′4″.
   - Using a scale of $\frac{3}{16}''=1'0''$, draw lines of: 8′8″; 15′1″; 21′6″.

4. Part 1: Starting at the bottom of a sheet of paper, draw five lines at a 60-degree angle to the left. Use a $\frac{1}{8}''$ scale, and draw the lines 5′ apart. Part 2: Starting on the bottom of the same sheet of paper, draw five lines at a 75-degree angle to the right. Use a $\frac{1}{8}''$ scale, and draw the lines which are 5′ apart.

5. Describe opportunities and constraints associated with computer-aided design programs.

REFERENCES

Bertauski, T. (2006). *Plan graphics for the landscape designer* (2nd ed.). New York: Prentice Hall.

Doyle, M. (1999). *Color drawing: Design drawing skills and techniques for architects, landscape architects, and interior designers* (2nd ed.). New York: John Wiley and Sons, Inc.

Lin, M. W. (1993). *Drawing and designing with confidence: A step-by-step guide.* New York: John Wiley and Sons, Inc.

Reid, G. (2002). *Landscape graphics: Plan, section, and perspective drawing landscape spaces.* New York: Watson-Guptill Publications.

Sipes, J. (2004). Digital tools for design/build. *Landscape Architecture, 94 (5),* 86–94.

Sipes, J. (2005). New options in planting design software. *Landscape Architecture, 95 (7),* 78–85.

ONLINE RESOURCES

Adobe Illustrator: http://www.adobe.com
Adobe Photoshop: http://www.adobe.com
AutoCAD: http://usa.autodesk.com/
Autodesk: http://usa.autodesk.com/
AutoSketch: http://usa.autodesk.com/
DynaSCAPE: http://www.gardengraphics.com

LANDCADD: http://www.eaglepoint.com/landscape
Microstation: http://www.axiomint.com
PlantARE: http://www.plantare.com
PRO Landscape: http://www.drafix.com
QuickCAD: http://usa.autodesk.com/
Sketchup: http://www.sketchup.com

CHAPTER 2

Landscape Graphics

A designer should make it a priority to master graphics and drawing skills. Poor quality drawings will not effectively communicate design ideas to a client. Also, if a designer with well-developed graphics skills draws too deliberately and slowly, he or she will be inefficient in design production. Time constraints, client expectations, and the ultimate use of the drawings will determine which graphics are most appropriate; and a designer should strive to develop a variety of graphics skills to meet these needs.

OBJECTIVES

Upon completion of this chapter you should be able to

1. describe the importance of quality graphics in landscape design.
2. describe the characteristics of quality black-and-white and color graphics.
3. identify and describe the basic symbols used in landscape design.
4. describe proper sheet layout and the scaling of sheets for design drawings.
5. describe the characteristics of quality lettering.
6. describe the basic applications of freehand graphics and hard-line graphics.
7. define the benefits and constraints of CAD and computer software in producing quality landscape graphics.

KEY TERMS

line and dot quality
line weight hierarchy
value contrast
texture

graphic symbols
section-elevation graphics

paraline drawing
perspective drawing

color-rendering
sheet layout

FIGURE 2-1

Precise graphics with high contrast and depth reflect strong professionalism (a); sloppy inconsistent lines combined with lack of textures create a drawing that lacks interest and professionalism (b). *(Steven Rodie)*

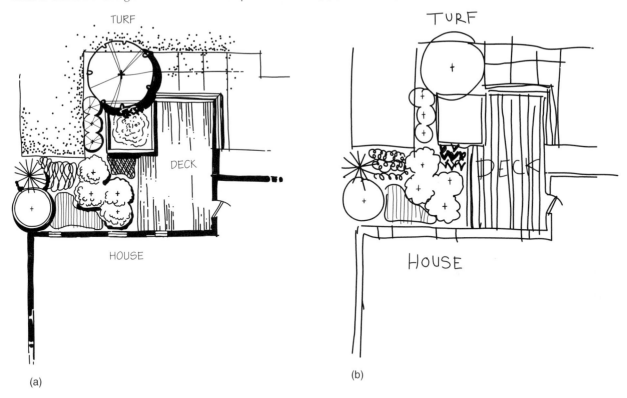

(a)   (b)

This chapter will provide an overview of why graphics are important, how graphics communicate information in several design process steps, what defines quality graphics, and specific information on drawing techniques. Mastering basic techniques will provide most designers the framework to build their own skills and graphic style. Anyone willing to put the necessary time into practice can develop quality graphics skills. Hand-drawn graphics are emphasized in this chapter, due to their relative simplicity, universal applicability, and on-going use in the landscape design industry. It is beyond the scope of this textbook to describe the complete variety of CAD software available for landscape graphics. Chapter 1 presents descriptions of numerous examples of CAD software. Although designers can develop quality graphics with either hand-drawn methods or CAD-based programs and tools, neither method in itself guarantees quality. The key to quality is the mastery of whichever tools are selected, practiced, and implemented.

 ## IMPORTANCE OF GRAPHICS

A designer's professionalism and credibility are reflected in his or her graphics (Figure 2-1). The general public, including clients, often view landscape design as an artistic profession. Therefore, there is an expectation that the designer's artistic abilities will be apparent in the design product.

FIGURE 2-2

Models provide a clear vision of a proposed design, but their high cost limits everyday usage. *(Courtesy of Ed May, Jason Beisner, Mark Ratzlaff, Michaela Fortina-Forst. Steven Rodie)*

Quality graphics clearly deliver information to the client and the landscape installer. Design drawings (together with written specifications in some cases) graphically describe how the project will be built. Small lapses in graphic quality can affect drawing legibility and accuracy. For example, poorly drawn lines without definite ends are difficult to measure accurately; sloppy leader lines used to identify plant symbols may connect to the wrong plants or be difficult to follow; and drawing information placed too close to a drawing edge may be lost in the photocopying process.

Clients sometimes have a difficult time understanding the bird's-eye view (plan view) used in most drawings. In an ideal setting, designers would have time to build intricate (and expensive) three-dimensional models to illustrate proposed design projects so the client could visualize proposed design features (Figure 2-2). As an alternative described in Chapter 1, computer programs can provide three-dimensional models of the landscapes that allow a client to virtually walk through proposed changes. When computers are not available, however, hand-drawn graphics on a sheet of paper are required, and the designer's ability to create depth and three-dimensional effect with simple graphics is essential. This is particularly important in residential landscape design, where most clients value extra assistance in understanding design plans (Figure 2-3).

An additional factor that complicates landscape graphics is that drawings will often serve multiple purposes. For many landscape projects, just one plan drawing is produced. This single drawing will serve as a sales tool for the designer, as an installation tool for the contractor, and as a reference for the client (Figure 2-4). Finally, quality graphics should get clients excited about landscape design. For the client, there is something very personal about seeing his or her landscape documented in an artistic drawing. While the hectic pace of residential design does not always allow time for a designer to develop a graphic masterpiece, many aspects of quality graphics can easily be incorporated into the everyday design routine.

FIGURE 2-3 A hand-drawn section-elevation (a) clarifies information found in the plan drawing (b). *(Courtesy of Karen Richards. Steven Rodie)*

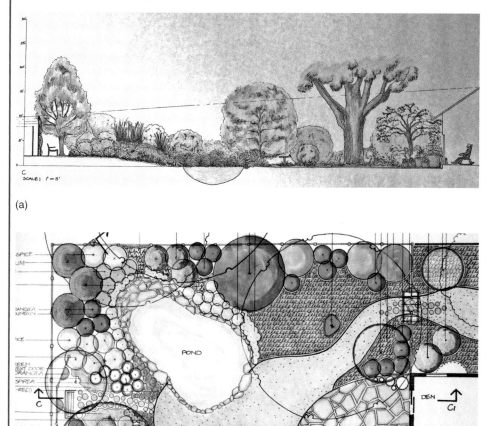

(a)

(b)

CHOOSING BETWEEN DRAFTED AND FREEHAND GRAPHICS

Beginning designers often struggle with freehand drawing and choose hard-line drawing (drafting; or using straightedges as drawing guides) instead because they feel it looks more professional. What these novice designers neglect to realize, however, is that slightly imperfect lines do not necessarily represent a poor quality drawing; in fact, they can augment a designer's drawing style.

Each type of drawing has its purpose. Designers use drafting for accuracy, and it is typical for engineering and architectural drawings. Straightedges and circle templates produce exact angles, line lengths, and shapes; the lines are typically dark and consistent. Although accurate, these renderings can lack artistic style and depth. Freehand drawings are typically less accurate; lines vary in their character and consistency. Freehand drawings are artistic and reflect the designer's personal style. Either method of drawing is acceptable to use at any stage of graphics production, provided the designer acknowledges the benefits and limitations of each technique.

FIGURE 2-4

The installation graphic (a) shows precise plant locations and spacing to assist the contractor during installation. The illustrative graphic (b) is more exciting to look at but lacks specific plant locations. Landscape graphics must typically compromise between these extremes in a drawing that is both interesting and precise. *(Steven Rodie)*

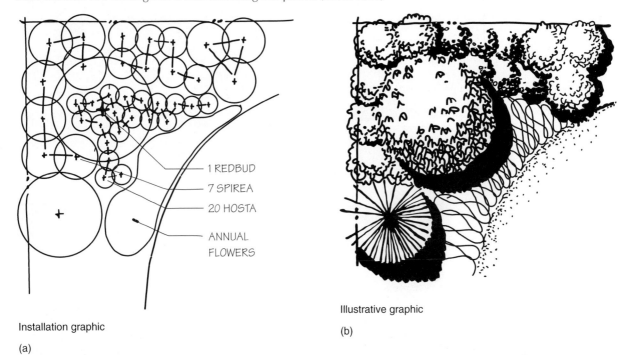

Installation graphic
(a)

Illustrative graphic
(b)

An alternative approach among many designers is to combine the methods. At each stage of design, as layers of information are combined and reworked, designers synthesize ideas. Hard-lining edges and using circle templates to draw specific plant sizes and groupings provide an accurate template to freehand trace for the final drawing. Method #1 requires two separate drawing layers over the sketch plan, but it keeps the final sheet clean of template and hard guidelines

Landscape Graphics Are Naturally Loose

Among the major design professions, landscape drawings tend to be the loosest. The informality of the renderings may stem from the many curving lines in a typical plan; it may be from the focus on fuzzy, leafy design elements; or it may be the constant changes that are inevitable in landscape designs. Regardless of the reason, there is an expected level of informality that landscape designers should be aware of (and have fun with) in their design graphics. As computer graphics have become more sophisticated, some landscape design offices have actually worked to produce computer-generated plant symbols and color-rendering qualities that appear hand drawn in an effort to meet this expectation.

FIGURE 2-5

Method combination #1 is summarized in a. It combines a freehand sketch (b), which is refined with a hard-line and template tracing of the freehand sketch (c) and finalized with a freehand tracing (d). *(Steven Rodie)*

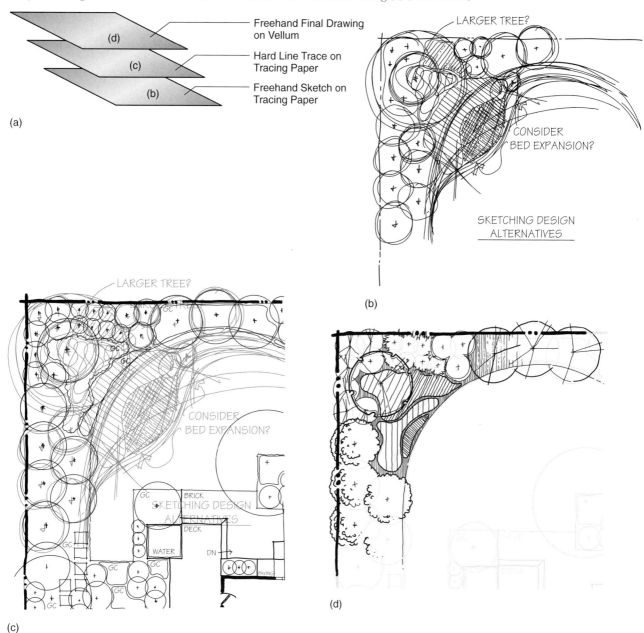

(Figure 2-5). Method #2 requires only one additional drawing layer, but it may require erasure of template and hard lines (Figure 2-6).

Designers can also combine methods by using computer graphics. For example, a computer-generated plan can be hand color-rendered for an accurate yet informal drawing style. In addition, a simple 3-D perspective drawing accurately generated with computer software could be traced freehand or with sketching software to produce a more complex yet artistic and informal drawing.

FIGURE 2-6

Method combination #2 (a) layers a sheet with blue or light pencil hard line and template tracing (c) over the free hand trace hard line drawing (b). The pencil guidelines are free-hand traced for the final drawing. *(Steven Rodie)*

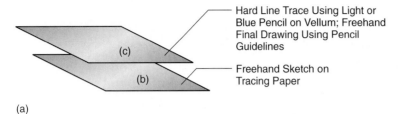

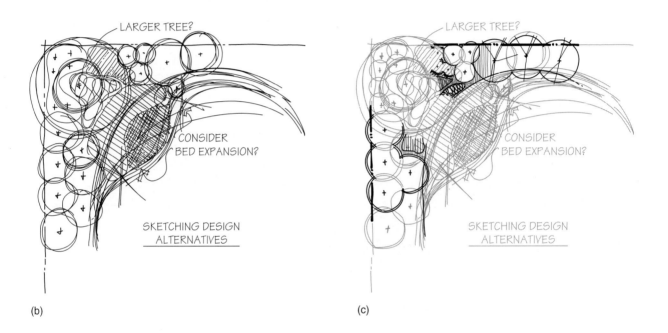

 ### GRAPHICS—BLACK AND WHITE MECHANICS

The following standards and techniques outline the basic mechanics for pencil and ink graphics. Learning these mechanics takes time and patience, and it may require a significant amount of doodling. Every designer may reach a slightly different proficiency and will likely prefer one method over another. The most important consideration is practice. Correct drawing techniques enable the designer to produce quality drawings efficiently. In many cases, poor graphics can take as long or longer to produce than quality graphics. The investment in establishing quality graphics skills will consistently yield dividends throughout a designer's career.

These graphic standards apply to all of the viewpoints typically used to represent a landscape. Plan view graphics represent the bird's-eye view of a site and are the most common way to convey design information. The section-elevation viewpoint provides a side view of the landscape with little or no depth. Paraline and perspective drawing typically look at a side view, but add depth (Figure 2-7).

FIGURE 2-7 Similar graphics quality standards apply to all graphic viewpoints, including plan (a), section–elevation (b; courtesy of Julie Catlin), paraline (c; courtesy of Alyssa Stapp), and perspective (d; courtesy of Karen Richards). *(Steven Rodie)*

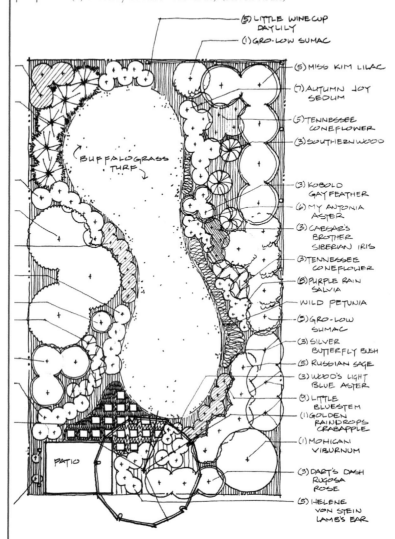

(a)

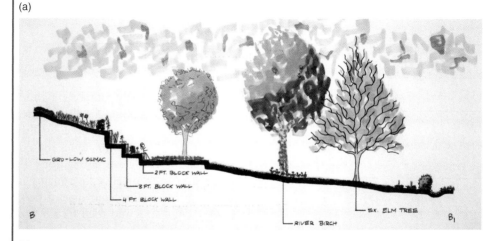

(b)

(continues)

FIGURE 2-7 | *Continued.*

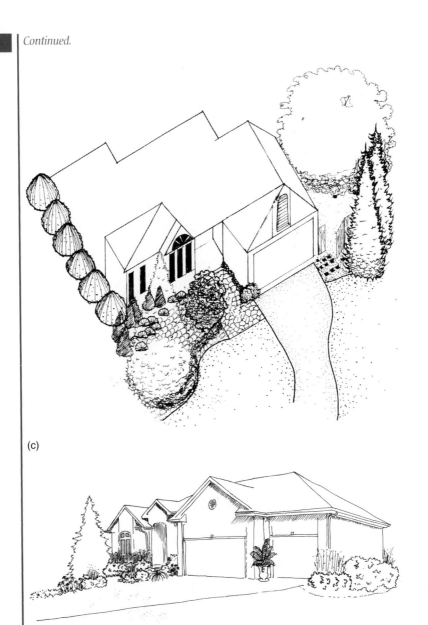

(c)

(d)

Line and Dot Quality

All pencil and ink drawings are basically composed of lines and dots. **Quality lines and dots** produce quality drawings. It sounds simplistic, but if a designer learns how to neatly draw consistent lines and dots, he or she will be able to produce quality drawings and sketches (Figure 2-8). Lines should have clear start and end points, and dots should have a clean point. Both should display some tonal variation. One of the best ways to practice line quality is to draw a series of line segments, varying the length, direction, and type of pen or pencil used. It also helps to draw boxes of varying sizes and then fill the boxes in with parallel lines (Figure 2-9).

Basic line quality can be summarized by the following characteristics (Figure 2-10).

- Each line has a distinct start and end point and does not trail off the paper.
- Lines are drawn neatly and deliberately without smudges.
- Lines must be drawn dark enough to reproduce effectively.
- Lines should start and finish with additional darkness (freehand lines may also exhibit a slight width increase) to emphasize the line endpoints. This requires a conscious effort to change the downward pressure when using a pencil. A pen, especially a felt-tip, will usually create this effect without additional pressure, if the line is drawn quickly.
- Where lines intersect, ends should slightly overlap. Drawings with a looser style will typically have more overlap. When freehand lines are continued end to end, or a complete circle is drawn, it looks better aesthetically to leave a small gap between the segments. This alleviates the inevitable misalignment of the adjoining ends.
- Dots should be dark, consistent in size, and deliberately produced with a vertically held felt-tip pen or soft pencil. Dots should not have tails.

One of the fears of drawing freehand is producing lines that are not straight and circles that are not circular. Experimenting with specific techniques such as the following will help develop proper technique (Figure 2-11).

Black and White versus Color

Black-and-white graphics are not as exciting as color, and they usually are not the best choice if a designer really wants to impress a client. For these reasons, designers sometimes elevate color above black and white in importance. Well-done color graphics make a difference when there are sufficient time and resources to complete them correctly. The vast majority of drawings, however, are black and white. Color applied to a quality drawing should be viewed as icing on the cake, rather than a fix for poor graphics.

- Try drawing straight lines in different directions on the sheet. Some people are better at straight lines drawn toward them whereas others are better at lines drawn away from them or at an angle.
- Try drawing lines by flexing just your wrist or fingers and contrast the results with an entire arm motion, keeping your wrist or elbow locked. Different line lengths may be more consistent with different approaches.
- For long lines, it may be more effective to look at the point you are drawing to rather than the point you are drawing from; this approach gives a target point to help visually guide your line to the specified endpoint.

FIGURE 2-8

A variety of well-drawn lines, combined with consistent dot patterns and quality, form the basis for all quality graphics. *(Courtesy of Alyssa Stapp. Steven Rodie)*

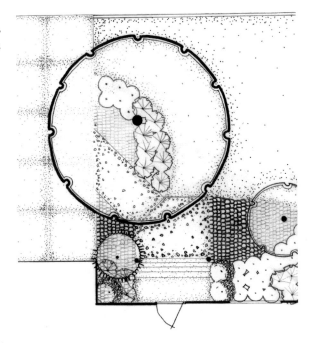

Consistent Inconsistency

Regardless of practice and patience, freehand lines and circles will not be perfect. The key to quality freehand graphics is not perfection—it is consistent inconsistency. Part of the beauty in works of art is human imperfection. As artists develop a unique style, consistent imperfections surface, and they can become an artist's trademark. Like artists, designers also develop drawing styles over time, and any consistent inconsistencies will develop into a unique drawing style.

- Tracing circles that were previously drawn with a template provides consistent practice for producing round circles.

Line Weight Hierarchy and Contrast

Line weight hierarchy is produced when the designer uses a variety of line widths in a drawing. Variable line widths create a visual hierarchy in a drawing and provide clues to the characteristics of drawing elements. The most consistent method to produce variable line widths is to use mechanical pencils or pens that have a specified width. A sharp, chiseled pencil tip or a felt-tip pen can create variable line widths, depending on the orientation of the tip and the applied pressure (Figure 2-12).

FIGURE 2-9

Practice drawing consistent lines and dots using a variety of pencil widths and lead hardness (a) and ink widths and tip types (b). Drawing within boxes or along lines provides a defined space to implement line quality principles. *(Steven Rodie)*

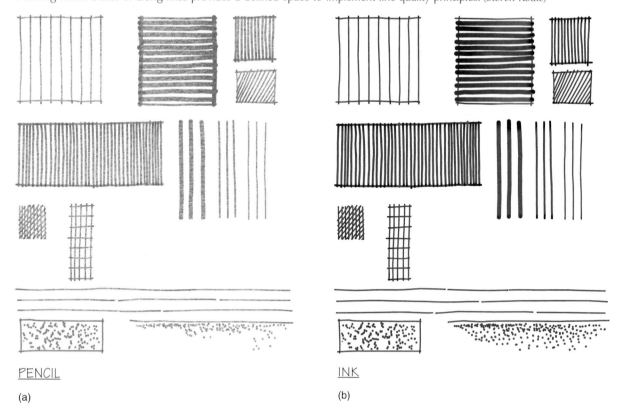

PENCIL

(a)

INK

(b)

FIGURE 2-10

Good/bad line quality ending, ink and pencil, slow and fast, overlap ends, leave gaps for lines and circles; area of good dots, some with tails. *(Steven Rodie)*

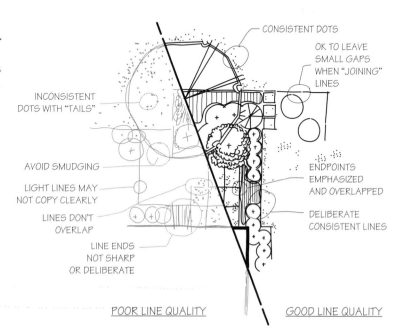

FIGURE 2-11

Designers should experiment to find the most comfortable directions and distances to draw lines (a). One of the best ways to practice consistent lines is to draw some light circles with templates and then freehand trace over the circles with heavier pencil or ink (b). *(Steven Rodie)*

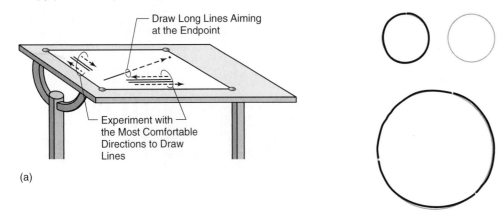

(a) (b)

Get a Grip on a New Habit

Line width variation is a powerful graphic tool. The concept of simply picking up a different pen or pencil to draw different line widths seems easy to understand. But in reality, beginning designers tend to stick with one writing tool, and transitioning to a multiple pen or pencil approach can be a challenge if the habit is not established early. The designer who lets this graphics habit take hold of him or her will be rewarded later with improved quality and efficiency.

The following guidelines summarize appropriate use of line width.

- A wider line indicates a higher relative importance of an element in a landscape drawing. For example, the outline of the house on a residential plan is typically the widest line, followed by a patio or driveway outline (Figure 2-13).
- Wide lines correlate to taller objects on the plan. This is one of the ways in which a two-dimensional landscape plan can indicate three dimensions. Using slightly wider lines for tree and plant mass outlines helps to elevate the plant symbols above the ground plane (Figure 2-14).
- The outer edges of all forms (decks, patios, roofs, etc.) should be slightly wider than the texture or pattern lines. For example, the line around the edge of a deck should be wider than the lines indicating board texture (Figure 2-15).

FIGURE 2-12

If mechanical pencils are not available, pencil line weight can be varied by chiseling the pencil point and varying the use of the flat surface and thin edge (a). Ink line weights are normally controlled by tip size, but tip wear and increased pressure will also affect width (b). *(Steven Rodie)*

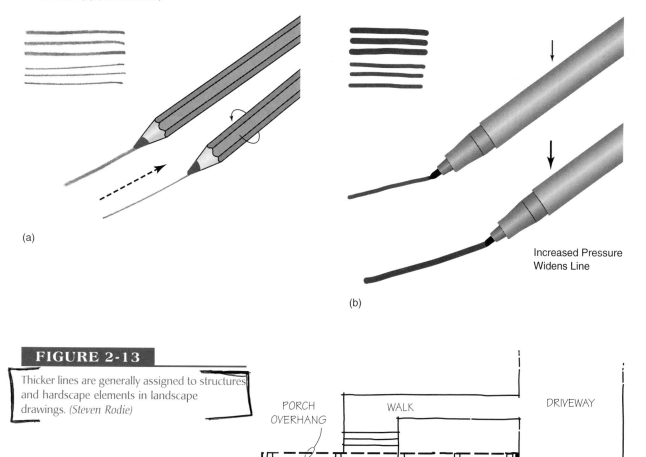

Increased Pressure Widens Line

(a)

(b)

FIGURE 2-13

Thicker lines are generally assigned to structures and hardscape elements in landscape drawings. *(Steven Rodie)*

Value Contrast

Value represents the relative amount of light and dark present in a drawing. Black-and-white landscape graphics can reflect the complete range of value: pure black, a wide range of grays, or pure white. Any difference in value between two adjoining areas will create contrast. The highest amount of **value contrast** occurs when white and black are adjacent to each other on the drawing. High contrast between adjacent areas differentiates them visually and generates interest.

FIGURE 2-14

Tall objects in a plan view drawing would hypothetically be closer to the viewer. The taller an object, the thicker the assigned line weight. *(Steven Rodie)*

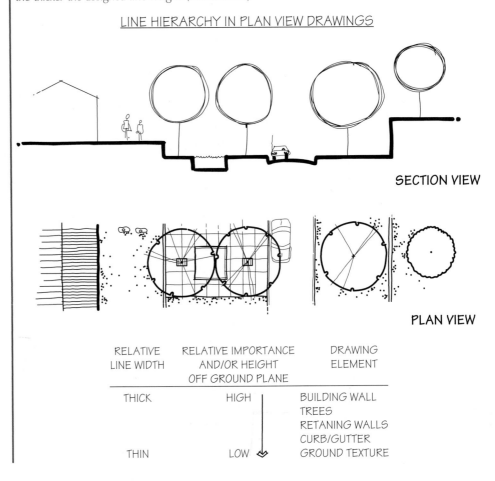

FIGURE 2-15

The deck, step, and patio textures are drawn with thinner lines than the edges of these areas. This creates a visual hierarchy and enhances the contrast between different ground plane surfaces. *(Steven Rodie)*

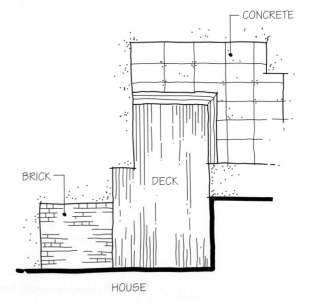

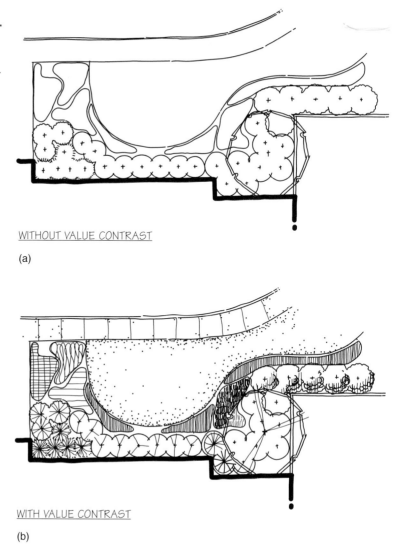

FIGURE 2-16

Graphics without value contrast (a) are less legible and interesting than graphics that contrast adjacent light and dark values (b). *(Steven Rodie)*

Drawings that contain high contrasts between elements are easier to understand than drawings that contain a limited value range (Figure 2-16). For example, applying a lighter value to an area inside of a tree circle symbol that contrasts with a darker outer area produces the illusion that the inside area is under the symbol (Figure 2-17).

Designers also create value contrast on landscape drawings through the use of shadows, one of the most useful and powerful additions of value contrast to a drawing. <u>Shadows significantly strengthen the three-dimensional qualities of the drawing.</u> They can also serve a functional purpose by reflecting the relative heights and shapes of objects on the plan (Figure 2-18). Several techniques for drawing effective shadows are shown in Figure 2-19 (pages 48–49).

All shadows on a drawing must orient in the same direction. They are normally placed in one of two ways: toward the bottom right of symbols to visually balance the dark visual weight, or in line with the expected shadow patterns that

FIGURE 2-17

Graphic depth is developed beneath a symbol by applying thinner line weights, pencil instead of ink, and omitting textures within the symbol. *(Steven Rodie)*

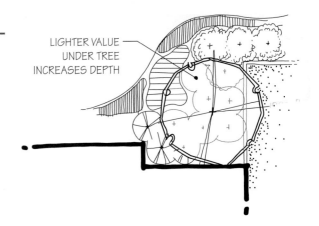

would naturally occur on the site (Figure 2-20, page 50). Black shadows provide the highest value contrast, but they can also hide significant ground plane areas that contain design information or elements, so their coverage area and direction should be considered before application (Figure 2-21, page 50).

Texture

Texture is an organized pattern of lines and dots that enhances value contrasts throughout the drawing and represents specific landscape materials and plants. Textures can be simple or complex, depending on their purpose. Effective textures can require a significant amount of time to plan and draw, so they should be given careful consideration in their implementation.

STRATEGIES FOR EFFECTIVE USE. Texture usage should incorporate the following strategies.

- Textures should be applied lightly; they can always be darkened but are difficult to lighten.

- A simpler pattern is best for most texture applications. Patterns must reflect a wide variety of light to dark values for maximum effect. Simple lines and dot patterns can be applied much more efficiently than complex patterns.

- Texture is basically lines and dots; therefore, better line or dot quality will result in superior quality for the entire texture area. Not every line or dot has to be perfect; consistent inconsistency will definitely enhance textured areas (Figure 2-22, page 50).

- Most texture patterns look best if they are consistently oriented on the drawing. If patterns change their orientation, be consistent and have a good reason to change it, such as to show site contours (Figure 2-23, page 51).

- A little bit of texture can create a distinct value contrast. A few dots (stipples) on one side of a line dividing a concrete surface from a turf area can provide enough textural difference. In addition, using texture gradation creates a sense of texture and surface depth without having to draw the texture for the entire area (Figure 2-24, page 52).

FIGURE 2-18

A plan view graphic without shadows (a) lacks depth. In contrast, the same drawing with shadows (b) reflects the relative shapes and heights of objects (pointed conifer tree, tall flagpole, etc). *(Steven Rodie)*

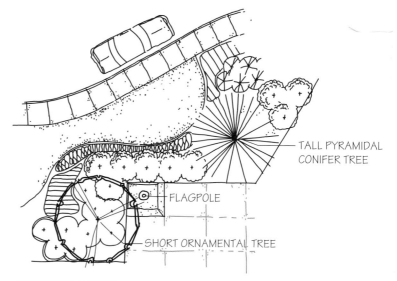

WITHOUT SHADOWS

(a)

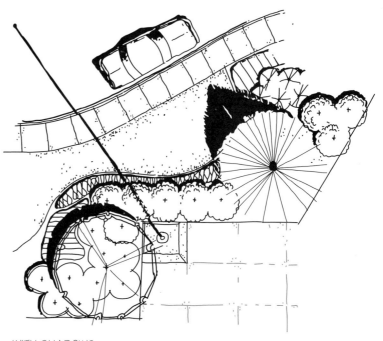

WITH SHADOWS

(b)

FIGURE 2-19

Shadow construction techniques include: Using an off-set circle as a guide for accurate shadows (a); applying a chisel-tip marker for quick shadows (b); showing relative heights and shapes of objects (c); and complementing the textured size of symbols with shadows (d). *(Steven Rodie)*

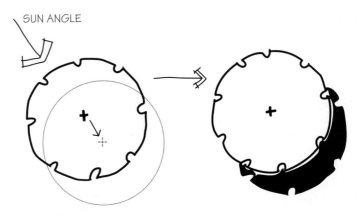

1) USE OFF-SET CIRCLE AS GUIDE

2) REFLECT EDGE PATTERN AND LEAVE WHITE SPACE BETWEEN SHADOW & SYMBOL EDGE

(a)

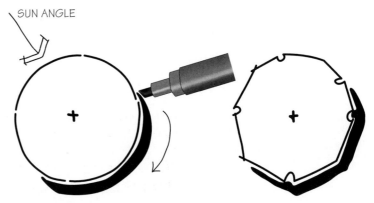

A CHISEL-POINT PEN OR MARKER CAN BE USED TO DRAW QUICK SHADOWS—FOLLOW THE BACK OF THE SYMBOL, KEEPING THE TIP PARALLEL TO THE SUN ANGLE

(b)

(continues)

FIGURE 2-19

Continued.

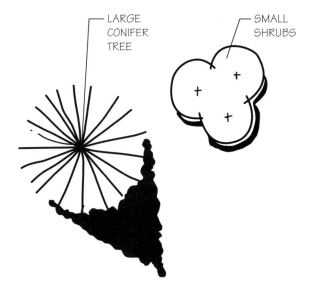

SHADOWS SHOULD REFLECT THE RELATIVE SIZE & HEIGHT OF THE OBJECT

(c)

SHADOWS SHOULD CORRESPOND TO TEXTURE ON "DARK" SIDE OF SYMBOL

(d)

FIGURE 2-20

All shadows should orient the same direction. They can be placed to reflect site sun angles (a) or to visually balance the drawing (b). *(Steven Rodie)*

SHADOWS REFLECT AFTERNOON SPRING OR FALL SUN IN CENTRAL U.S.
(a)

SHADOWS PLACED AT LOWER RIGHT OF SYMBOL TO BALANCE VISUAL WEIGHT OF DARK AREAS
(b)

FIGURE 2-21

Shadows should be sized or rendered to minimize the loss of important information under the shaded area. *(Steven Rodie)*

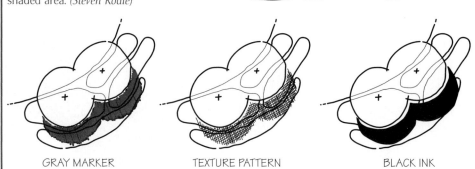

GRAY MARKER   TEXTURE PATTERN   BLACK INK

FIGURE 2-22

Simple symbols and textures drawn freehand can create effective graphics in spite of slight variations in application. *(Steven Rodie)*

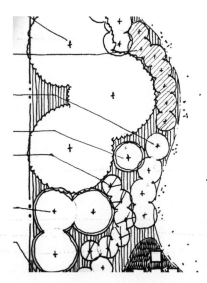
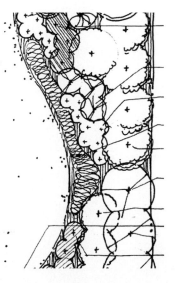

FIGURE 2-23

Flat areas of similar material (groundcover or mulch) should have a consistent graphic alignment to enhance graphic simplicity (a). Texture symbols can vary when reflecting slopes or other groundplane characteristics (b). *(Steven Rodie)*

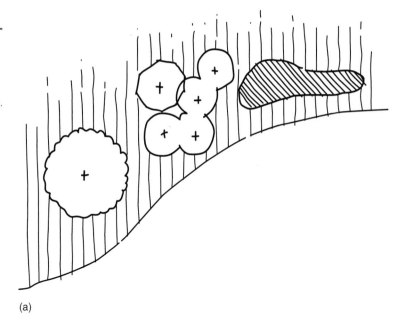

(a)

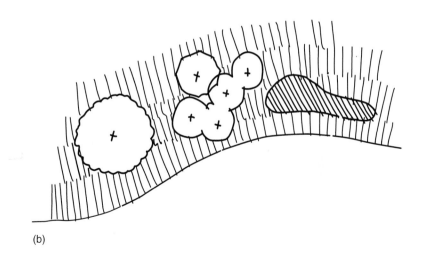

(b)

LESS IS MORE. Texture patterns can take a significant amount of time to produce on a landscape drawing. Before deciding on texture pattern placement, look at the drawing as a whole, and locate the largest areas requiring the same texture and the areas that are adjacent, which require a contrasting texture. Based on this comparison, choose the lightest textures for the largest areas, because light textures are quickest to draw. If plain white is the chosen texture, the largest area is already finished! Try to minimize the use of dark, densely patterned textures, except for smaller areas. On most drawings, turf and paved surfaces are usually light texture. It pays to strategize before putting pen to paper.

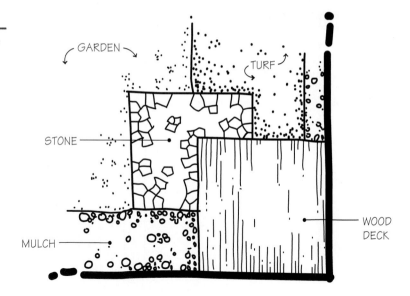

FIGURE 2-24

Texture focused along object edges effectively differentiates adjacent surfaces and saves drawing time. Similar symbols can be used in adjacent areas (garden and turf) to enhance drawing simplicity if varied slightly. *(Steven Rodie)*

GRAPHIC SYMBOLS AND DEFINITIONS

Graphic symbols are an essential component of landscape graphics. There are a variety of standard universal symbols that every designer should use for consistency in drawing information. While some of these symbols overlap other design professions, many are unique to landscape design and represent specific plant types or other information. Designers interested in developing a personal library of graphic symbols should study a wide range of landscape drawings and learn symbols that suit their style and drawing needs. Although a variety of symbols provides more drawing flexibility, a core group of symbols should be developed and practiced.

Plan View Graphics

A plan view drawing presents a view that looks straight down on the landscape. This is the most commonly used format for design because it documents the entire landscape in one diagram that shows scaled sizes, distances, and relationships for all ground plane elements. While quality graphic techniques, combined with clearly drawn symbols and textures, can improve client understanding of design features, it is not uncommon for the nondesigner to have difficulty interpreting plan view graphics.

STANDARD SYMBOLS. The following conventional symbols are used on landscape drawings (Figure 2-25). They represent a form of shorthand that saves drawing time and reduces drawing clutter. Should it be necessary, the designer can use a legend to define symbols that appear on the drawing (Figure 2-26).

canopy: The branched portion of a tree typically measured parallel to the ground at the widest distance between the ends of opposite branches; the edge of the canopy is referred to as the tree drip line.

PLANT SYMBOLS. Plan view plant symbols represent a variety of information on landscape drawings. They can be very abstract and simple (an open circle) or very realistic and complex (almost every leaf or branch is portrayed). They can be drawn to reflect a vegetated **canopy** (growing season) or a branched form without foliage (winter months) (Figure 2-27).

FIGURE 2-25

Consistent use of standard symbols enhances design clarity and professionalism. *(Steven Rodie)*

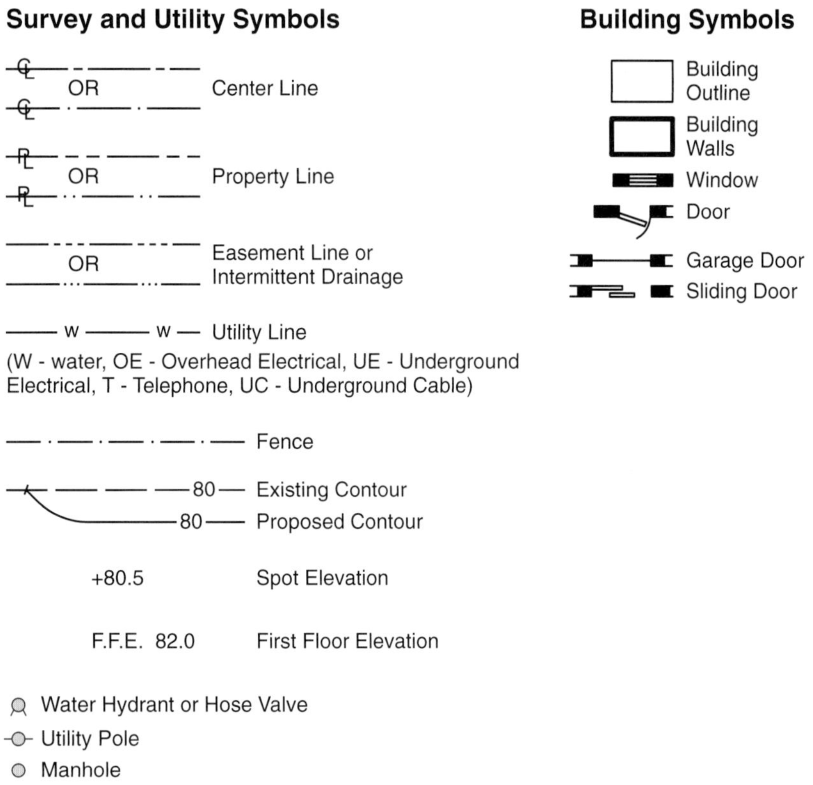

Survey and Utility Symbols

- ₵ OR ₵ — Center Line
- ₽ OR ₽ — Property Line
- OR — Easement Line or Intermittent Drainage
- — w — w — Utility Line
  (W - water, OE - Overhead Electrical, UE - Underground Electrical, T - Telephone, UC - Underground Cable)
- — · — · — · — Fence
- — — 80 — Existing Contour
- ——80—— Proposed Contour
- +80.5   Spot Elevation
- F.F.E. 82.0   First Floor Elevation
- Water Hydrant or Hose Valve
- Utility Pole
- Manhole

Building Symbols

- Building Outline
- Building Walls
- Window
- Door
- Garage Door
- Sliding Door

FIGURE 2-26

A legend can efficiently define symbol graphics and eliminate repetitive labeling. *(Courtesy of Thomas Ogee. Steven Rodie)*

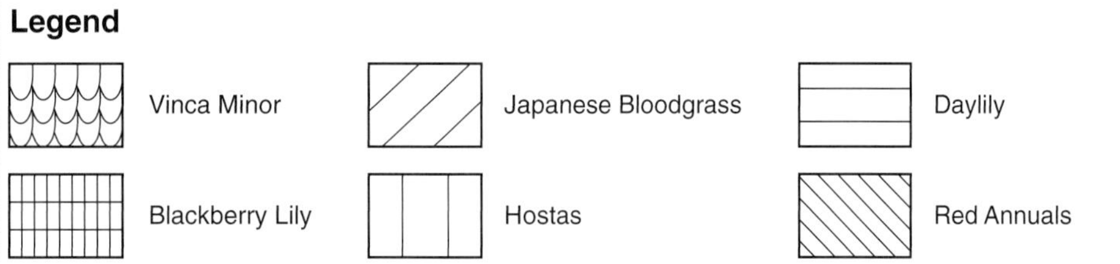

Legend

- Vinca Minor
- Japanese Bloodgrass
- Daylily
- Blackberry Lily
- Hostas
- Red Annuals

deciduous plant: A plant that loses its foliage (leaves or needles) annually.

Symbol Style. Symbols with rounded or relatively smooth edges tend to represent plants with relatively smooth or foliage textures, such as **deciduous plants**. Symbols that contain rounded textures or branching patterns inside the edge also represent deciduous plants (Figure 2-28). Symbols with jagged, irregular edges or edges composed of short, straight lines represent plants with pointed or needled foliage (Figure 2-29).

Symbol Size. Symbol size should correlate with expected mature spread of each plant so the design accurately represents the mature landscape. Tree symbols are an

FIGURE 2-27

Plant symbols should be suited to the required drawing style and information. *(Courtesy of Karen Richards. Steven Rodie)*

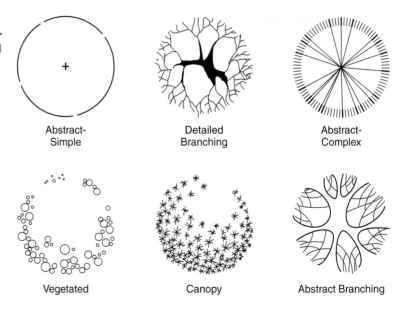

FIGURE 2-28

Deciduous symbols reflect leafy and relatively soft vegetation qualities. *(Courtesy of Karen Richards. Steven Rodie)*

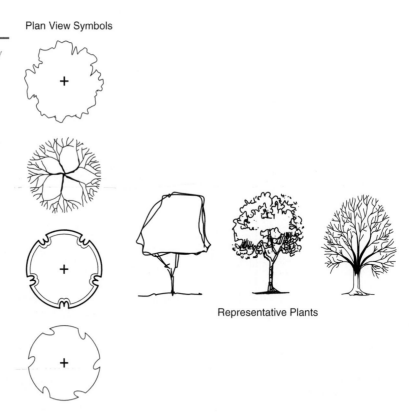

exception. They are typically drawn smaller, two thirds to three fourths of the expected mature canopy size. This more accurately represents the spread of these plants in the first 20 to 30 years of growth, and it reflects the smaller mature size indicative of many trees in stressful urban environments (Figure 2-30).

Landscape Graphics 55

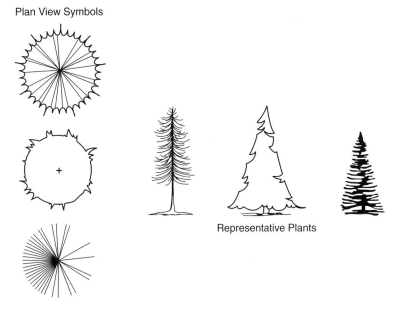

FIGURE 2-29

Coniferous and broadleaf evergreen symbols reflect needle-like or spiny vegetation qualities. *(Courtesy of Karen Richards. Steven Rodie)*

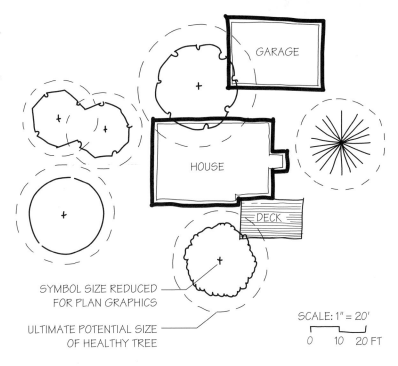

FIGURE 2-30

The dashed lines approximate the ultimate size of mature healthy trees. The tree symbols indicate realistic anticipated tree sizes during the life of the project. *(Steven Rodie)*

Symbol Pattern and Spacing. Using correctly sized symbols, a designer can accurately gauge the correct spacing for plants. Plant symbols should graphically indicate whether plants are expected to overlap in a design or remain separate (Figure 2-31).

FIGURE 2-31

Plan graphics should clearly show the intended plant spacing to confirm massing as well as identify overcrowding. *(Steven Rodie)*

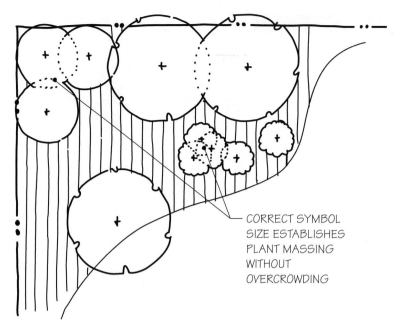

CORRECT SYMBOL SIZE ESTABLISHES PLANT MASSING WITHOUT OVERCROWDING

FIGURE 2-32

Overlapped circles clearly show individual plants (a); illustrative symbols with texture show outlines of overlapped plants (b); bold outline graphically denotes plants as a group while light inner lines define specific overlap (c); and a squared version of (c) reads as a group of sheared or hedge plants (d). *(Steven Rodie)*

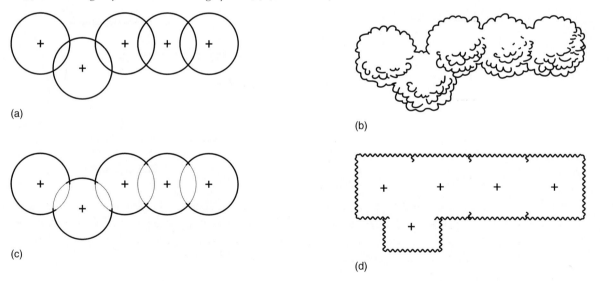

The designer can draw overlapping plant symbols in several ways, depending upon how realistic or illustrative the drawing needs to be. Symbols can be drawn with center points and outlined edges clearly shown. Additional texture can be added to produce an illustrative symbol that appears more realistic but omits the individual plant locations. Or a bold outline can be added to the grouping to

FIGURE 2-33 A variety of textures can be used to represent groundcover or perennials areas (a). Edges of bubbles can also be symbolized to represent groundcover textures (b). Contrasts in adjacent patterns and values enhance drawing quality (c). *(2-33a and 2-33c courtesy of Karen Richards. Steven Rodie)*

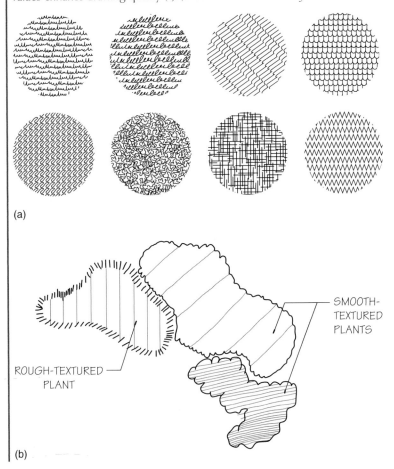

(continues)

emphasize the plant mass while retaining the individual plant locations. Finally, squaring off the group symbol reflects a sense of formality or higher maintenance, such as a hedge (Figure 2-32). Each approach emphasizes different information and should be applied according to the needs of the designer.

Bubbles and Textures for Groundcovers, Perennials, and Turf. Groundcovers, herbaceous perennials, and annuals are typically planted in masses and are often relocated during installation. For this reason, most are shown as bubbles on landscape drawings, with an edge or interior texture that correlates to the plant texture (Figure 2-33). When individual plant locations are required for a detailed garden or intricate plant layout, simple symbols to represent a plant are used (Figure 2-34).

herbaceous perennial: A plant with no persistent woody stem above the ground that lives for three or more growing seasons.

FIGURE 2-33

Continued.

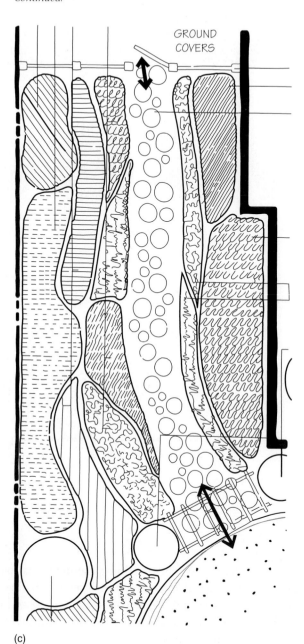

(c)

stippling: An area of consistently applied fine dots to create drawing textures and shading patterns; also the activity to create such a pattern.

hatching: A series of fine, parallel or crossed lines drawn with a relatively consistent spacing and length used to indicate a ground texture or shading pattern; hatching is also the activity to create such a texture or pattern.

contour line: A line on a map that connects all points of the same elevation in a given area; most base maps indicate contours at a 1-foot or 5-foot interval.

Turf areas are often left blank to generate value contrast with the textures of adjoining planting beds and hardscape. If a turf area requires texture, simple **stippling** or **hatching** is used. Another option that requires less time is to emphasize texture only near turf edges to increase contrast. In sloping turf areas, it may be appropriate to graphically accentuate **contour lines** to visually enhance relative slope steepness (Figure 2-35).

FIGURE 2-34

Individual plants within groundcover or perennial flower areas can be shown as small circles or as plant centers (+) within areas. *(Steven Rodie)*

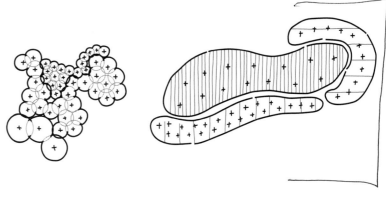

FIGURE 2-35

Using stipples or hatching creates effective turf texture that reflects the steepness and orientation of slopes. *(Steven Rodie)*

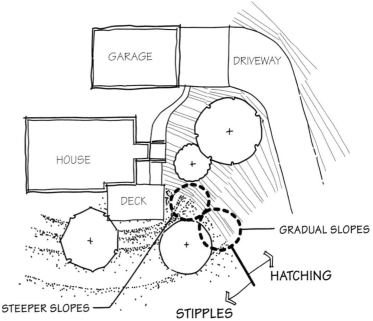

PLAN VIEW HARDSCAPE AND WATER TEXTURES. Hardscapes can cover large areas on landscape drawings. Selected textures for these surfaces should reflect material and character (Figure 2-36). Concrete and asphalt are relatively smooth and can be shown with simple stippling (Figure 2-37). Wood, brick, and stone are normally drawn to reflect individual boards or paving units, but the scale of the drawing will determine the exact detail. Large-scale drawings may only require a hint of character but finely detailed drawings usually require a higher attention to detail (Figure 2-38).

Designers often draw flowing water as a light value in contrast with darker streambanks and vegetation, and they sometimes draw informal narrow lines parallel to the bolder edge lines to create a contrast in line weights while hinting at surface ripples along the banks. Pools of water may be left light or may contain a variety of textures to simulate ripples (Figure 2-39).

FIGURE 2-36

Plan graphics can effectively reflect the character and texture of hardscape surfaces. *(Courtesy of Karen Richards. Steven Rodie)*

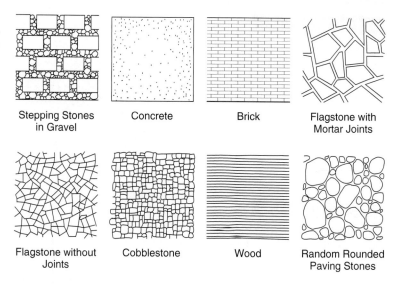

FIGURE 2-37

Subtle contrasts in texture patterns and density can identify a surface contrast between adjacent concrete and asphalt, the two most common paving materials. *(Steven Rodie)*

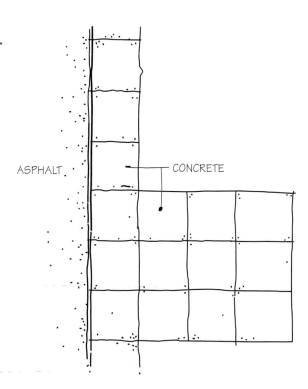

SYMBOL AND TEXTURE SUMMARY. The following guidelines should be followed when developing graphic symbols and textures.

- The simpler the better; a simple graphic takes less time to draw and is easier to read.
- Every designer should have a basic well-practiced collection of symbols. Although it can be tempting to develop a different symbol for every plant

FIGURE 2-38

Small scale close-up views require more detail than large-scale drawings. *(Steven Rodie)*

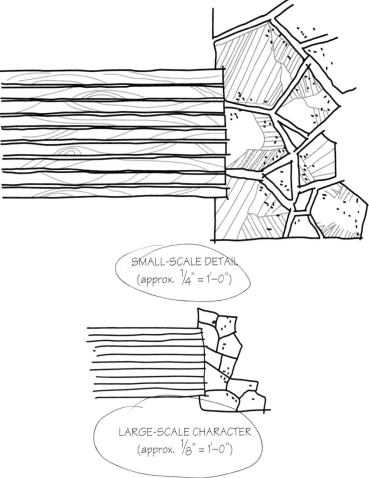

FIGURE 2-39

Streambank texture and value contrast, together with water surface patterns, define water feature character and movement. *(Steven Rodie)*

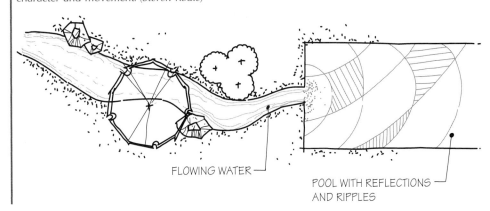

species on a drawing, too many symbols (especially for plants) can lead to a cluttered drawing (Figure 2-40).

- Avoid being overly realistic, and remember the graphic quality guidelines. Practice a graphic style that is comfortable to draw and that maintains an appealing abstract quality.

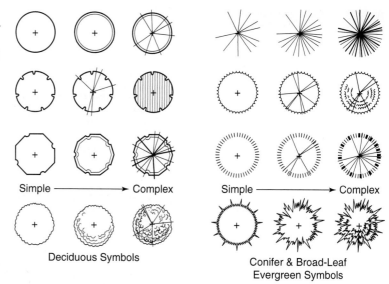

FIGURE 2-40

A basic selection of plan-view plant symbols includes a selection of four basic deciduous and coniferous symbols each drawn at three detail levels. *(Steven Rodie)*

Additional Graphic Methods

A variety of drawing methods are available to restructure two-dimensional plan information into formats that are more easily understood by clients. Three common methods include section elevations, paraline drawings, and perspective drawings (Figure 2-41). Additional time is required for each method, so the designer needs to be selective in choosing the approach, the number of additional drawings, and the specific viewpoints.

SECTION-ELEVATION GRAPHICS. **Section-elevation graphics** are powerful tools in landscape design communication. Spatial relationships and the feel of a landscape space are much more evident in a section elevation than they are in a plan view (Figure 2-42, page 65). Section elevations clarify **topographical** and landscape features. Even though they are most often used later in the design process, section elevations can also provide critical assistance to the designer in earlier stages by clarifying complex landscape areas. The more comfortable a designer becomes with the production and use of section-elevation graphics, the stronger his or her design awareness and communication skills will become.

topography, topographical: The three-dimensional characteristics of land, including hills and valleys, drainage and slopes; synonymous with landform.

Definitions and Guidelines. Basic section-elevation definitions and guidelines include the following concepts (Figure 2-43, pages 66–67):

- A section represents the cross-section of a landscape at a specific location. Information contained in a section is limited to landscape elements and features that are bisected at the section location.
- Elevations depict building facades and landscape features.
- A section elevation combines section information with landscape elements and features located slightly in front of or behind the section. Section elevations

FIGURE 2-41 A plan view garden design (a) is illustrated with three drawing methods, including section-elevation (b), paraline (c), and perspective (d). *(Steven Rodie)*

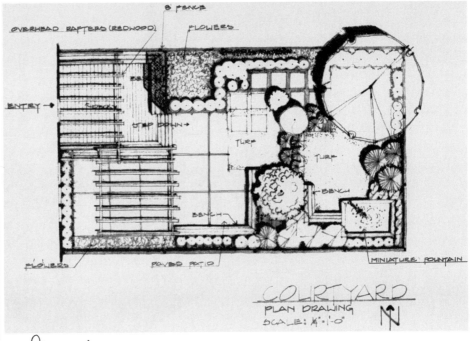

(a) Plan view

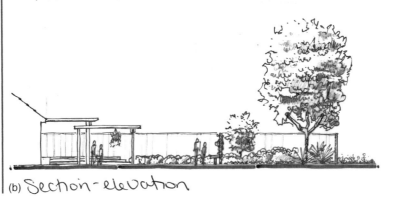

(b) Section-elevation

(continues)

tend to be more useful than sections because they contain more information and depth.

- A designated location and direction of view are required for any section elevation to provide orientation for the drawing.
- Section elevations can be estimated (Figure 2-44, page 67) or carefully constructed (Figure 2-45, page 68), depending on the purpose and the required clarification of plan information.
- Vertical heights in section elevations are typically drawn at the same scale as horizontal distances so that drawings do not become distorted. In addition, all vertical dimensions in a section elevation should use the same scale.

FIGURE 2-41 Continued.

(c) Paraline

(d) Perspective

FIGURE 2-42

Although the plan graphic for this backyard (a) provides a lot of information, the sloping terrain and spaces created by the tree canopy are not evident until a section-elevation is constructed (b). Note location of section-elevation on plan at C-C_1. *(Courtesy of Julie Catlin. Steven Rodie)*

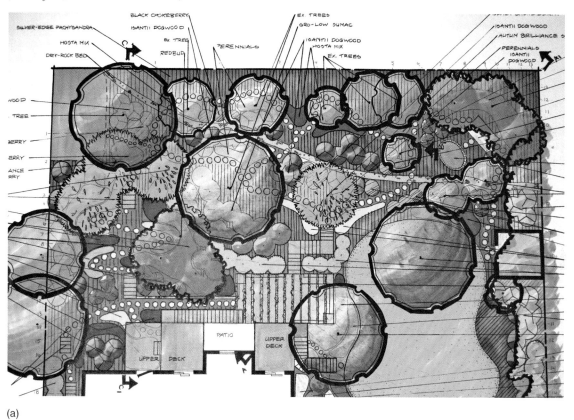

(a)

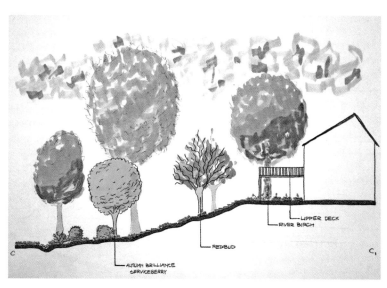

(b)

FIGURE 2-43 A vertical slice through a landscape (a) establishes the following graphic information: a section line (b); an elevation (c); a section-elevation (d); and a length and view orientation on the plan drawing (e). *(Steven Rodie)*

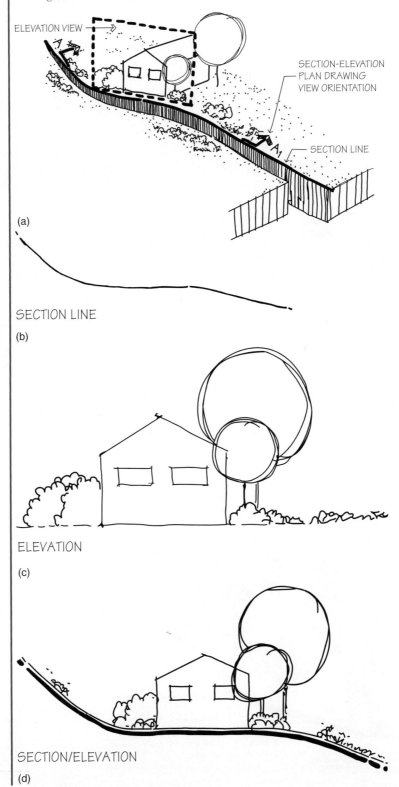

(continues)

FIGURE 2-43

Continued.

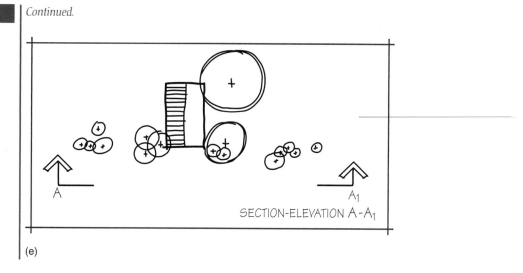

(e)

FIGURE 2-44

Estimated dimensions from the plan view drawing can be used to quickly develop a reasonably accurate sketch. This method works well for small flat landscape spaces where tree canopies and overhead structure heights can be easily estimated. *(Steven Rodie)*

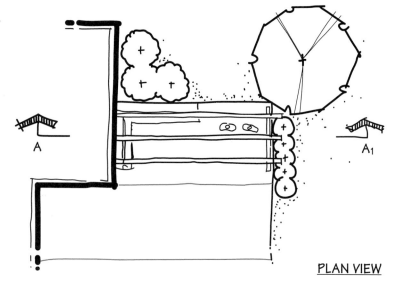

FIGURE 2-45 Plan view measurements and land elevation information are projected onto a grid of scaled height and width dimensions. This information provides a framework for accurately developing a detailed section-elevation. *(Steven Rodie)*

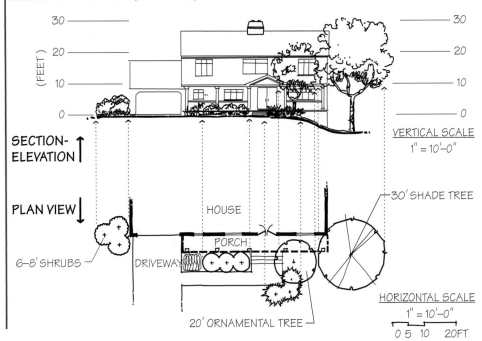

Grid of scaled hgt. + width dimensions

Styles. Section-elevation graphics represent a familiar view of the landscape, so less symbolism is required to convey information. For instance, a pine tree can simply be drawn as a pine tree in an elevation graphic; it does not need a special circular symbol to identify its character as it does in plan view.

Section-elevation graphic styles vary from the abstract to the detailed and specific (Figure 2-46). The challenge in drawing section-elevation graphics is to avoid being overly realistic and to maintain an abstract style. The graphic quality principles for section elevations are exactly the same as for plan graphics (line quality, line weight hierarchy, appropriateness, and variety of textures, etc.). Value contrast is especially important because not only are adjacent elements differentiated, but depth of view is also developed (Figure 2-47). Finer lines for background elements, or the combined use of ink (foreground) and pencil (background), will create visual depth.

Two additional style elements should be included in all section elevations. The designer should add a bold line just under the landscape scene to provide a strong visual base for the drawing and to emphasize the specific topographic features in that particular area (Figure 2-48).

In addition, the designer should add renderings of people to section elevations to provide a sense of scale and to suggest that the landscape design will be a people-friendly space. There are easily as many people symbols as there are graphic symbols for trees (Figure 2-49).

Incorporating people symbols into section-elevation drawings follows the same graphic principles as those outlined for plants and other elements. In addition,

Landscape Graphics 69

FIGURE 2-46

Abstract section-elevation graphics (a) can be drawn quickly while detailed graphics (b) can be used for more refined presentations. *(Steven Rodie)*

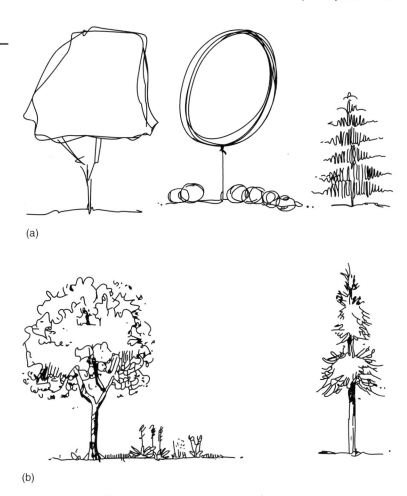

FIGURE 2-47

Background objects should be drawn lighter to enhance graphic depth. A consistent sunlight/shadow orientation also enhances graphic quality. *(Steven Rodie)*

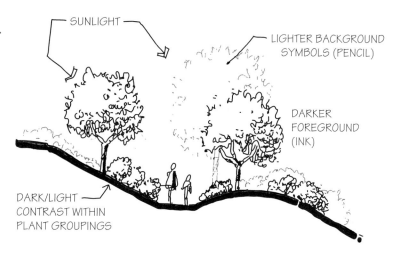

FIGURE 2-48

A bold section line provides a visual base for a section-elevation and reinforces land form slopes and features. *(Steven Rodie)*

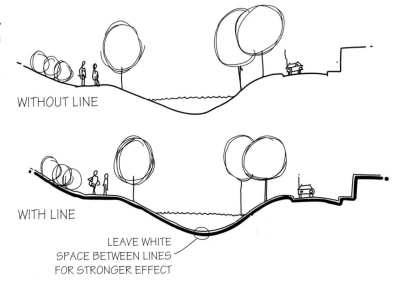

FIGURE 2-49

People can be simple or complex, and add interest, scale and life to drawings. *(Courtesy of Karen Richards, Steven Rodie)*

the designer should minimize details such as faces, which are difficult to draw effectively, and the people should represent a realistic height (Figure 2-50).

Develop an Abstract Style. First attempts at section-elevation graphics do not seem to go as smoothly for most designers as plan graphics. It may be that we are trained from childhood to draw elevation views of trees, flowers, people, and other natural features in a particular way. In most cases, our drawing memory becomes imprinted with a cartoonish style of drawing rather than an abstract interpretation of real shapes and textures. Enhanced observational skills developed as part of sketching and drawing exercises will improve interpretation and graphic quality (Figure 2-51). In addition, the simple exercise of observing and tracing quality abstract styles from other designers will help undo established drawing patterns and broaden a designer's capability.

FIGURE 2-50

Keeping people at a reasonable distance from the viewer eliminates the need for facial and character detail. All adult figures should be approximately the same height (5 to 6 scale feet), and all heads should align with a scene horizon line regardless of figure distance from viewer (a). Rule-of-thumb proportions should be used to keep figures relatively consistent (b). *(Steven Rodie)*

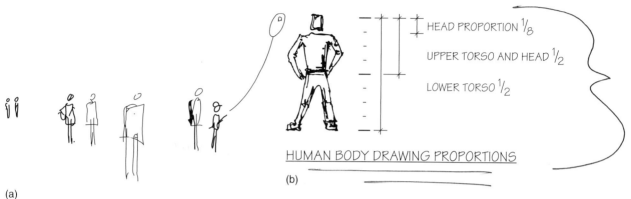

FIGURE 2-51

Moving beyond a cartoon style (a) to an artistic abstract or detailed style (b) strongly enhances drawing quality and professionalism. *(Steven Rodie)*

FIGURE 2-52

Paraline drawings are developed using oblique frameworks to establish consistent ground plane and vertical lines (a). Once the framework is selected, the plan view drawing is placed under the framework, and vertical projections are used to develop object heights and widths (b). The completed drawing (c) shows completed object shapes and textures. *(Steven Rodie)*

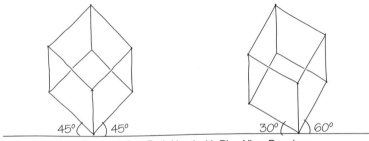

(a) Plan Oblique Frameworks - Both Used with Plan View Drawings

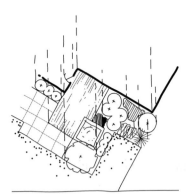

(b) Plan View Objects Projected Vertically for Height (Same Scale As Plan View)

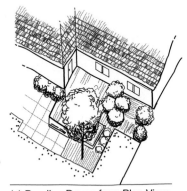

(c) Paraline Drawn from Plan View
- All Vertical Lines Remain Vertical
- All Parallel Lines Remain Parallel

PARALINE DRAWINGS. Designers use **paraline drawing** techniques to project a plan-view drawing into a three-dimensional drawing. Since all vertical lines and parallel lines remain parallel, there is some distortion in the drawing, especially when circular or rounded objects are represented. The enhanced view of spatial character and relationships more than compensates for the accuracy limitations, since paraline drawings are relatively quick to produce. The basic framework for developing a paraline drawing is shown in Figure 2-52.

PERSPECTIVE DRAWINGS. **Perspective drawings** are similar in some ways to section-elevation and paraline drawings. They typically provide a side view of the landscape rather than a plan view (Figure 2-53). But they also have two major differences. First, perspective drawings accurately reflect depth and space through the construction of converging lines and vanishing points; and second, perspective drawings typically take more time to develop, especially if they are hand drawn accurately. Many CAD programs are capable of quickly generating simple to complex perspective views of plan information, so the availability of perspective drawings as an important visualization tool is becoming commonplace.

Considerations for perspective drawing include the following:

- Because they typically require a greater time investment than section-elevation or paraline drawings, the production of perspective drawings needs to be planned carefully. With practice, designers can efficiently produce perspective

FIGURE 2-53

Perspective landscape views provide the most realistic interpretation of design ideas. (Steven Rodie)

drawings, especially if they use a loose, informal graphic style or CAD software.

- Perspective drawings are normally drawn in one- or two-point perspective (Figure 2-54). The choice of perspective, as well as specific location and height of the viewpoint, is based on the types of information that the drawing is intended to convey.

- Designers develop perspective drawings from a predrawn framework of lines that denote the horizon line and distances into the view (Figure 2-55).

- Designers can use computer design programs to develop a three-dimensional wire frame of the landscape, over which a hand-drawn perspective drawing can be quickly sketched. This normally saves time compared to manually drawing the framework by hand. The frame can also be rotated and used for additional sketch viewpoints.

Analysis and Concept Graphics

Analysis and concept graphics serve an important role in design process communications. Designers use symbols such as bubbles, arrows, dashed-line patterns, and asterisks to condense information on drawings. Distinct graphics that clearly convey important information are more effective at defining abstract design ideas than written text. Although designers use many of the symbols interchangeably, the definitions in Figure 2-56 reflect the most common usage.

COLOR BASICS

Appropriately applied color adds clarity and refinement to landscape drawings. It helps identify landscape elements through associated colors (blue water, gray concrete, brown decking, etc.), differentiates similar elements (dark green plant mass

FIGURE 2-54

One-point perspective (a) has one vanishing point on the horizon line and tends to produce a centered viewpoint; two-point perspective (b) uses two vanishing points and typically provides a corner viewpoint. *(Courtesy of Karen Richards. Steven Rodie)*

FIGURE 2-55

The sample one-point perspective chart indicates the vanishing point, horizon line, and relative measurements across and into the viewpoint. *(Steven Rodie)*

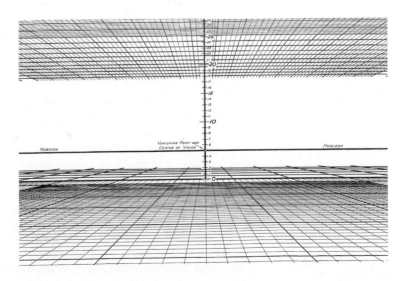

Landscape Graphics 75

FIGURE 2-56

Basic graphic design symbols include pedestrian and automobile circulation (a); wind and views (b); edges, screens, and barriers (c); focal points/areas of interest and buildings (d); and designated assessment or use areas (e). *(Steven Rodie)*

Circulation:
Automobile
Pedestrian

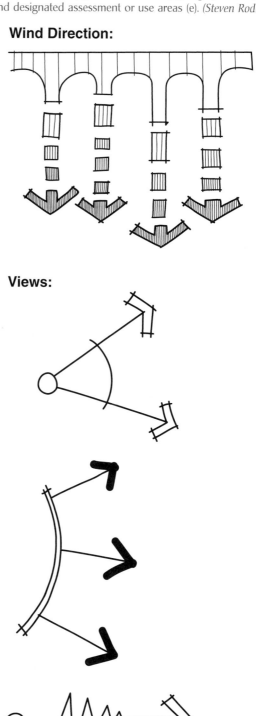

(a)

(b)

(continues)

FIGURE 2-56

Continued.

Edges, Screens, & Barriers:

Focal Points/Areas of Interest:

(c)

Buildings & Structures:

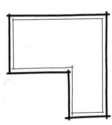

Designated Assessment or Use Areas:

(d)

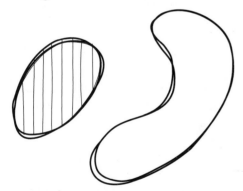

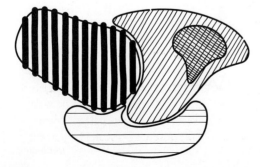

(e)

adjacent to a light green mass), heightens the quality of drawings, and adds interest. Color may also enhance the marketability of the design by making it more attractive.

Color application does come with trade-offs, however. **Color-rendering** can take significant time, and mistakes are difficult to fix. Also, color does not change design quality; a superior design will stand on its own. In fact, ineffective use of color can reduce design clarity and degrade the professionalism of an otherwise stellar design.

The basics of color-rendered landscape drawings are discussed below, including methods, effective use, color use in computer drawings, and applications for different drawing types.

Common Media

Colored pencils and markers are the two most common media for color-rendering. Other media that can be used include pastels, watercolor pencils, and water colors. Each type of media has its benefits and limitations. For example, compared to markers, colored pencils take longer to apply effectively over large areas but provide subtle color layering and gradation. Markers can quickly cover large areas of drawing, and they provide vibrant colors, but they can be very unforgiving in their application and attempts to correct errors. The two media combine well and are often used together on the same drawing.

Watercolor pencils and water colors generate a very artistic style, but they may be limited in their application to drawings produced on photocopy or plotting papers. Pastels can quickly cover large areas, and they work well when blended or used for subtle coloring, but they typically require the use of a fixative to protect the finished drawing from smudging. Once again, these media can be combined with other media to develop specific styles and effects.

COLOR LAYERING WITH MARKERS. Markers provide a flexible approach to developing the color gradation and depth that define the three-dimensional feel of well-rendered landscape drawings. The simplest approach to layering involves application of wet marker over dry applications of the same color (Figure 2-57). The changes are subtle, but a sense of depth is produced. Using a range of several **complementary colors** will produce more dramatic results. The key to this approach is to apply the light colors first, followed by the darker colors, and to allow the colors to blend while they are still wet.

complementary colors: Colors that are located directly across from each other on the color wheel and that create a high contrast when viewed together, such as red and green or yellow and purple.

MURPHY'S LAW AND MARKER USE. Markers quickly cover large areas with consistent color; they dry quickly, are transparent so they do not hide drawing information, and come in a wide array of colors. For all these advantages of markers, they also have disadvantages, and mistakes are inevitable. Proper planning can reduce the likelihood of experiencing an "Anything that can go wrong, will go wrong" moment.

Always print an extra copy or two of the drawings you are planning to color. Having a reserve copy provides a backup for at least one fatal coloring mistake, and a second copy provides some doodling space to see how color combinations might go together. Always have on hand fresh markers for the colors used to cover the largest areas of the drawings. Markers in their final death throes always fail halfway through a large area, and no amount of handiwork can fix the resulting streaked lines and

FIGURE 2-57

Light to dark color gradation is possible with one marker if each layer of color is allowed to dry before succeeding layers are applied (a); multiple colors create a stronger gradation, but must be selected to complement one another in the composition (b). *(Steven Rodie)*

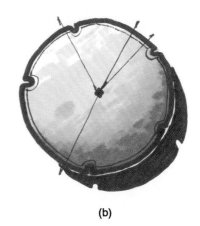

(a)   (b)

inconsistent color. The authors are well aware that marker problems most often occur after the copy shop and art store close and immediately before a major deadline.

Principles for Using Color Effectively

The key to effective use of color is the same for all other graphic techniques: experiment and practice! There are many variables in color application, including marker or pencil brand, marker wetness or dryness, type of paper, speed of drawing, and the designer's skill level. The best way to develop confidence in applying color is to establish some basic skills, know what standards to use as a benchmark, and practice.

The following guidelines summarize basic color application.

- Create a swatch library of different colors on commonly used paper types. Use this as a reference for color comparisons and effects (Figure 2-58).
- Use color to highlight a particular area rather than applying color to the entire drawing. Leaving large areas around the perimeter of the drawing mostly unrendered focuses the color on key landscape areas and saves time (Figure 2-59).
- Apply dark colors over light colors where overlaps occur.
- Apply color to large areas first. If a mistake is made in a large area, it is usually more noticeable and difficult to correct; the designer can start over without wasting efforts in other completed areas.
- Strive to create a contrast between the ground plane (usually light tone) and the plants and trees that would have height in the real landscape (darker or bolder tones).
- Combine marker and colored pencil to take advantage of the speed of marker application and the detail and luminescence of pencil. Apply marker first; then add pencil to enhance texture, emphasize details, and produce highlights (Figure 2-60). Applying pencil first will seal the paper, and the marker ink will not soak into the paper correctly.

FIGURE 2-58

A simple color chart with named marker applications provides a quick check of expected marker colors on a particular paper type. *(Courtesy of Karen Richards. Steven Rodie)*

FIGURE 2-59

Significant areas of this drawing have been left uncolored, which helps to focus attention on the garden design and saves considerable rendering time. *(Steven Rodie)*

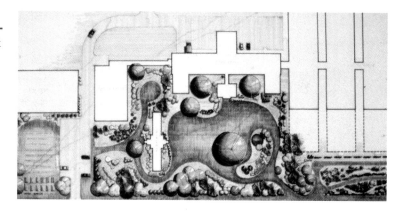

- Colored pencil is slow to apply but is easier to control than marker. It can also be layered to produce additional color depth (Figure 2-61). Colored pencil is less bold than marker and is not as vibrant and dramatic when viewed from a distance (as in a design presentation) or when photographed (Figure 2-62, page 82).
- Apply all color in structured patterns to avoid smudged or scribbled patterns. Straight, parallel lines for large open areas work best (Figure 2-63, page 83).

FIGURE 2-60

Marker alone provides bold colors and contrasts (a); colored pencil layered over marker creates highlights, depth, and subtle gradation not possible with marker (b). *(Steven Rodie)*

(a) Marker alone

(b) Colored pencil layered over marker

FIGURE 2-61

Light applications of colored pencil can be layered to create depth and a wide range of subtle color contrasts. *(Courtesy of Sanford Neuharth, Steven Rodie)*

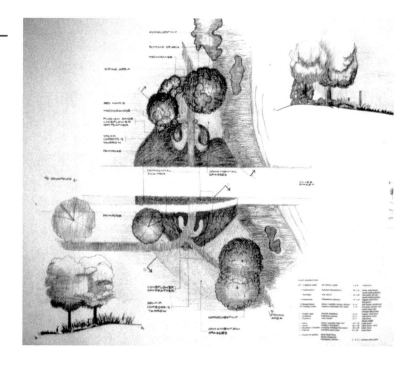

- Developing a light-to-dark value change, combined with shadows, is the classic technique to develop depth in a plan view drawing (Figure 2-64).
- Symbols that have layered or blended color take on a three-dimensional feel (Figure 2-65, page 84).
- A subtle technique to add more contrast is to leave some white space in the drawing. This technique can also be used to represent sunlight on plant and paved surfaces (Figure 2-66, page 84).
- Creative color palettes can add artistic value to a landscape drawing but can also be more difficult to interpret. When in doubt, use a realistic palette, such as shades of green for landscape plants, grays for concrete, and browns for mulch.

Applications

Designers can apply color to drawings at any stage of the design process. Site analysis and functional diagrams normally include a variety of bubbles and open-structured symbols that can be quickly filled with color to enhance legibility (Figure 2-67, page 85). Final design presentation drawings are the most common drawings that receive color and generally require the greatest amount of time and detail (Figure 2-68, page 86). In most cases, copies of drawings, rather than originals, will be colored.

Sketch drawings can also be enhanced with color to quickly add additional clarity and visual interest. On tracing paper, marker can be added to the back of the sketch (to avoid smudging ink or pencil lines), and colored pencil can be added to the front, creating a temporary drawing with a more polished look (Figure 2-69, page 87).

FIGURE 2-62

A landscape design rendered with colored pencil (a) and marker (b) illustrates the contrast in final drawing quality. *(Courtesy of Alyssa Stapp, Steven Rodie)*

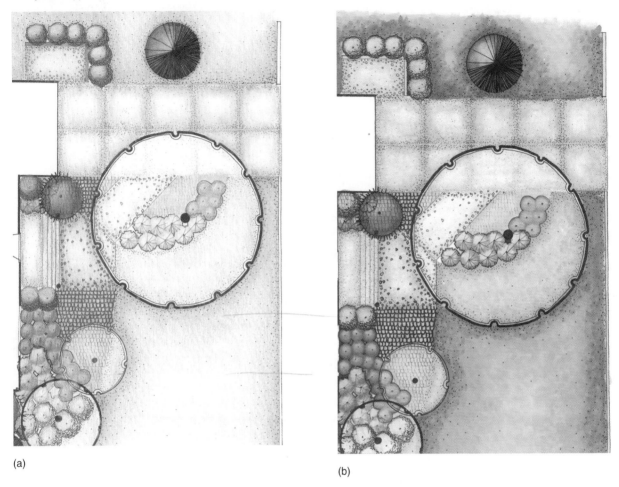

(a) (b)

FREEHAND DRAWING AND SKETCHING

Freehand drawing is a skill that every designer should develop. Regardless of how talented many designers have become with computers, freehand drawing is a skill worthy of learning. This is especially true for residential design, where many smaller companies have not made the significant investment for design software and staff training.

Designers who sketch well quickly draw what they carefully observe (shapes, value contrast, textures, etc.) without overanalyzing. Sketching sometimes requires a conscious effort not to think too much; this intuitive technique is hard to learn but easy to practice. The most basic premise in developing sketching and drawing ability is being able to see what you are looking at. As simplistic as it sounds, it is the most critical element in drawing success.

FIGURE 2-63

Marker (a) should typically be applied in a consistent pattern (especially over large areas), allowing for color spread depending on paper type. Pencil (b) can be applied in linear patterns, or lightly blended to create a gradation. *(Steven Rodie)*

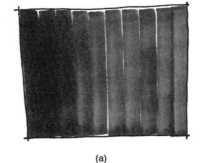
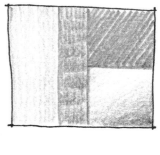

(a)   (b)

FIGURE 2-64

A gray-scale version of a color rendering illustrates the light (ground plane) to dark (tall trees) value hierarchy that creates depth in a color drawing. *(Steven Rodie)*

Art does not reproduce what we see; rather, it makes us see.

- Paul Klee, Swiss Expressionist Painter

The Right Frame of Mind for Accurate Sketching

Developing sketching ability goes hand in hand with the ability to carefully observe character and detail in any object or setting. Designers who sketch well have learned how to successfully use their right-brain mode of thinking (Edwards, 1999). In essence, they sketch what they have actually observed (shapes, value contrast, textures, etc.) rather than what the analytical left side of their brain has told them they have observed. For example, left-brain reasoning typically places the eyes two thirds of the way from the chin to the top of the head. Right-brain thinking allows us to see and draw the true facial proportion of 95 percent of human faces—the eyes are half way between the chin and top of the head (Figure 2-70, page 87).

Artists are generally born with right-brain assessment capability. With practice and application, all designers can effectively assess, draw, and design from a right-brain approach.

FIGURE 2-65

Color layering in plant symbols, together with textures and shadows, creates a sense of three-dimensional height for each symbol. *(Courtesy of Lindsay Hanzlik. Steven Rodie)*

FIGURE 2-66

The streaks of white space in the turf area and along symbol edges create highlights and contrast. Marker bleeds and spreads on paper, so white space must be planned for. *(Courtesy of Julie Catlin. Steven Rodie)*

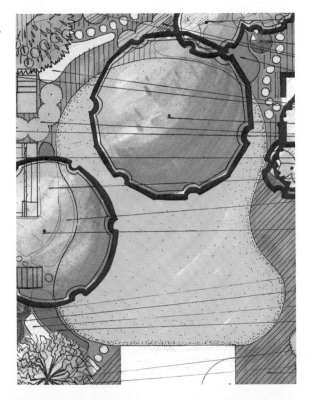

FIGURE 2-67 Color was applied to the back of the symbols on this tracing paper site analysis overlay. *(Steven Rodie)*

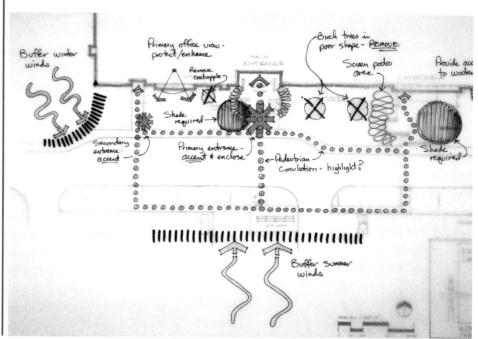

 DRAFTING

Drafting is best used when drawing accuracy is required for defined straight lines, accurate angles, and circular circles. Due to the precise and accurate nature of drafted drawings, it is critical to maintain graphic quality. All of the graphic quality rules apply to drafting, especially line quality and line width. Drawing against a straightedge allows downward pressure on the pencil or pen that should result in consistent dark lines with defined endpoints. At corners, line ends should have a slight overlap. Holding the pen or mechanical pencil perpendicular to the paper will produce consistent widths. To minimize pencil smudging, avoid using soft leads, except when required for wide, dark lines; and apply soft pencil toward the end of the drawing process.

 LETTERING

Written text (lettering) expands the meaning of symbols and adds additional information not represented in typical graphic layouts. Lettering also conveys basic information, such as the names of clients and designers, dates of completion, and project addresses. Because lettering consistency affects drawing quality and text legibility, standard lettering practices are used, and the result is the classic lettering style evident in the design professions. Hand lettering adds a significant amount of design style to a drawing, and when done well, it reflects an attention to detail and professionalism.

There are several alternatives to supplement or replace hand lettering, including using a typeface that looks like lettering in a word-processing program for

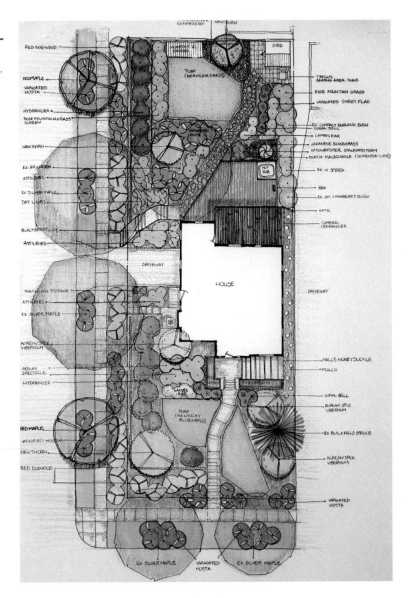

FIGURE 2-68

Final presentation drawing includes a combination of marker and colored pencil rendering. *(Courtesy of Cynthia Corrigan. Steven Rodie)*

information attached to the drawing, adding printed text to a drawing using dry transfer letters, or using an adhesive lettering device such as Kroy Lettering Systems.

Often, designers use a combination of hand lettering and printed text to convey design ideas. Less formal drawings (site analyses, functional diagrams) require quick notes and written documentation, which are done efficiently using informal hand lettering. In contrast, master plans and section elevations normally contain blocks of typed information (plant lists, design concepts), while individual plant labels and other smaller blocks of information are hand lettered.

Criteria for All Lettering Styles

Quality lettering should be legible and consistent. Creative interpretations of letter shapes can help establish an overall drawing style, but too much embellishment

FIGURE 2-69

Marker was applied to the back of this tracing paper design, which eliminates ink line smudging and creates softer colors. Colored pencil was applied on the front, which creates light contrasts and enhances color gradations. *(Steven Rodie)*

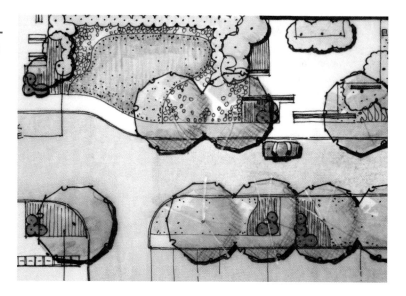

FIGURE 2-70

Comparing left-brain and right-brain views of a typical human face illustrates the difference in perceived relationships of facial characteristics. Accurate observation leads to right brain visualization and sketching ability. *(Steven Rodie)*

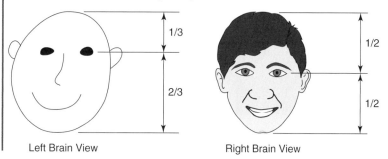

is distracting and potentially unreadable. Letters should be consistent in size and spacing, dark enough to copy well, and parallel to the edge of the sheet. Suggestions for developing good lettering skills (Figure 2-71) include the following:

- Use horizontal guidelines to establish letter heights; tools such as the Ames® Lettering Guide can speed up the guideline drawing process. Vellum with preprinted grid lines also provides lettering guidelines, but sizes beyond the grid pattern may be required.

- The most common lettering height is $\frac{1}{8}''$; it is large enough to read, small enough to not overwhelm drawing graphics, and a comfortable size to produce.

FIGURE 2-71

A composite lettering example illustrates the basic characteristics of consistent quality lettering. *(Courtesy of Karen Richards. Steven Rodie)*

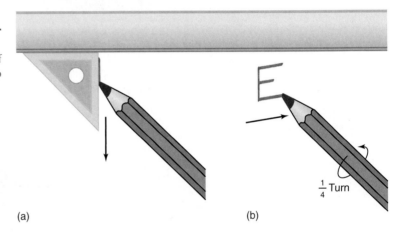

PLANT SCHEDULE

QTY	BOTANICAL NAME	COMMON NAME	SIZE	COND.
3	PINUS STROBUS	EASTERN WHITE PINE	5'–6'	B & B
5	PYRUS CALLERYANA	BRADFORD PEAR	1-½"	B & B
1	QUERCUS PALUSTRIS	PIN OAK	2"	B & B
3	QUERCUS RUBRA	RED OAK	2"	B & B
12	RHUS AROMATICA	FRAGRANT SUMAC	5 GAL	CONT.
5	CORNUS SERICEA	REDTWIG DOGWOOD	5 GAL	CONT.
3	TAXUS CUSPIDATA	JAPANESE YEW	15"–18"	B & B

GENERAL SPECIFICATIONS

1. PLANTS SHALL BE NURSERY GROWN, AND SHALL MEET THE REQUIREMENTS OF AMERICAN STANDARDS FOR NURSERY STOCK ADOPTED BY THE AMERICAN ASSOCIATION OF NURSERYMEN.

2. PROPERLY IDENTIFY EACH SEPARATE PLANT.

3. ALL NURSERY STOCK SHALL BE WATERED UPON DELIVERY AT SITE, AND SHALL BE PROTECTED WITH SHADE CLOTH FROM WINDS AND FULL SUN UNTIL INSTALLED.

FIGURE 2-72

The keys to quality lettering are using a guide for all vertical lines drawn with the edge of a chiseled pencil lead (a) and rolling the pencil to use the broad chisel for consistently wide horizontal lines (b). *(Steven Rodie)*

(a) (b)

Other sizes should be practiced and developed. The larger the lettering, the more difficult it becomes to maintain consistency.

- Always use a vertical guide. A small triangle is used above or below the T-square or parallel bar to align vertical lines. Although awkward initially, after this method is practiced it is quick and efficient.
- A pencil or mechanical pencil with a chiseled point and lead in the HB to 2H range produces dark lines with less smudging.
- Draw vertical lines downward with the chiseled edge against the guide. Alternating back and forth between the edge and face of the chisel produces the strong line weight contrast while the edge and face of the chisel are automatically maintained (Figure 2-72).
- Vertical and horizontal lines should be dark and legible, and they should have defined ends.
- You need to practice; good quality lettering can be produced as long as lettering is consistently inconsistent.

SHEET MECHANICS AND LAYOUT

At each stage of landscape design, information must be legibly and consistently arranged into a particular drawing. The layout of this information on individual sheets and as a combined package affects the quality of the design.

General Layout

Landscape drawings should follow a general format similar to the layout of a book page. The binding edge is to the left, and important reference information (such as titles, sheet numbers, etc.) is located along the bottom or right edge of the sheet. For drawings that are part of a multiple-sheet design package, this is especially important (Figure 2-73). One-of-a-kind drawings can be formatted with greater flexibility. Sheets can be arranged with a portrait or landscape format; most multiple-sheet sets have a landscape format (Figure 2-74).

White space between elements and around the edges of the sheet is an important visual factor in **sheet layout**. There should be a blank margin of at least $\frac{1}{2}''$ around the edges of the sheet and enough space between sections of sheet information that the layout appears balanced and not overcrowded (Figure 2-75). Actual arrangement of information will depend on the scale of the drawing, the size of the sheet, and the preferred orientation of the property information.

Title Block

The title block contains all of the pertinent sheet information, including the sheet title, client name and property address, name of designer, drawing date, and sheet number (if a multiple set). Additionally, the designer may leave room for revision notes, signatures, the drawing scale, and north arrow. The title block is generally located in the lower right-hand corner of the sheet but may also be located along the bottom or right edge of the sheet (Figure 2-76). A hierarchy of text sizes is used to indicate the sheet name and client, followed by the designer and additional information. Text can be informally hand lettered or produced with computer software. Borders, logos, and additional graphics add a professional touch to the title block (Figure 2-77, page 92).

North Arrow and Drawing Scale

The designer should include a north arrow on every sheet that includes a plan view graphic. A drawing scale should be included for every sheet that includes a plan or section-elevation graphic. They can be included in the title block or placed together relatively close to the graphic for clearer reference. Similar graphic styles are often used for the arrow and scale to visually tie them together, or they can be physically combined into one graphic (Figure 2-78, page 93).

Like maps, most sheets are formatted for a north orientation that points upward on the drawing. However, a client who faces south to view his or her house from the street may be more comfortable with that same view orientation on the drawings, which places south on the top of the sheet. Regardless of how the drawing is oriented, be sure to confirm that north is correct.

FIGURE 2-73

Sheet layout should emulate bound book pages, locating important information along the bottom or right page edges for easy reference (a). Multiple sheet sets should remain consistent in this format (b). *(Steven Rodie)*

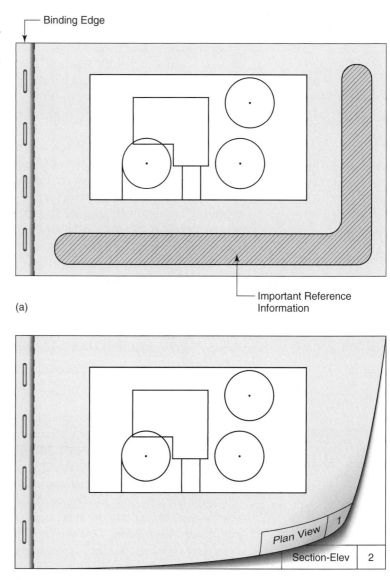

The Smudgier, the Better

Many designers consider tracing paper and a soft pencil to be the ultimate tools for initial design development. The thick, rounded lead makes it difficult for the designer to be precise about lines, and the tracing paper helps create a sense of noncommitment to the lines being drawn so the designer can stay open to new ideas. Computer drafting has changed the way many designers use tracing paper and pencil. At one time, wearing graphite smudges on arms and hands was a sign that a designer was taking the time to creatively explore a variety of design alternatives. Such marks are not evident when designing in front of the cool glow of an 18-inch flat panel monitor. It is not for everyone, but if you are not feeling loose in your design process, you might try the old standbys of tracing paper and a soft, thick pencil.

Landscape Graphics 91

FIGURE 2-74

Select a drawing orientation that best suits information formatting; sheet sets are typically produced in a landscape format. *(Steven Rodie)*

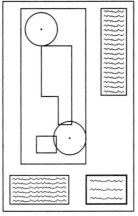
Portrait

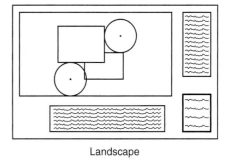
Landscape

FIGURE 2-75

White space around all sheet elements adds to sheet legibility and composition. *(Steven Rodie)*

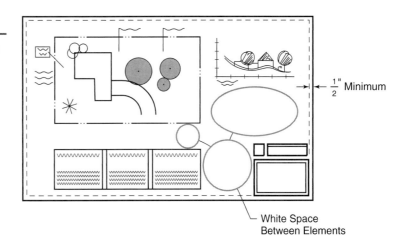

FIGURE 2-76

Title blocks are generally located along the bottom or right edge of the sheet. *(Steven Rodie)*

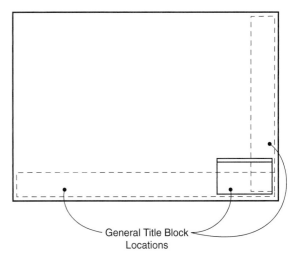

FIGURE 2-77

Title block content and layout should follow basic graphic guidelines. Designs can be simple (a) or more complex (b and c). *(2-77b courtesy of Julie Catlin and 2-77c courtesy of Julie Catlin and Karen Richards. Steven Rodie)*

CONCEPTUAL LANDSCAPE PLAN
LUTHERAN FAMILY SERVICES OF NEBRASKA, INC
124 S. 24TH ST, SUITE 230 OMAHA, NE
AUGUST 27, 2004

PLAN COMPILED & DEVELOPED BY:
STEVEN RODIE, ASLA

(a)

CATLIN'S CUSTOM LANDSCAPE

MASTER PLAN
DESIGNED FOR: FIENHOLD RESIDENCE
2812 FRANKLIN AVE
COUNCIL BLUFFS, IA
DESIGNED BY: JULIE CATLIN - LANDSCAPE DESIGNER
1426 S. 134TH STREET OMAHA NE
PHONE: (402)201-4440
MAY 3, 2005

(b)

Base Map

Norby Residence
5801 Deerhaven Drive,
Lincoln, Nebraska
Prepared by: Karen Richards & Julie Catlin
Date: 09-26-05

(c)

FIGURE 2-78

North arrow and scale examples include separate (a) and combined features (b). *(Steven Rodie)*

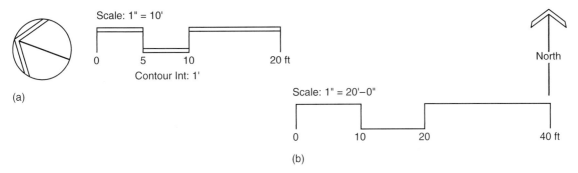

FIGURE 2-79

Blocked lettering and consistent non-overlapping leader lines reduce drawing clutter and enhance legibility. *(Courtesy of Karen Richards. Steven Rodie)*

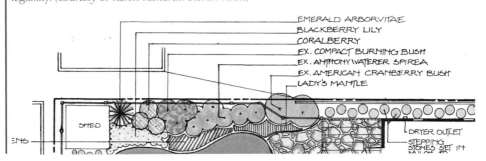

Additional Information and Formatting

Additional information that must be incorporated into various landscape drawings might include the following: keys or legends required to enhance graphic clarity, plant lists, and general notes. All of these items should be located near the title block so they are easy to find and reference. Whenever possible, lettering should be blocked into logical groupings to provide a more structured and legible format. Plant labels should also be blocked and directly connected to plant symbols with clear leader lines (Figure 2-79). Whenever possible, plant names (not symbols) should be used in plant labels. While direct labeling eliminates potential confusion, sometimes it is not possible due to the number of plants or lack of space.

GRAPHICS APPLICATIONS FOR SPECIFIC DRAWING TYPES

The complete landscape design process requires the designer to work on at least five different types of drawings or sheets: base map, site inventory and analysis, functional diagram, preliminary design, and final design. General graphics considerations for each landscape drawing are summarized in the following sections.

Base Map

Base maps are normally drafted on vellum or Mylar for accuracy and longevity. The size of the property, layout of the house, and other information must be correct from the start of the design process, or errors can quickly mount. Drafted lines will best ensure the accuracy required for this sheet. For most projects, this sheet serves as the bottom layer for the design process. Once the design is layered, processed, and finalized on tracing paper, it is transferred to this sheet for the final design drawing.

Prior to drafting the base map, thought should be given to selecting an appropriate scale for the plan and sizing the sheet size for the property. The typical drawing scales of $\frac{1}{8}'' = 1'0''$ or $1'' = 10'$ will generally fit an average-sized residential lot on an $18'' \times 24''$ sheet, with room to spare. The base map should include a title block, drawing scale, north arrow, line border (if desired), and additional space for notes and plant information. If the information will not format correctly, a smaller scale ($\frac{1}{16}'' = 1'0''$ or $1'' = 20'$), a larger sheet ($24'' \times 36''$), or both can be used to provide additional space.

Site Inventory and Analysis

Designers typically draw site inventory and analysis information in an informal, freehand style with pencil or pen. Graphic symbols include a variety of analysis and concept graphics. Since it is primarily used by the designer to summarize site conditions and preliminary design decisions, it is normally drawn on a tracing paper overlay of the base map (Figure 2-80). This saves time and produces the first sheet of the design overlay process. The site analysis forms the bottom layer of the process, so the use of bold graphics keeps critical site information visible as the next layers are developed. Color can help clarify information and create a more presentable drawing if the information is shown to the client. The inventory and analysis can be presented on separate sheets if the site conditions are complex, but they are normally combined on one sheet. Designers should consider using the sheet as a note pad to summarize all important site information and annotating drawing graphics wherever necessary to ensure that important information is included (Figure 2-81).

Functional Diagram

Functional diagrams represent the first stage of creative design development and should reflect a loose, freehand graphic style (Figure 2-82). Graphics may also include thumbnail sketches or quick section elevations, whatever best conveys design ideas at this early stage of development. It is preferable to draw this information on a sheet of tracing paper or vellum so that it can be layered over the site analysis (Figure 2-83). Keeping this drawing stage informal allows the designer to quickly develop design alternatives that can be explored and presented to the client without a significant time commitment. Clients also tend to be more open in their critique when they know that the drawings are in process and appear temporary rather than refined and complete (Figure 2-84).

Preliminary Design

Preliminary design drawings incorporate a variety of graphic symbols and techniques. Much of the drawing remains freehand, but drafted lines can be used to refine

FIGURE 2-80

A typical site analysis drawn on a tracing paper overlay. *(Courtesy of Lindsay Hanzlik. Steven Rodie)*

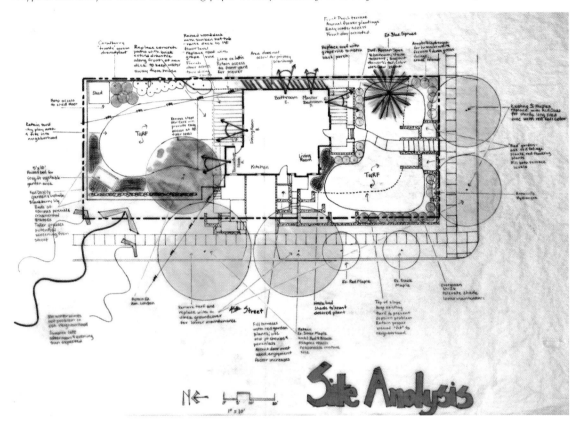

FIGURE 2-81

Annotating the site analysis with notes and observations provides important information for future design ideas. *(Courtesy of Lindsay Hanzlik. Steven Rodie)*

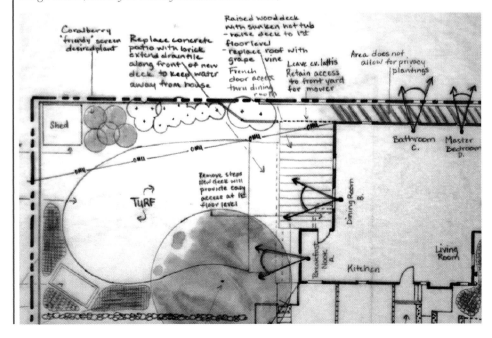

FIGURE 2-82

A section of a typical functional diagram showing the freehand graphic style. *(Steven Rodie)*

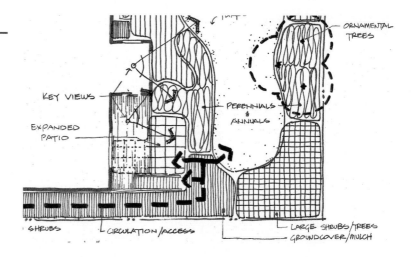

FIGURE 2-83

The functional diagram overlay should develop from the site analysis overlay information. *(Steven Rodie)*

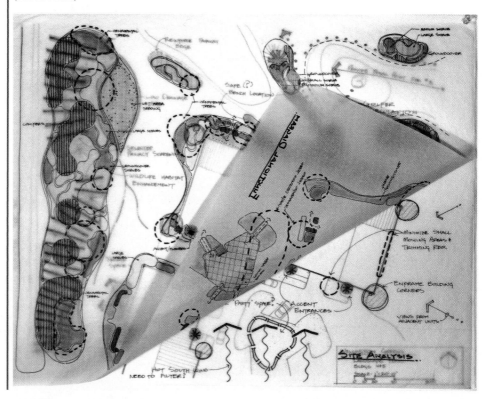

building edges, walkways and streets, and other straight landscape elements. Designers combine standard plant and material symbols with freehand bubbles and shapes representing undifferentiated plant masses and planting beds. Since client review of the preliminary design is typical, the graphics should strike a balance between presentation quality and efficiency because some changes may still occur prior to final design (Figure 2-85).

FIGURE 2-84

A functional diagram should clearly illustrate the basic design elements in a format that reflects the "in-process" nature of the design. *(Steven Rodie)*

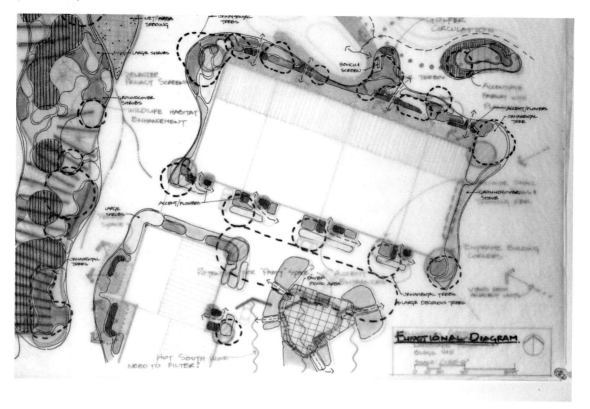

Final Design

Final design graphics are typically presentation quality. Graphics have been drafted or freehand traced from drafted lines (maybe even a combination of the two); quality lettering has been added; symbols and textures have been finalized, and all additional information has been added to sheets prior to final copies being made. Color-rendering should be completed on clean drawing copies so that originals can be kept and copied in the future (Figure 2-86).

DRAWING SETS. Whenever multiple sheets are required for a design project, they should appear consistent in their format and style. Using the same title block (with sheet number and title changed) for each drawing in the set, together with consistent borders and binding edges, enhances the packaged look for a multiple-sheet project (Figure 2-87).

FIGURE 2-85 A preliminary design drawing reflects accurate locations and sizes for plant materials and bedlines. Preliminary design also defines plant characteristics and roles in the landscape, and may include selected plant species. *(Courtesy of Karen Richards. Steven Rodie)*

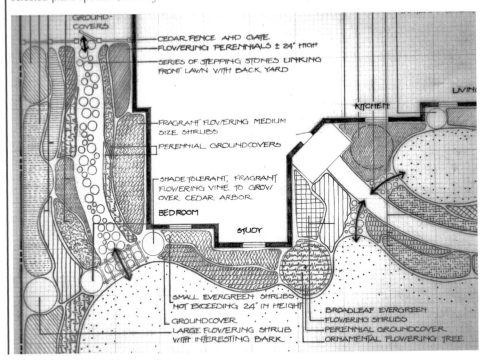

FIGURE 2-86

Final design graphically incorporates all pertinent information; color rendered and black and white copies are produced for the client while the original drawings remain with the designer. *(Courtesy of Karen Richards. Steven Rodie)*

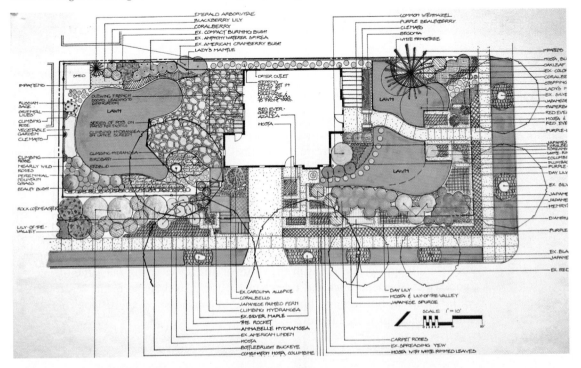

FIGURE 2-87 Multiple sheet sets (a and b) should appear as a package, reflecting similar layout, title block, and orientation characteristics. *(Courtesy of Casey Hughes. Steven Rodie)*

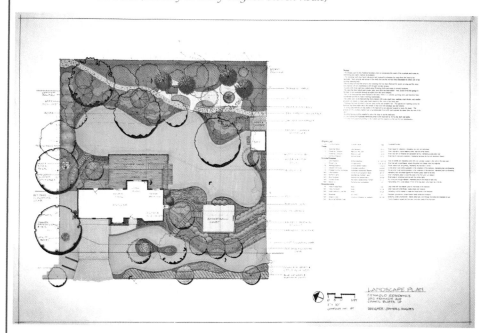

(a)

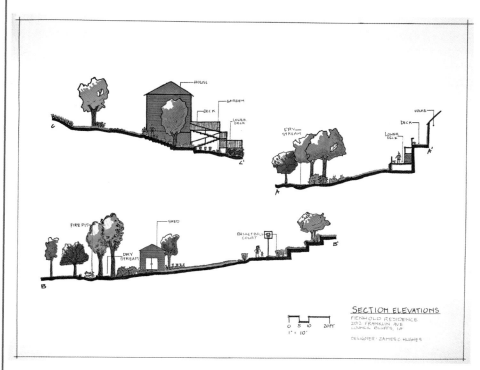

(b)

Is It OK to Borrow Someone Else's Graphics and Style?

One of the best ways to learn how to draw well is to borrow graphic symbols and ideas from other designers. There is not such a thing as copyright infringement when it comes to an effective, quality symbol for an evergreen tree or a high-contrast groundcover texture that takes half the time to draw as other textures. Literally tracing quality line work helps develop a sense of the differences in line weight and the variability in patterns that define graphic composition. Through time, every designer will develop a unique, personal style, regardless of how much he or she borrowed.

If you borrow graphics, make sure they are quality graphics, and make sure you like them and are willing to practice them. Ultimately, you should attempt to make them better. The designer who borrows from you will then have an even better graphic to build from.

Pablo Picasso realized how artistic skill and ideas are developed when he said, "Bad artists copy. Good artists steal."

COMPUTER AND CAD GRAPHICS

As summarized in Chapter 1, computer-aided drafting and design (CAD or CADD) software provides a wealth of graphic tools for use by landscape designers. Benefits are numerous, including increased production efficiencies, exciting three-dimensional and photographic products, and the ability to quickly edit and revise graphics for evolving design alternatives. In addition, however, CAD also has limitations of which designers need to be aware.

Benefits of Computer-Aided Design

One of the most important graphics benefits in CAD use is the increased efficiency of repetitive drawing tasks and base map production. The drafting requirements for base maps are a good fit for CAD. Direct use of numerical site measurements can automatically produce accurate property line layouts and house foundations without manual scaling. Information can be drawn on electronic layers and made visible or hidden. Text and drawings can also be moved as blocks; site information, title block, notes, or plant lists can all be arranged to best fit the sheet size. Title block templates can be revised and reused, and north arrow and scale graphics can be quickly copied into every drawing.

Photo-realistic simulations provide dramatic before-and-after comparisons of landscape ideas (Figure 2-88). Depending on complexity and detail, they take comparatively little time to develop. Relatively sophisticated software packages can assign three-dimensional attributes to every two-dimensional element in a drawing and automatically create three-dimensional images from various landscape views.

FIGURE 2-88

A photo of an existing landscape (a) and a computer rendering (b) showing a landscape concept. *(Bryan Kinghorn)*

(a)

(b)

Computer Graphics Trade-Offs

When computer-aided design is used in place of hand drawing, it is important to consider possible pitfalls and trade-offs. These considerations depend on the type of software and the designer's skills and expectations.

Designing on a computer often reflects a different process and context than designing with pencil and paper. Drawing context can be difficult to establish on

Acknowledging Efficiencies

Some advertisements for landscape designers promise a beautiful landscape design with before and after photographs and a detailed estimate delivered via laptop computer all in one site visit. Designers also advertise complete design production using e-mailed photos. Do these approaches seem realistic? It all depends.

Experienced designers with photo imaging skills working on relatively small projects may be able to deliver reasonable quality. But any designer who designs without a site visit cannot have the same valuable information gained from actually viewing the property. Designs that require minutes instead of hours, regardless of computer efficiencies, must compromise something in the process. Reality often dictates a quick turnaround for design, and some clients are only interested in achieving a landscape look rather than a complete design. The designer and the client must acknowledge the limitations of whatever process is used. Landscape design services are much like many other professional services: You get what you pay for.

the computer because the entire drawing may not be in view. Zooming in and out provides the close-up scale necessary to incorporate details, but context can be easily lost.

Computer capabilities for sketching and shaping curved lines continue to improve dramatically, but some designers may still get a better feel for these when they are done by hand with pencil and paper. Using a stylus rather than a mouse enhances the sense of drawing, but many drawing commands are still input with a keyboard and mouse.

It is not uncommon for users of computer graphics packages to overindulge in available graphic choices. Color-rendering can lack the three-dimensional depth and color gradations that simple color marker and pencil can deliver. Software also makes it easy to use too much color. Some CAD packages have hundreds of symbols available to represent plants, groundcover textures, paved surfaces, etc. Using a different symbol for every species of plant and ground plane surface can make a drawing appear too busy and difficult to interpret.

Adding proposed landscape elements to a house photo is simple in concept but is not always successful. Photo manipulation is as much of an art form as photography itself. Some software packages allow the user to digitally enhance an image; and, if done correctly, the resulting image may show a true representation of the designer's ideas. But a poorly done result can be unnatural looking or even misleading. Realistically enhanced landscape photographs tend to create an expectation level that is more defined than the level generated through an artist's sketch of the same landscape. Designers who use before and after images should be sensitive to potentially increased expectations (Sheppard, 2001).

CAD Graphics Summary

Computer-generated graphics represent a significant and growing portion of current drawing production in the landscape design industry. Current software can effectively address virtually all of the quality graphic standards and products discussed in this chapter, including:

- Line weights and line types can be specifically selected for appropriate drawing uses and line hierarchy.
- Line and dot quality is ensured, as are proper joining of lines and development of shapes and specific line patterns.
- Choices for developing value contrasts and textures that reflect materials and ground surfaces are almost limitless.
- Color palettes can be specifically selected; color shading, gradation, and transparency and opacity can be adjusted to create drawing interest and depth.
- Lettering fonts, sizes, and hierarchies can be creatively applied to drawings.
- Hundreds of plant symbols are available, and customized symbols can be added to existing databases.
- Sheet layout and sizing can be quickly completed; and title blocks, scales, north arrows, etc. can be created and pasted into drawing files for efficient base map production.
- Basic plan view information can be assigned heights so that accurate three-dimensional modeling of landscape areas can be produced.
- Software is available for drafted and sketched drawings, depending upon the accuracy and drawing quality desired.
- Software is also available to produce drawing elements (lines, symbols, lettering, etc.) in a drafting mode (for accuracy) that appear to be hand drawn (for artistic enhancement).

Regardless of software capabilities, designers should consider even the most advanced CAD equipment as a drawing and production tool that requires thoughtful, practiced use if quality graphics are expected in final drawing production.

OTHER GRAPHICS APPLICATIONS AND CONSIDERATIONS

Several additional issues are worth noting in a discussion about landscape graphics. Landscape photography and the ability to electronically manipulate images should be considered as valuable a skill as landscape sketching. One example is to use site images as backgrounds for conceptual design sketches. Other times, a designer may design directly on a photographic print (or copy) using tracing paper (Figure 2-89). As noted previously, this technique can also be performed with photo-imaging software on a digital picture.

Finally, every designer should strive to maintain a current personal portfolio. Although portfolio formats and content will vary with personality and career, including a variety of projects will show the depth and breadth of the designer's experience. Electronic formats and tools, many of which are described in Chapter 1, allow designers to store and share their portfolios digitally. With current technology, it is interesting to note that an entire career of design work can be captured onto a single digital storage device.

FIGURE 2-89

A photo of an existing landscape (a) and an informal tracing paper overlay (b) showing relative plant sizes during the growing season and other design enhancements. *(Steven Rodie)*

(a)

(b)

SUMMARY

Quality graphics are at the core of the entire design process. They improve communications between designers, expand client awareness of design products, and serve as a tool for designers to better assess design solutions. Graphics should evoke professionalism and enhance the process as well. Creating quality graphics should not be a burden. In fact, graphic quality and efficiency go together and when properly executed require no extra effort.

KNOWLEDGE APPLICATION

1. Describe a minimum of six important reasons to produce quality graphics in landscape design.
2. Describe the quality characteristics for both black-and-white and color landscape graphics. Then review three examples of each graphic type, and rate the quality of all six examples.
3. Define the character differences between the two basic plant symbol categories. Using this information, identify and practice a minimum of a dozen symbols reflecting each category. Vary the style from simple to complex, and practice both copied symbols as well as your own symbols.
4. Describe proper sheet layout, and develop several distinct layout variations that all follow the layout guidelines.
5. Describe the similarities, differences, and basic applications of engineer's and architect's design scales.
6. Describe the characteristics of quality lettering. Review samples of technical lettering, and judge the quality using the stated characteristics.
7. Describe the basic procedures that ensure quality color graphics.
8. Describe the basic differences between freehand graphics and hard-line graphics, and explain how and when each type is used.
9. Define the benefits and constraints of CAD and computer software graphics tools when compared to traditional methods of hand drawing.

LITERATURE CITED

Edwards, B. (1999). *The new drawing on the right side of the brain.* New York: Penguin Putnam.

Sheppard, S. (2001). Guidance for crystal ball gazers: Developing a code of ethics for landscape visualization. *Landscape and Urban Planning, 54*(25), 173–188.

REFERENCES

Bertauski, T. (2006). *Plan graphics for the landscape designer* (2nd ed.). New York: Prentice Hall.

Doyle, M. (1999). *Color drawing: Design drawing skills and techniques for architects, landscape architects, and interior designers* (2nd ed.). New York: John Wiley and Sons, Inc.

Lin. M. W. (1993). *Drawing and designing with confidence: A step-by-step guide.* New York: John Wiley and Sons, Inc.

Reid, G. (2002). *Landscape graphics: Plan, section, and perspective drawing landscape spaces.* New York: Watson-Guptill Publications.

CHAPTER 3

Defining Landscape Design and the Foundations That Shape the Design Process

It is not easy to define landscape design. It is a complex concept that includes problem solving, application of universal design principles, integration of human landscape preference, and visual and oral communication; and it ultimately results in a well-designed landscape. Defining landscape design is occasionally a nebulous exercise because the successful application of these processes, theories, and principles will vary with every project and every designer. Ultimately, a quality landscape can be defined as a product where appropriate design solutions are reached.

OBJECTIVES

Upon completion of this chapter you should be able to

1. define landscape design.
2. describe some basic misconceptions about landscape design.
3. define the universal design process.
4. describe the benefits of applying the landscape design process.
5. describe how the landscape design process supports quality design.
6. describe the theories that help explain human landscape preference and provide basic examples of each theory.

KEY TERMS

design process:
  accept situation
  analyze
  define desired
    outcomes
  generate ideas
  select
  implement
  evaluate
Prospect-Refuge Theory

landscape preference:
  coherence
  complexity
  legibility
  mystery

restorative characteristics:
  fascination
  escape
  extent
  compatibility

This chapter will discuss the complex subject by providing a definition of landscape design, discussing common misconceptions about landscape design, outlining the design process and benefits, and describing the relationships of people and the outdoor environment.

A LANDSCAPE DESIGN DEFINITION

The theoretical foundation for landscape design combines a universal problem-solving process with human landscape preferences and the restorative qualities of nature. The landscape design definition used in this textbook connects this foundation to the common factors in residential landscape settings.

> *Landscape design is the art and science of organizing and enriching outdoor space through the placement of plants and structures in agreeable and useful relationships with the natural environment.* (Adapted from the Nebraska Master Gardener Handbook, *1994*)

Aesthetic and Functional Components of Design

aesthetic(s): Characterized by a heightened sense of beauty; the study of beauty and the human psychological responses to it.

A designer must address **aesthetics** and function simultaneously; one cannot exist without the other in quality design. Aesthetic considerations include the application of design principles such as order, unity, rhythm, and balance to the landscape. Functional considerations include addressing maintenance requirements and creating comfortable outdoor living spaces with effective circulation patterns. Functional considerations also include problem-solving related to issues like challenging growing conditions and drainage issues.

Additional Components That Define Landscape Design

The human backside is a dimension architects seem to have forgotten.
- William Whyte

In addition to the broad categories of aesthetic and functional components, landscape design is further defined by other essential factors. These factors include the following:

LANDSCAPE DESIGN BLENDS ART AND SCIENCE. Design training should include the study of art as well as the environmental, physical, and biological sciences. Emulating a prairie or forest canopy in a landscape design requires an understanding of natural environments. Calculating the pressure requirements in an irrigation line or properly designing a retaining wall requires knowledge of the physical environment. Finally, ensuring that trees are handled, planted, and maintained properly is a key role of the designer that requires biological information.

OUTDOOR SPACE IS A FOCUS OF LANDSCAPE DESIGN. Well-defined landscape space creates quality living areas that are preferred in their character and useful in their layout, and that provide high levels of human restoration (Figure 3-1).

hardscape: The structural elements of a landscape composed of nonliving materials, such as paving, decks, walls, fences, arbors, and pergolas.

LANDSCAPE DESIGN COMBINES PLANTS WITH STRUCTURAL COMPONENTS. Landscape design is not limited to plant materials. **Hardscapes** (walls, patios, fences, etc.) that complement plants are critical to the success of a design.

FIGURE 3-1

This landscape contains several garden rooms that vary in ceiling heights and sense of enclosure, a variety of interesting views, easy access to all areas, and numerous choices in sun/shade and wind exposure that meet the preferences of most every visitor. *(Steven Rodie)*

FIGURE 3-2

The walls in this landscape setting are a critical structural feature as well as an important color, texture, and plant materials backdrop. *(Steven Rodie)*

Hardscapes help to define and personalize outdoor rooms and make the space more useable (Figure 3-2).

microclimate: The climatic conditions (wind, solar exposure, precipitation, humidity, temperature, etc.) within a defined landscape area that are unique to that space.

THE NATURAL ENVIRONMENT IS THE CONTEXT FOR LANDSCAPE DESIGN. Designers must consider drainage, solar exposure, soils, and **microclimatic** conditions in design decisions. In addition, landscapes suited to the preferences and restorative requirements of the homeowner make the outdoor living spaces more valuable to them (Figure 3-3).

 ## COMMON MISCONCEPTIONS ABOUT LANDSCAPE DESIGN

There are a variety of common misconceptions that cloud the understanding of landscape design. These false perceptions are not limited to the general public; in fact, many designers harbor (and even support) these myths. Overcoming these misconceptions will help shape a better understanding of design and designers.

FIGURE 3-3

The natural context for design may not always be as evident as in the setting for this garden, but natural processes and conditions are present in every landscape setting. *(Steven Rodie)*

Enlightening Clients about What Design Is and Is Not

Landscape design involves significantly more than the dictionary defines or what a majority of the public tends to understand. It incorporates functional considerations and problem-solving as well as aesthetics and art. It is a process, not just a product. Changing the paradigms of people who know design by its dictionary definition is an important and ongoing role for the landscape designer. With better understanding comes enhanced valuation of professional design services and a higher likelihood that designers will be well compensated for their skills.

Misconception: Landscape Design Is a Product
Reality: Landscape Design Is Both Product and Service

project scale: The relative size of a project; large-scale typically encompasses a large area (e.g., park or subdivision), whereas a small-scale project encompasses a small area (residential backyard).

Regardless of a **project's scale**, quality design should be recognized as a service that reflects a professional process and is worthy of compensation. The nondesigner often sees design as a product. Whether it is a commissioned painting or the rough sketches that eventually lead to a newly constructed home, design is represented by a drawing or work of art—a product. Not commonly acknowledged is the process used to achieve a successful design product that met the client's needs, lifestyle, and budget (Figure 3-4).

Misconception: Good Designers Do Not Have to Think about Design
Reality: Good Design Requires Time and Energy

Good design is hard work and should take time. It is multilayered and requires effective communication, planning, and forethought. The real world, however, demands that design happen quickly. In fact, many assume there is a direct

FIGURE 3-4

The completed design (c), developed in concert with a site analysis (a) and functional diagram (b), is an integral design process step rather than a product. *(Courtesy of Kathy Heupel. Steven Rodie)*

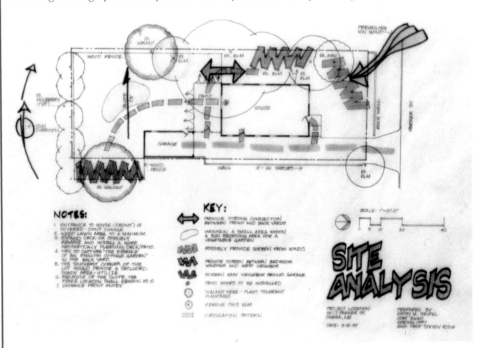

(a)

(b)

(continues)

FIGURE 3-4 *Continued.*

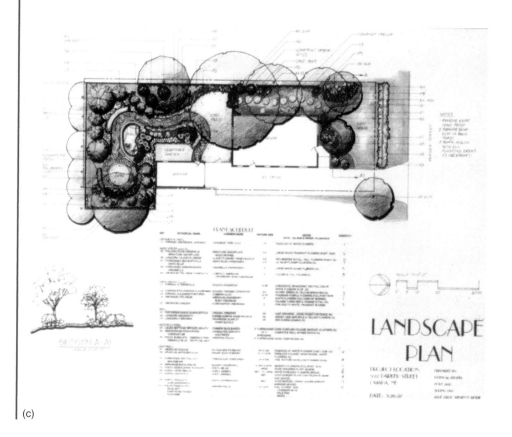

(c)

Art Is Hard Work

Grant Wood, the famous Iowa artist, stated the following about his artistic method (Roberts et al., 1995): "The public does not realize, perhaps, the amount of work that goes into one painting before I begin to set it down on canvas. In my last picture, I spent two months—fourteen hours a day, including Sundays—sketching, making notes, rejecting ideas." Other artists have similar stories to tell. Quality design, including the effective combination of colors, textures, shapes, and contrasts, does not occur without skills, patience, and hard work.

correlation between the quality of a designer and how quickly he or she can develop creative and workable design schemes. Most designers combine efficiency and speed only after many years of experience. A designer's ability to balance deadline demands with the time required to produce quality designs is an important consideration.

Misconception: Design Is a Rare Talent
Reality: Design Is a Learned Skill

More than a rare talent, good design is a set of skills that can be taught and learned. Years of practice, intuitive thinking, and analytical assessment synthesize into a solid core of characteristics needed for a successful designer. Many students (and designers) mistakenly feel they lack either the intuitive or analytical skills needed to produce sound design. What the student usually lacks, however, is confidence in his or her ability. This can be overcome through training, practice, and discipline.

Left Brain and Right Brain

The left and right sides of the human brain have been shown to process information differently (Sperry, 1976). The left side of the brain tends to process information rationally and analytically, and the right side tends to depend on insights, visualized information, and feeling. Since skilled designers and artists tend to be identified as a right-brain group, beginning designers can be intimidated because they lack confidence in their right-brain capabilities. Two key considerations level the playing field for all designers: By default, the design process requires a significant left-brain component of analytical rational assessment; and right-brain skills can be learned and improved upon. Quality design comes from quality thinking, and both sides of the brain are required in the design process.

Misconception: Good Designers Have an Obvious Signature Style
Reality: Personal Style Is Universal

While experienced designers will often develop a noticeable style in their work, it should not overshadow the project. Much like the personal nuance displayed in hand-drawn architectural lettering, naturally developed design style should subtly enhance a project's aesthetics. Signature features used to enhance the design might be reflected in plant selection, a favorite hardscape material, or a unique solution to a common landscape problem. Adding the same obvious specimen plant or accent feature to every landscape may "sign" the landscape canvas for the designer but will likely detract from overall landscape style in the process.

Misconception: Good Designers Work Best Alone
Reality: Good Designers Appreciate the Benefits of Additional Viewpoints

Although many designers work alone due to time or other constraints, design collaboration almost always enhances the process and ultimately the outcome. Adding further viewpoints and experience to the design process can significantly improve design quality. Seeking additional input from others involved in the design—such as clients, contractors, and other designers—provides learning opportunities for everyone.

Designing is an intricate task. It is the integration of technological, social, and economic requirements; biological necessities; and the psychophysical effects of materials, shape, color, volume, and space: thinking in relationships.
- László Mholy-Nagy, Hungarian Artist and Photographer

Misconception: Quality Design Is Unique
Reality: Quality Design Is Grounded in Creative Adaption

Quality design ideas that have stood the test of time can be an inspiration for new projects. Because functional, aesthetic design is timeless, very few design ideas are truly unique. Rather than going back to the drawing board, savvy designers look for similarities that exist between their project and previous projects under similar circumstances. Good design creatively adapts proven solutions dealing with site conditions, client desires, and design scale (Figure 3-5).

Misconception: Good Designers Do Not Need to Follow the Basic Aesthetic Design Principles
Reality: Good Designers Know the Basic Design Principles Well and Know How to Stretch Them

Design principles represent universally accepted, time-tested guidelines that define aesthetically pleasing landscapes. Top designers know how to stretch these principles without breaking them. They also know when the strength of an entire design should hinge on the pure implementation of a single design principle. A student of design needs to remember that principles are not laws or rules. Like other principles, they can be diluted, intensified, and corrupted. For the landscape designer, they can be dramatically manipulated into a near limitless array of compositions.

 ## DESIGN AS A PROCESS

A fundamental **design process** discussed by Koberg and Bagnall (1981) provides a framework from which to build a meaningful landscape design definition. This process applies universally to all design, including the arts, engineering, and landscapes. It acknowledges outcomes and products but goes a step further and emphasizes process and functional issues as key components. Koberg and Bagnall define design as "a process used to accomplish change," and as a "problem-solving journey." These seven stages are included in the process:

- accept situation
- analyze
- define desired outcomes
- generate ideas
- select
- implement
- evaluate

Regardless of how this process is adapted, two things are certain about this process. First, design is cyclic, not linear. Each new design will build on previous successes and failures. Second, the foundation of design rests on effective problem-solving skills and the desire to accomplish beneficial change. This dynamic process integrates aesthetics and creativity and is a collaboration between the designer, client, and contractor.

FIGURE 3-5

The proven design success reflected in the sightlines and focal points at Versailles (a) was adapted in the inspiration for the design for Washington, DC. (b). *(a–Landscape Interpretations, Thomson Delmar Learning. b–Library of Congress Web site.)*

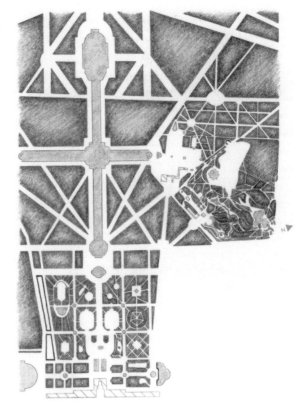

(a)

(b)

Defining Landscape Design and the Foundations That Shape the Design Process 115

FIGURE 3-6

The design process can take on many configurations depending upon designer/client personalities and project scale. Whether linear (a), complex (b), or freeform (c), the evaluation of project successes and shortcomings typically leads to another iteration of design process. *(Steven Rodie; adapted from Koberg and Bagnall)*

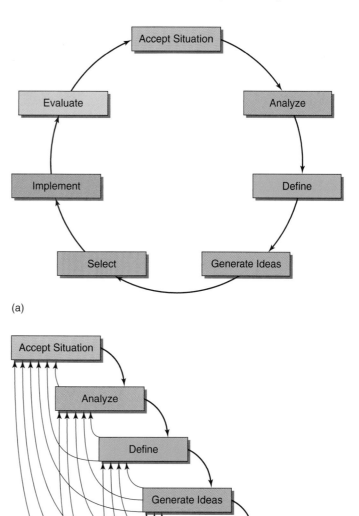

(continues)

FIGURE 3-6
Continued.

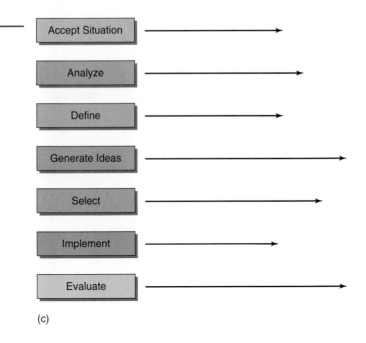

(c)

Each step of the design process has many components, and each step can be applied in a variety of ways. The specific sequencing of these steps varies as much as the designers and projects to which they are applied (Figure 3-6). A case study illustrating the design process is included at the conclusion of this section.

Accept Situation

Before a problem can be solved or a change implemented, the designer needs to fully identify and understand what constitutes the problem or the need for change. Oftentimes, what appears to be a problem is simply a symptom of the problem. If a dead plant is replaced with the same plant in a landscape, the symptom (dead plant) is treated, but the problem is left unsolved (too much shade, etc.). The adage "understanding the problem is half the solution" is especially true when it comes to landscape design.

For instance, most homeowners would perceive a continually wet area in a landscape as a problem that requires expensive drainage installation or site grading. If the wet area is instead viewed as an opportunity to incorporate unique and relatively inexpensive moisture-loving plants, then it becomes an asset without further costs (Figure 3-7).

Analyze

Fact-finding and information-gathering comprise a critical step in the design process. Analyzing a site includes the land inside of the property lines and its relation to all adjacent properties. The wants and needs of the homeowners must also be taken into account.

FIGURE 3-7

A wet area can be a landscape constraint requiring a concrete channel solution (a) or a landscape opportunity for diverse moisture-tolerant plants and a unifying design element (b). *(Steven Rodie)*

(a)

(b)

Define Desired Outcomes

Goals and objectives should be developed for every design project. A goal is the big-picture view of the design. Statements such as "We want low maintenance," or "I want a comfortable private garden," are design goals that reflect broad, desired design parameters but do not specify outcomes. Goals act as the umbrella under which specific measurable outcomes can be organized. Measurable outcomes for the above goals might include: "Replace all turf on steep slopes with adapted suckering woody shrubs," or "Ensure that lattice, fencing, or broadleaf evergreen shrubs are located in direct visual line between the garden patio and the windows and patios of neighboring homes." The goal will most likely be achieved when all of the measurable outcomes have been met.

Generate Ideas

The ability to generate creative ideas and solutions that successfully address goals and outcomes is an obvious factor in quality design. Creativity enhances every step of the design process and is particularly important when formulating ideas. Brainstorming a variety of ideas and formulating fresh approaches take time. Both are necessary for effective, creative problem-solving.

Select

At some point, the designer must develop a design from all the ideas previously generated. Typically, the final design is a synthesis of multiple ideas. This step can be difficult to complete, especially if numerous ideas all successfully address a design problem. In this step, the designer must verify that the selected design reflects an accurate response to the problems, analyses, and project outcomes. Due to installation costs, maintenance, and client interests, compromises are often made between competing priorities and available resources.

Implement

Design implementation is the next progression in the design process, and even though it seems straightforward, breakdowns in communication can occur as drawings are read and concepts discussed. For example, a mediocre or poor quality result can occur when the design is poorly implemented. The most critical issue in this step is to ensure that someone who truly understands the design (usually the designer or client) is directly involved with implementation.

Evaluate

Evaluation tends to be the forgotten step in the design process. Moving on to the next design leaves little time to go back and evaluate the previous project. At times, it may be difficult to evaluate a project when the designer already knows it did not go as planned. Evaluating previous projects provides the opportunity to assess overall design, plant selection skills, and installation.

I've Got the Plan but You've Got the Shovel

Designers vary in their approaches to design installation. Some develop trusted relationships with landscape contractors, whom they depend on for correct project installation. Others must literally place the plants and help dig the holes themselves. Some are open to contractor feedback concerning technique or materials; others are adamant about precise implementation of the design. Regardless of approach, two considerations must be taken into account during the implementation stage.

First, the communication link between designer and installer is critical. The designer typically has to communicate design specifications and design concepts to nondesigners who have a significant role in the success or failure of the project. As a designer, it is important to develop a trusted network of professionals who can bring the design to life. Second, designers who have installed landscapes themselves have an advantage in implementing quality design. Installation and maintenance experience can result in a realistic design process on paper; and, in turn, problems are less likely once installation begins.

CASE STUDY

Applying the Design Process

The following sequence shows how the process is reflected in a simple backyard project. The homeowners may envision themselves more as problem-solvers than designers, but the design process is clearly reflected in their approach, solutions, and conclusions.

Accept Situation

The homeowners have a small backyard and spend little time in it. They initially assumed the yard was unused because the patio lacked privacy. After they constructed a lattice screen, the patio remained unused, and they began to assess the real reasons for the lack of use. They enjoyed gardening, but acknowledged the limited time they had to take care of their entire yard. What they needed was gardening space to grow flowers or vegetables, not additional privacy.

Analyze

They knew from previous gardening experiences that full sun and good soil were very important. An old garden area left from the previous owners measures $10' \times 20'$. They calculated that the new garden should be triple this size. In addition, such a change would mean less lawn maintenance. The existing garden is in full sun and is well drained, while the adjacent areas where additional garden could be added vary in sun exposure. The flowers they like grow well in shade, so the sun and shade patterns will likely dictate where the vegetables and flowers should be planted. They enjoy watching the rabbits and other wildlife in the yard and also know that the new garden area is far from the faucet on the house.

(continues)

CASE STUDY
(continued)

Define Desired Outcomes

The goal is to enhance their backyard gardening experience. Desired outcomes include: increased garden area, space for both flowers and vegetables, reduced lawn area that is still easy to mow and water, keeping the rabbits but not allowing them to eat the entire garden, and limited additional maintenance for the entire garden.

Generate Ideas

The homeowners thought about a variety of ideas. Some were very straightforward: buy a longer garden hose to reach the new garden area; put the vegetable area in the sun and the flowers in the shade; put a fence around the garden to keep the rabbits out. Some ideas were unusual: plant some garden outside the fence just for the rabbits; try to find some shade-loving vegetables and sun-loving flowers since interplanting was supposed to encourage beneficial insects; move the entire garden area out to the middle of the yard; build a vegetable trellis so the plants would be easier to enjoy from the patio; and plant the old garden areas with wildflowers.

Select

It turned out that a variety of ideas worked best. The garden was expanded in sunny areas to maximize vegetable quality; a trellis was installed in the old garden for flowering vines instead of vegetables; organic mulch was used throughout the garden to reduce weeds; and a bean variety was found that would tolerate some shade.

Implement

Garden installation went smoothly for the most part. Typical implementation challenges occurred. A portion of the garden had to be moved after the buried television cable was severed; the rabbit fence was not rabbit-proof after all; and part of the garden ended up next to the patio, where cooking herbs could be reached more easily from the kitchen.

Evaluate

The new garden area met almost all of the desired outcomes. Gardening area was increased; there was space and good growing conditions for both flowers and vegetables; the reduced lawn area was easier to mow and water; summer watering was available when needed; and the rabbits did not eat the entire garden (just a big corner). The homeowners believed they met their goal of enhancing their backyard gardening experience. They also wished they would have made it five times the original size and not believed the "rabbit-proof" fencing labels. Next spring, they will be deciding what their true needs are for more garden space and investigating ways to discourage the rabbits.

How a Landscape Designer May Have Enhanced the Process. Although the homeowners were able to complete the design process on their own, they likely would have benefited from collaborating with a landscape designer. Designers are able to apply their multiple design experiences to each new project, allowing them to view a new project in the context of previous work. In some cases, they may have already worked on a project with similar opportunities and constraints.

An effectively applied design process, particularly if a designer is involved, has numerous benefits, including:

- Landscape design requires a framework to create successful designs. Establishing a framework for the process provides a roadmap for the entire design project—initial conception through evaluation.
- Applying the design process ensures that desired outcomes are reached.
- The design process identifies the most effective use of available resources in meeting priorities.
- The design process provides a basis for effective communication between designer, contractor, and homeowner.
- Implementing the design process ensures that maintenance issues will be considered as design priorities and that outcomes are established.
- The design process supports the phasing of larger projects.

 ## RELATIONSHIPS OF PEOPLE AND THE OUTDOOR ENVIRONMENT

Research about the human response to landscapes provides direction for quality landscape design. Responses correlate to a variety of factors, including the preference for landscape settings based on behavioral or survival mechanisms and the ability to understand a landscape. Additionally, studies have examined the restorative value of landscapes. This section will address how each of these factors is directly applicable to residential landscapes.

Landscape Preference

In the past 50 years, a variety of research has focused on identifying preferences for landscape features and characteristics. Two areas of preference assessment have evolved that correlate to quality landscape design. Behavioral Assessment considers our historic and genetic desire to be in landscapes that benefit survival. Humanistic Assessment connects preference to our ability to understand and make sense of landscapes. These basic theories are subtle, but salient, features of quality design.

Behavioral Assessment. Behavioral Assessment contends that evolutionary adaptation and biological need are the major factors influencing landscape preference. Two theories within this assessment are relevant to landscape design: Habitat Theory and Prospect-Refuge Theory. Habitat Theory proposes that humans prefer landscapes that provide ideal habitat for survival. In essence, our biological needs have shaped our preferences and adaptations to particular

FIGURE 3-8

Water is a universally preferred landscape element, whether it be an active (a) or contemplative (b) feature. *(Steven Rodie)*

(a)

(b)

Water is the driving force of all nature.
- *Leonardo Da Vinci*

landscapes. **Prospect-Refuge Theory** hypothesizes that humans most prefer landscape settings where they can see everything going on (prospect) without being seen (refuge) (Appleton, 1975).

The following examples demonstrate these theories. People highly prefer landscapes that include water, especially water that appears drinkable. The sight and sound of water are key reasons for the popularity of backyard ponds and water gardens (Figure 3-8).

People strongly prefer a combination of shade and sun over either one exclusively. Favorite landscapes provide choices in how they are experienced, regardless of weather conditions, and they provide locations that feel comfortable for all landscape users (Figure 3-9).

FIGURE 3-9

The two structures create a range of outdoor seating conditions to meet variable human preferences: The structure on the left provides a small, open, exposed seating area while the structure on the right provides a more shaded, protected outdoor seating space. *(Steven Rodie)*

FIGURE 3-10

The walls, floor, and ceiling of this outdoor space provide a strong sense of enclosure while focusing views to the sunny garden in the distance. *(Steven Rodie)*

Most people prefer backyard decks and patios that create a sense of refuge from the world while still offering a view (Figure 3-10). In public settings, favorite places to talk or sit are typically located slightly off the beaten path but still close enough to be near the action (Whyte, 1980) (Figure 3-11).

FIGURE 3-11

Seating spaces should provide a comfortable setting with a distinct sense of enclosure and accessibility, regardless of whether the seating is adjacent to (a) or at a distance (b) from the nearest access. *(Steven Rodie)*

(a)

(b)

HUMANISTIC ASSESSMENT. Humanistic Assessment stipulates that we learn many **landscape preferences** as we assimilate our landscape surroundings. We prefer landscapes that physically and emotionally engage people, have distinct patterns, and seem familiar. Human engagement and understanding of the landscape are significantly enhanced when four characteristics are evident (Kaplan et al., 1998). These characteristics are as follows:

Coherence: The landscape can be visually organized and logically divided into physical units that the viewer already understands (Figure 3-12).

Complexity: The landscape holds viewer attention with a balance of texture, color, form, and size variety (Figure 3-13).

FIGURE 3-12

The strong, flowing lines and structural elements in this view combine to create a visually and physically coherent landscape. *(Steven Rodie)*

FIGURE 3-13

The broad variety of colors and textures heightens landscape complexity, while the distinct bedline and repetition of specific plants limits complexity to a comfortable level. *(Steven Rodie)*

FIGURE 3-14

The informal stone path provides legible garden access, and the open gate welcomes the viewer and creates a memorable garden landmark. *(Steven Rodie)*

FIGURE 3-15

A landscape that contains a sense of mystery and engages the viewer to "find out more" is typically highly preferred. *(Steven Rodie)*

Legibility: A legible landscape is readable and safe. A visitor feels welcome to explore and is equally confident he or she will not get lost (Figure 3-14).

Mystery: Examples of mystery include an unusual landscape feature or the promise of something just around the corner. Mystery is the highest predictor of landscape preference (Figure 3-15).

Coherence and complexity are most often reflected in the visual (two-dimensional) character of a landscape, while legibility and mystery correlate to the feel and spatial (three-dimensional) qualities of a landscape. A designer's ability to layer these human preference predictors with traditional design principles can significantly enhance the quality of designs that are produced.

FIGURE 3-16

Fascination with other living creatures in the garden (a), as well as the continual changes of intricate plant characteristics (b), can be an important restorative garden element. *(Steven Rodie)*

(a)

(b)

Everybody needs beauty as well as bread, places to play in and pray in, where nature may heal and give strength to body and soul.

- John Muir

Restorative Value of Landscapes

People restore themselves mentally and physically in natural settings. Natural views and access to the outdoors have been shown to enhance healing in hospitals, reduce neighborhood violence, reduce blood pressure and stress, and improve accuracy in simple task completion (Berto, 2005; Hartig et al., 2003; Ulrich, 1984).

Restorative settings that connect people to nature typically display four important characteristics: fascination, being away or escape, extent, and compatibility (Kaplan et al., 1998). These **restorative characteristics** are defined below.

Fascination: Fascination requires focused energy and effort. Well-designed landscapes contain a wealth of natural attributes that provide fascinating mental and visual diversion. Examples include water patterns and sounds; changing light qualities; the characteristics of leaves, flowers, and wildlife; wind and light movement; and the changing seasons (Figure 3-16). Quality design provides an opportunity to make the most of fascination by incorporating interesting elements in accessible locations.

Escape: Before being restored, people need to escape whatever stresses them. Escape can occur mentally (looking out a window) or physically (moving away from the stressful environment). While a walk through the woods or a park is a classic example of escape, a carefully designed backyard landscape can also provide a sense of escape (Figure 3-17).

Extent: Although escaping to a place removed from the source of stress is part of restoration, a boundless destination will provide deeper restoration through a sense of extent. For example, if a person feels as if he or she could walk

FIGURE 3-17

Walking in a naturalistic garden setting provides restorative physical and mental escape from everyday stressful environments. *(Steven Rodie)*

FIGURE 3-18

Multiple garden paths and destinations enhance the sense of extent in a restorative garden setting. *(Steven Rodie)*

Look deep into nature, and then you will understand everything better.
- *Albert Einstein*

forever in his or her thoughtfully designed garden, restoration is more likely to occur on a regular basis (Figure 3-18).

Compatibility: A restorative landscape setting must be compatible with personal needs and expectations. For example, even though a wilderness backpacker may get wet and cold on a weekend hike, restoration is still very likely because the hiker expects to experience those conditions. Similarly, proper relaxation in a backyard requires basic comforts to mitigate undesirable temperatures and exposure to wind, rain, and sun (Figure 3-19).

Designers who incorporate personal restoration objectives, in combination with landscape preference factors, greatly enhance their potential for design success and value.

FIGURE 3-19

This shelter enhances landscape compatibility by protecting garden visitors from weather extremes *(Steven Rodie.)*

SUMMARY

This chapter provided four perspectives on the definition of landscape design. First, a definition of landscape design rooted in theoretical foundations and applicable to residential landscape design was presented. Secondly, landscape design was defined by what it is not; misconceptions were explained and debunked. Next, landscape design was described as a goal-driven process with distinct stages, and clear benefits associated with the process were highlighted.

Finally, the relationship between people and nature and its influence on design decisions were discussed.

As you develop a broader appreciation and understanding of landscape design, it is important to remember that quality design is a process, a service, and a product that all require effective oral and written communication. It is learnable, hard work and a blend of tried-and-true principles with old and new ideas.

KNOWLEDGE APPLICATION

1. Consider the landscape design misconceptions presented in the text. From your experience, select the three that are most common or important, and describe their meaning to you personally.
2. Identify and define the seven steps in the universal design process.
3. Apply the universal design process to a personal problem-solving exercise.
4. Describe the benefits in applying the landscape design process.
5. Compare and contrast the two assessment approaches used to identify human landscape preference, and define the core theories or factors in each approach.
6. Identify and describe the four qualities of natural elements that maximize human restoration in a landscape setting.

LITERATURE CITED

Appleton, J. (1975). *The experience of landscape.* London: John Wiley.

Berto, R. (2005). Exposure to restorative environments helps restore attentional capacity. *Journal of Environmental Psychology, 25*(3), 249–259.

Hartig, T., Evans, G., Jamner, L., Davis, D., & Gärling, T. (2003). Tracking restoration in natural and urban field settings. *Journal of Environmental Psychology, 23*(2), 109–123.

Kaplan, R., Kaplan, S., & Ryan, R. (1998). *With people in mind: Design and management of everyday nature.* Washington, DC: Island Press.

Koberg, D. & Bagnall, J. (1981). *The all new universal traveler: A soft-systems guide to creativity, problem-solving, and the process of reaching goals.* Los Altos, CA: William Kaufmann, Inc.

Roberts, B., Dennis, J., Horns, J., & Parkin, H. (1995). *Grant Wood: An American master revealed.* Rohnert Park, CA: Pomegranate Artbooks.

Schoneweis, S. (Ed.). (1994). *The Nebraska master gardener handbook.* Lincoln, NE: The University of Nebraska-Lincoln, Institute of Agriculture and Natural Resources.

Sperry, R. (1976). *A unifying approach to mind and brain: Ten year perspective.* In M. A. Conner & D. F. Swaab (Eds.), *Progress in brain research* (pp. 463–469). New York: Elsevier.

Ulrich, R. (1984). View through a window may influence recovery from surgery. *Science, 224*(4647), 420–421.

Whyte, W. (1980). *The social life of small urban spaces.* Washington, DC: The Conservation Foundation.

REFERENCE

Lewis, C. (1996). *Green nature/human nature: the meaning of plants in our lives.* Urbana, IL: University of Illinois Press.

The Principles of Design: Description and Application

CHAPTER 4

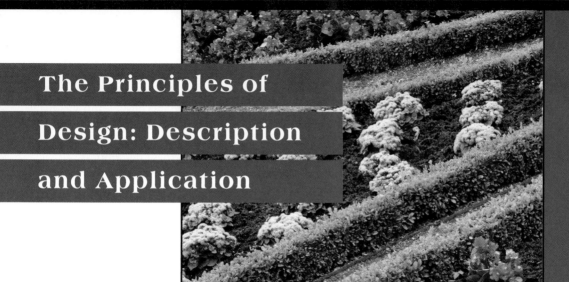

Most homeowners appreciate a beautiful and functional landscape. They desire a place for the kids to scamper about, a backyard that provides privacy in an otherwise cramped subdivision, or perhaps a front porch that allows them to watch neighborhood happenings on long summer evenings. Whatever the client's goals, skillful and creative application of design principles will help achieve them.

OBJECTIVES

Upon completion of this chapter you should be able to

1. describe the elements of art.
2. apply examples of the elements of art to landscape design and give examples.
3. describe overarching design principles.
4. describe the aesthetic principles of design.
5. provide examples of aesthetic design principles in a landscape context.
6. provide examples of how aesthetic design principles affect a person's experience in the landscape.
7. describe how aesthetic design principles influence landscape preference.
8. define the concept of functional design principles.
9. provide examples of how functional design principles affect a landscape design.

KEY TERMS

line	complementary colors	negative space	unity
form	value	order	balance
hue	chroma	repetition	proportion and scale
analogous colors	positive space	rhythm	emphasis

To explain how designers skillfully and creatively apply design principles, this chapter includes broad and theoretical topics as well as specific examples of issues that must be addressed to create a quality landscape. Specifically, this chapter will describe how we learn design; the elements of art; and landscape design principles, including overarching principles, aesthetic principles, examples of applied aesthetic principles, and functional principles. Along with these descriptions, numerous examples are provided to illustrate how each can be applied to residential landscapes.

Landscape design is a blending of art, function, and science. The art of design, through the manipulation of line, form, color, texture, and space, determines what the garden visitor sees and feels. Design function addresses circulation patterns, access, and spatial definition and use. The science of horticulture primarily influences plant selection, based on identifying and developing appropriate growing conditions and maintenance considerations. When the art, function, and science of design are creatively combined and implemented, they come to life in a living, multidimensional landscape.

Thomas Church, in his 1955 book *Gardens Are for People,* wrote: "Horticulture excellence in the garden can never compensate for fundamentally bad layout." Good landscape design maximizes the best natural features of the site and minimizes the less-desirable features. A well-designed landscape also unifies the entire property. A poorly designed landscape lacks a sense of cohesiveness and can appear jumbled or even chaotic. Each landscape will be different because of unique characteristics of the site and house, and most importantly, the client's preferences.

 ## HOW WE LEARN TO DESIGN

As Chapter 3 describes, landscape design is a process, and it is something that can be learned. Although not everyone learns exactly the same way, in general people learning about design pass through a series of phases. For most people, there are four phases: awareness, perception, imagination, and expression.

Awareness

Observing the world around us is a common experience. Although they are casual and seemingly untrained, these experiences generate ideas. Looking at any landscape —be it commercial, residential, or natural and untouched—generates ideas. Imagine, for example, a blue house with a white picket fence. Every day on your way to work, you walk by this house and think to yourself how nice the landscape always looks. What you notice most is the color. The planting color combinations complement each other and the house, and there is something colorful in the landscape throughout the entire year.

Perception

Seeing is an intense and focused form of observing. It brings details such as texture and subtle color differences into focus. Imagine, again, the blue house. One particularly crisp fall morning, you notice a shrub in the landscape that has gorgeous fall color. On this single shrub, there are deep oranges, rich burgundies, and bright yellows. Upon closer examination, you realize it is fothergilla *(Fothergilla major).*

Imagination

During this phase, past experiences are brought into association with present experiences. Connections are made, and new ideas are generated. Return to the blue house with the white picket fence. A couple of weeks after noticing the striking fothergilla, you are working on a landscape renovation project for a new client, who is excited about incorporating plants that provide more than one season of interest. During the brainstorming process, fothergilla comes to mind. It has fragrant white flowers in the spring, deep bluish-green foliage in the summer, and, of course, the striking fall color you remember seeing.

Expression

I am enough of an artist to draw freely upon my imagination. Imagination is more important than knowledge. Knowledge is limited. Imagination encircles the world.
- *Albert Einstein*

These first three steps lead to the last step in the learning process: expression. Creating and forming ideas into an actual design are ways to express what was observed, perceived, and imagined. A successful design will meet the client's needs and exhibit a sense of whatever style to which the designer and client have agreed. It is a synthesis of the designer's experiences, the client's desires, and interpretation of the unique needs of the site. Using the blue house example a final time, you decide fothergilla would be a great addition to your client's landscape, and you include three masses of fothergilla in the preliminary design. When you present the concept to the client, he or she is delighted with how many four-season plants you have incorporated into the design, including the fothergilla.

 ## ELEMENTS OF ART

All forms of visual art (e.g., painting, sculpture, architecture) are derived from line, form, color, texture, and space. These elements are combined and used in a variety of ways, and they influence how we experience the artist's vision. In landscape design, as with other forms of visual art, these elements are the foundation of good design.

Line

Lines can be horizontal, vertical, diagonal, or curved. It is how they are used individually or in combination that gives an object, including a landscape, dimension. Lines can appear graceful or crude, bold or delicate. Some common examples of how line is used in the landscape include: accentuating an object or drawing attention to a focal point, altering a viewer's perspective, or making a large space seem smaller or a small space seem larger. In the case of accentuating a focal point, using a walkway with a direct route to the focal point will naturally draw a person physically and visually to that area. In contrast, a walkway that meanders, where the focal point is not always in view, will create a mood of surprise, privacy, or even suspense. In some cases, lines are not specifically developed, but instead appear as the difference between contrasts in texture or color, or as shadow lines where changes in elevation occur (Figure 4-1).

Form

Shape refers to two-dimensional objects, while **form** is associated with three-dimensional objects. Both are made from connected lines, and the way these lines are arranged determines the shape or form. For example, the shape of a bedline might

FIGURE 4-1

The contrast in texture between this groundcover and patio creates an obvious curvilinear line and mimics the line of the soldier course of pavers along the edge. *(Cynthia Haynes)*

FIGURE 4-2

The rounded forms of these hostas and daylilies reinforce the gentle curve of the bedline and give the space a flowing feeling. *(Ann Marie VanDerZanden)*

be a long gentle curve. When plants with rounded forms are planted in a pattern that follows that curve, the landscape takes on a flowing and natural feel (Figure 4-2). In contrast, if the bedlines are linear and geometric in layout, and plants with columnar forms and tight, upright branching habits are planted along those bedlines, the landscape takes on a rigid and formal feel (Figure 4-3).

COMBINING LINE, SHAPE, AND FORM. A brainstorming part of the landscape design process involves completing numerous form composition studies. These studies explore different ideas of how the landscape could be arranged, based on the client's needs. Different forms (squares, rectangles, circles, etc.) are used to represent areas in the landscape, such as a patio, lawn, or planting beds. The areas are given different forms in order to establish an attractive and functional layout. Functional diagrams and form composition are discussed in more detail as part of the design process in Chapter 5.

FIGURE 4-3

The upright forms of these plants fit in the linear and narrow shape of this bed, although the effectiveness of these plants as a screen is debatable. *(Ann Marie VanDerZanden)*

FIGURE 4-4

The Munsell color system includes a 12-color color wheel (a) and the concept of a color solid (b) to illustrate the complex characteristics of color. *(4-4b courtesy of Gretag Macbeth. Gretag Macbeth)*

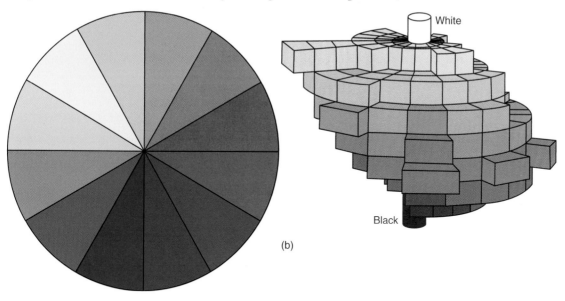

(a)   (b)

Color

Color is perhaps the most complex art element, and it tends to be the most-used (as well as over-used) element. Although there are a variety of color systems that organize the dimensions of color, the Munsell Color-Order System is specifically referenced in this text due to its simple structure and applicability to student use. The system defines color three dimensionally through both a color wheel (Figure 4-4a), as well as color solid (Figure 4-4b). It defines specific hues and the related values and chromas associated with each hue.

FIGURE 4-5

Warm (a) and cool (b) color combinations have different visual impacts in the landscape. *(Ann Marie VanDerZanden)*

(a)

(b)

Based on the Munsell System, color has three dimensions: hue, value, and chroma. **Hue** is the name of a color, such as red or blue. Although the total number of hues varies by color system, most systems are based on 10 to 12 basic hues. Hues are also categorized according to cool and warm designations. Warm colors include reds, yellows, and oranges (Figure 4-5a), while cool colors include blues, greens, and purples (Figure 4-5b). This distinction is important for designers, because warm hues tend to look closer to the viewer, while cool hues tend to appear farther from the viewer. Applying this rule of thumb allows a designer to decrease or increase the perceived dimensions of space within a landscape.

FIGURE 4-6

Analogous colors blend into one another, creating a mass of similar color. *(Cynthia Haynes)*

Hues can be further categorized as analogous or complementary. **Analogous colors** are those adjacent to each other on the color wheel. For example, yellow-orange is analogous to yellow and to orange (Figure 4-6). Analogous colors tend to blend into each other and can be used to create a continuous spectrum of colors. Alternatively, the highest contrasts between hues are generally between **complementary colors**, which are located opposite one another on a color wheel (Figure 4-7). This visual relationship explains why red flowers tend to appear so vibrant when they are backed by green foliage and why purple/yellow or blue/orange flower combinations appear so striking.

Value describes how light or dark a color is, and it is sometimes called brightness, lightness, or luminosity. This light or dark quality is based on the amount of light reflected back by the colored object. Hues that are light (as a result of having a lot of white added to them) have a high value and are referred to as tints. In comparison, hues that have black added have a lower value and are called shades. For example, a light purple would have high value in comparison to the hue in its pure state (Figure 4-8).

Chroma is the third dimension of color; it is sometimes called intensity, saturation, or purity. It is a measure of the actual hue content. Pure hues have the most chroma, while grays have the least chroma. A color with high chroma is vibrant looking. When black or gray is added to that same color, it will have a low chroma and look somewhat grayish (Figure 4-9).

Landscape design students who carefully observe color in all three dimensions and experiment with its use, in graphic rendering and plant compositions, can enhance their overall design skills and their understanding of color as an important design element. For additional information, *Color Drawing* by Michael Doyle is an excellent resource (Doyle, 1996).

Color is an important landscape design consideration for both the plants and hardscapes. Faint differences in the value or chroma of green among plants, for example, can be enough to make a planting composition interesting. Well-executed Japanese gardens typically reflect strong usage of color in an understated way.

FIGURE 4-7

Complementary color pairs contrast with each other for a bold color statement in the landscape. *(Cynthia Haynes)*

FIGURE 4-8

This series of purple color dots shows a range of values from low value, where the color is pure, to high value, where white has been added to the hue. *(Joseph E. VanDerZanden)*

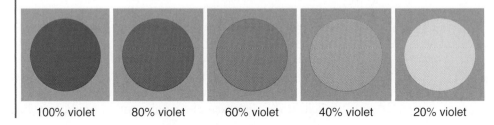

FIGURE 4-9

These yellow color dots show differences in chroma. High chroma is vibrant (far left), whereas low chroma is dull and grayish (far right). *(Joseph E. VanDerZanden)*

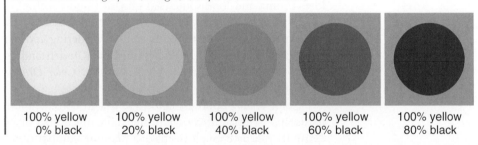

FIGURE 4-10

Using shades of green and capitalizing on shadows within the landscape creates a beautiful and simple garden. *(Ann Marie VanDerZanden)*

Color Basics: Understanding Value and Chroma

The difference between value and chroma can best be seen in looking at color as black-and-white imagery. If several values (tints and shades) of one hue are seen in black and white, they will vary from light to dark. If several chromas of one hue are seen in black and white, no noticeable difference will be apparent because the color intensity is changing, not the light or dark appearance.

These gardens include green in varying shades that combines with light and shadow to create a beautiful and peaceful retreat (Figure 4-10). Another consideration in a landscape is that color value and chroma can change throughout the day, depending on the sun's location and intensity. Color is also influenced by the growing season as leaves evolve from a fresh, light-green in the spring to deep, gray-green in the heat of summer to amber, orange, or maroon in the fall.

Texture

Texture is a surface characteristic that can be seen and even felt. Texture can be loosely divided into three categories: coarse, medium, and fine (Figure 4-11). All of the various surfaces and elements that make up a landscape—plants, decks, patios, walkways, water features, and garden art—have their own textures. Contrasting textures add interest to a design, but too much textural contrast can result in a chaotic scene. Because texture is typically a more muted or understated art element, especially compared to color, it tends to be used less often and less effectively than the other elements. However, it can, and should, play an important role in virtually every landscape composition.

FIGURE 4-11

In this planting composition the large hydrangea has coarse texture, the silver-leafed Brunnera has medium texture, and the small variegated hosta has fine texture. *(Cynthia Haynes)*

Space is the breath of art.
- *Frank Lloyd Wright*

Space and light and order. Those are the things that men need just as much as they need bread or a place to sleep.
- *LeCorbusier*

Space

This art element surrounds us; we move through it every day. In designing landscape space, it is important to distinguish between positive and negative space. **Positive space** is occupied or filled space, while **negative space** is unoccupied or empty space. Space can be defined two dimensionally as a shape (a natural limestone patio defines a positive space adjacent to the negative space of an undefined surrounding planting area; Figure 4-12a) or three dimensionally as a form (a mass of plants defines positive space as a rounded shape adjacent to the negative space of undefined stone mulch; Figure 4-12b).

In landscape design, it is important to effectively use the element of space as a dominant feature to help organize a landscape and provide focus to specific areas. A good example of a dominant negative space is the lawn area of a home landscape. The lawn area creates a relatively large, uniform, green, negative space. It typically does not have a defined shape when compared to the distinct bedlines of the planting areas and the obvious forms of the different plant masses. A good designer will consider the potential shapes and forms this negative space can exhibit (Figure 4-13). Considering adjacent positive and negative space simultaneously results in greater physical and visual contrast between the two, which in turn leads to a stronger delineation of space and a more interesting design.

LANDSCAPE DESIGN PRINCIPLES

The principles of landscape design are the building blocks used by designers to create beautiful and functional landscapes. For discussion in this text, the principles have been divided into four categories: overarching principles, aesthetic principles, application of aesthetic principles, and functional principles.

FIGURE 4-12

The limestone walk and patio define a positive two-dimensional space surrounded by the empty or negative undefined planting area (a). In the planting bed the shrub forms function as positive space, while the stone mulch work as negative space (b). *(Ann Marie VanDerZanden)*

(a)

(b)

Overarching principles provide a general context for quality design and include the concepts of simplicity, blending of form and function, and designing to reflect local elements of architecture, history, nature, and a sense of place. Aesthetic principles are developed when the elements of art are applied to a landscape. Applied aesthetic principles reflect the impact of integrating aesthetic principles into a landscape design setting and the impact they have on how people experience the space. The fourth category of functional principles provides guidance on creating landscape designs that fulfill client desires for comfortable and maintainable outdoor spaces, while effectively solving landscape problems.

OVERARCHING

Overarching design principles provide a framework that can be used to guide initial design development. These broad concepts direct general design considerations but do not determine specific decisions made by a designer and client.

FIGURE 4-13

The frontyard of this landscape was designed with bold curvilinear bedlines to create a distinct contrast between the positive space of the planting beds and the dominant negative space of the turf area. *(Thomas W. Cook)*

FIGURE 4-14

This landscape is defined by simple shapes, simple bedlines, and simple plant forms. *(Ann Marie VanDerZanden)*

Simplicity

Regardless of the scale of a landscape, simplicity should be an overriding design consideration. About the concept of simplicity, Church (1955) wrote: "It is not wise to be overambitious in designing the garden. Too many things going on in a small area produces a restless quality which will leave the onlooker dissatisfied. The house dominates the area and must be allowed to dictate the general lines of the garden."

Simplicity in a landscape can be created both physically and visually. Physical simplicity refers to a design in which the actual shapes are simple (Figure 4-14). Because the house tends to dominate a residential landscape design, consider simple bold, rectilinear lines that mimic the house or strong audacious curves rather than thin, wiggly bedlines that resemble a snake lying in the grass. In addition to simple bedline shapes, include plants with simple forms, such as rounded or oval, as opposed to severely pruned topiary or intricate espaliers.

FIGURE 4-15

A hedge is created by planting arborvitae in a linear arrangement, and the spireas in front were installed using an off-set spacing arrangement. *(Ann Marie VanDerZanden)*

FIGURE 4-16

The trees on the right also read as a group of three (2 + 1), even though there is space between them. *(Steven Rodie)*

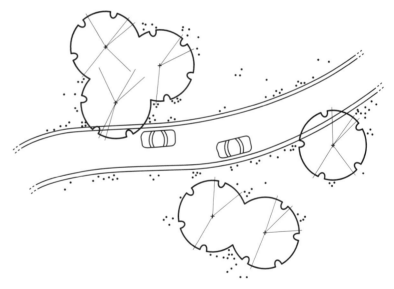

Visual simplicity can be achieved when plants are grouped or massed together. Cohesive plant masses, rather than a series of individual plants, enhance simplicity. A common design blunder in creating a mass is to position plants in a row like fence posts. Although some formal designs or functional screening applications warrant a linear arrangement, a more pleasing approach is to use off-set spacing to create an informal yet uniform grouping (Figure 4-15). Generally, plants should be consistently spaced so they overlap slightly and form a single mass at maturity. Somewhat longer distances between massed, or individual, plants can maintain simplicity while introducing additional interest (Figure 4-16). Normally, plants should only be used individually if they are serving as a focal point or filling an important functional role in the landscape such as a shade tree.

Beauty of style and harmony and grace and good rhythm depend on simplicity.

- Plato

The basic landscape structure, including the major components (turf areas, patios and paving, plant masses, etc.) and their shapes, should reflect visual simplicity. Within this backdrop, however, a significant variety of ornament can be incorporated to add interest to the design while at the same time not diluting the strength of the overall visual structure.

Blending Form and Function

Form and function must both be addressed in the design process. "Form follows function" is one popular cliché that supports this principle, and allowing the functional needs of a design problem to drive the aesthetic qualities has long been an equation for successful design. In contrast, some designers are initially driven by aesthetics, and they follow up with functional problem-solving. Perhaps the most ideal perspective is offered by Frank Lloyd Wright, one of the most influential architects of the twentieth century: "Form follows function—that has been misunderstood. Form and function should be one, joined in a spiritual union." Joining form and function, rather than addressing each consecutively in the design process, ensures a successful design.

Reflecting Local Elements of Architecture, History, Nature, and Sense of Place

Quality design typically draws inspiration from its surroundings. Architectural building character and nearby natural landscape features combine to establish an important context for landscape design. In addition to developing a visual fit for a proposed landscape, it is also important for the designer to consider the spatial qualities it has and their relationship to the surrounding site.

Any intelligent fool can make things bigger and more complex.... It takes a touch of genius—and a lot of courage to move in the opposite direction.

- Albert Einstein

ARCHITECTURE. The architecture and design details of a house should be carefully considered as initial landscape concepts are developed. For example, a sharply peaked roofline, or prominent dormers, might establish an angular theme to bedlines or a pyramidal theme to plant form. In contrast, a round window or other rounded architectural feature may inspire circular paving patterns, curvilinear bedlines, or curved wall corners. A brightly colored front door may prompt using that same color in sculpture or flowers to further accent that focal point. In many cases, portions of several different architecture themes may be combined for a single landscape. Regardless of which elements are selected, they should be strategically implemented so they complement the architecture and unify the landscape.

HISTORY. The historic character of local and regional landscapes can also provide design inspiration. On a local level, older, established residential neighborhoods may have a dense tree canopy from large street trees and homes with front porches that needed screening. In other historic neighborhoods, there may be a strong sense of formality due to well-maintained hedges along property lines and expansive turf areas. Considering these features is especially important when a designer is renovating an existing landscape in such neighborhoods. On a regional basis, elements such as rural stone fences, brick-paved streets, or small frontyard grottos (religious figure displays) may offer themes, textures, and colors that can be incorporated into a residential landscape and help the new design blend with the historic character of the area.

FIGURE 4-17

The native stone and alpine plants in this landscape reflect the surrounding landscape of the Colorado Rockies. *(Steven Rodie)*

Less is more.
- Ludwig Mies van der Rohe

NATURE. Every landscape project has a physical context and location, and nature is an integral element. Natural landscape character is created by types of plants and plant communities, types of stone, and inherent visual patterns found in that location. A landscape designer who is sensitive to these characteristics can develop comparable design themes in a residential setting.

Examples of natural landscape character include locales identifiable by certain native plants such as flowering dogwoods in the forests of the Eastern United States, little bluestem across the Midwest prairie states, or aspen in the Front Range and mountains of Colorado. Regional field or quarried stone also provides a strong tie to local landscape character, as do the regional patterns of geology, water, and wind (Figure 4-17).

Obviously, a prairie or forest requires ample space to be fully developed. The flow of wind in prairie grasses or the vividness of spring bloom in a forest understory, for example, is not possible with just a few plants. The scale of residential landscapes often poses significant limitations on how this large-scale natural character can be implemented, but carefully selecting representative elements can result in a condensed version of some natural features.

Simplicity and repose are qualities that measure the true value of any work of art.
- Frank Lloyd Wright

SENSE OF PLACE. Sense of place is hard to define and quantify, but it is critical to the successful design of landscape space. Human beings are drawn to spaces that have a special ambiance. The draw may be generated from combinations of many factors such as excitement (a popular outdoor festival), drama (overlooking spectacular scenery at the edge of precipitous cliffs), spirituality (an ancient cemetery), or nature (the smells, sounds, and feel of walking through a dense forest). Most people have visited places that have remained vivid and timeless in their minds. Clients who can mentally or physically revisit these places can provide guidance on how to re-create these spaces in their home landscape. Sense of place is typically a by-product of quality design. Yet, it can be an elusive design objective without a focused effort to create such an atmosphere within the framework of aesthetic and functional design goals.

LIMITATIONS. Not all of the characteristics described previously will be appropriate for every residential landscape. For example, a client may not be willing to take on the high level of maintenance associated with the formal landscape style of an established historic neighborhood. Planting a frontyard prairie or forest may reflect regional nature, but it might conflict with the historic character of an old neighborhood or sense of monotonous style in a new subdivision neighborhood. Sensitive application of these principles, in combination with the other design principles, is essential to quality design.

AESTHETIC LANDSCAPE DESIGN PRINCIPLES

In the creation of a healthy environment, nature's collaboration is not only important, but also indispensable.
- Eliel Saarinen, LA, p. 372

Landscape design principles are the same for all types of landscapes. They are applied in many different ways, depending on the site, the client's wishes, and, of course, the designer's knowledge and preferences. Similar to the elements of art, these principles represent the primary concepts that influence landscape design. Even though the principles are typically applied concurrently in a design, the ability to break them apart into identifiable components is an important skill to develop when getting started in the landscape design profession. More importantly, it helps a designer understand what he or she is seeing and assists the designer in explaining and selling his or her design to a client.

Depending on the latest garden design book or television show, experts will describe anywhere from three to twelve landscape design principles. This text will focus on seven principles: order, repetition, rhythm, unity, balance, proportion and scale, and emphasis (Figure 4-18). The principles collectively determine how a landscape is organized (generally influenced by order, repetition, and rhythm) and how it looks and feels (generally influenced by unity, balance, proportion and scale, and emphasis). Many times, however, their impact on design will overlap or combine, providing innumerable design applications. For example, the rhythm developed in the strong curves and varying widths of a pathway provides structure and spatial definition to an adjacent turf area, but the curves also provide visual interest and can be felt by the viewer as he or she moves along the path. Table 4-1 summarizes the different principles, how they overlap, and examples of how they appear in the landscape.

Study nature, love nature, stay close to nature. It will never fail you.
- Frank Lloyd Wright

Design principles are just that—principles. They are not rules that require precise adherence, but instead reflect a framework of universal concepts that prove effective in creating designs. Ineffective application of these principles can lead to visual boredom or chaos. Creative designers will use these principles as a starting point, and they will stretch them, reconfigure them, turn them inside out as they strive to implement new ideas that successfully address both aesthetic and functional landscape needs.

Order or Design Framework

Order can also be thought of as a design's overall framework. To start, order can be achieved by using a consistent design theme such as formal, informal, or structured informality (Figure 4-19). All three layouts contain a lawn area, and in each case the lawn area is open and uncluttered, but the overall framework of each layout varies dramatically. Importantly, in each layout the expanse of lawn is useful as a recreation area; it is easy to mow and creates a dominant space within the landscape.

Grouping plants and hardscapes so there are physical connections between the elements will also enhance order and create a cohesive whole (Figure 4-20).

FIGURE 4-18 Combining multiple design principles will create a cohesive, functional, and enjoyable landscape for the client. *(Steven Rodie)*

(a)

(b)

(c)

Order is further reinforced by grouping plants together in masses (as illustrated in Figure 4-19), rather than scattering individual plants around the landscape. Plants grouped into odd numbers are more natural looking than even-numbered groups because odd-numbered groups resist being visually split into two equal halves. For example, a pair of plants tends to look like two individual plants side by side, rather than a single mass. Once a grouping contains more than 10 plants, or if the plants will grow together quickly, the odd-numbered grouping tenet is less critical.

TABLE 4-1

Summary of the design principles and examples of how the principle can be implemented in a landscape.

PRINCIPLE	TYPICAL IMPLEMENTATION	ENHANCED AND STRENGTHENED BY:
Order (Design framework)	• Use well-defined bedlines and a consistent bedline theme • Plan for masses of landscape plants rather than all plants being separate • Establish defined rooms in the landscape by creating walls and ceilings with plant masses, tree canopies, and hardscape elements (fences, pergolas, walls, etc.)	Balance: tends to make order more visible and feel stronger
Repetition	• Repeat one or more of the basic art elements (form, color, texture, size) • Examples: planting red flowers at all entrances to the house or using the same groundcover along the front edge of all perennial flower beds • Remember: Too much repetition leads to boredom; too little leads to visual chaos	Rhythm: created through repeated patterns and elements
Rhythm	• Edges or outlines of elements (landscape bed edge, various plant heights, etc.) create rhythm when they move back and forth or up and down • Varying widths of ground features (turf area or sidewalk) create rhythm • Rhythm is heightened when variations are broad and strong (a broad "S" curve is more aesthetic than a series of short, tight "S" curves)	Repetition
Unity	• Line up edges between patterns and spaces • Example: a bedline "crossing" a sidewalk and continuing on the other side will appear connected across the sidewalk	Repetition and emphasis
Balance	• Symmetrical—place identical design components on both sides of a doorway, front walkway, window, etc.; one element out of place will create imbalance • Asymmetrical—place a small tree at one front house corner and three large shrubs at the other front house corner; the smaller shrubs will visually weigh as much as the one larger tree and balance the entire house view	
Proportion and scale	• Plants at house corners should be approximately $\frac{2}{3}$ to 1 times the height to the eave (shorter is not visually effective; larger plants will dwarf the house) • An intimate patio space should have a low ceiling created by a small tree with a low canopy or a pergola at indoor ceiling height (8 feet)	
Emphasis	• Place a unique color, shape, texture, or form (or combination of these elements) in select landscape locations to contrast with the surrounding elements • The more unique the element, the more attention it will attract	

The Principles of Design: Description and Application 149

FIGURE 4-19

Examples of common design frameworks include formal (a), informal (b), and structured informality (c). *(Lynn Kuhn)*

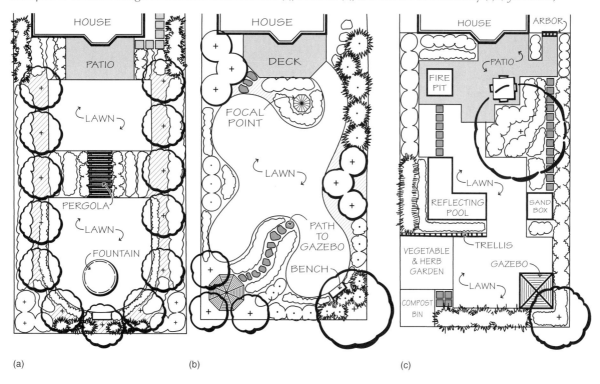

FIGURE 4-20

A cohesive whole is created in this landscape by: the repeated line of the bed and walkway, the white wood structure over the entry ties and white trim on the house, and the boulders and plants arranged to reinforce the bedline. *(Ross and Beth Brockshus)*

FIGURE 4-21

Repetition can be achieved when plants or groups of plants are repeated. In this case combining Groups A and B results in a relatively large plant mass (Group C). *(Elizabeth Crimmins)*

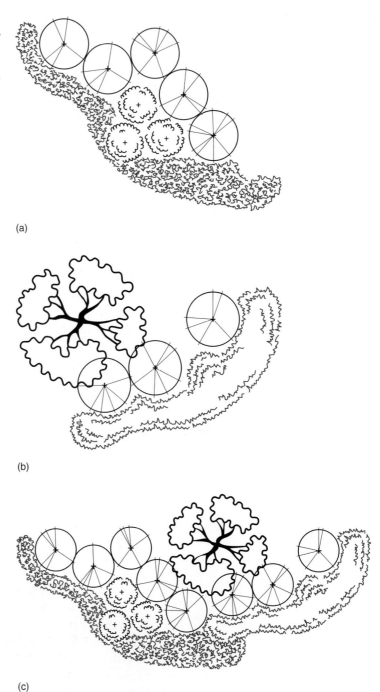

(a)

(b)

(c)

Repetition

Repetition is created in a landscape when anything is repeated, including color, form, texture, a particular plant, a mass of similar plants, or a plant composition. Repeating plants or groups of plants tends to be an obvious way to achieve this principle. A grouping can consist of just one type of plant, or it may be a collection of a few different species. For example: Group A consists of five large shrubs, three medium shrubs, and a mass of herbaceous groundcover (Figure 4-21a). Group B has a small ornamental tree, three large

FIGURE 4-22

The fence posts and finials in this landscape create a distinct vertical rhythm. *(Steven Rodie)*

shrubs, and a mass of an evergreen ground cover (Figure 4-21b). The next step is to combine the two different plant groupings and to repeat this combined grouping within the landscape (Figure 4-21c). This type of repetition results in a sense of unity between the different plant species that make up the design.

Beginning designers often make the common mistake of incorporating too many different plant species in a layout. For smaller landscapes, it is generally wise to use fewer plant species than in larger landscapes. Too much plant diversity results in a landscape that looks in disarray. When an individual plant is placed by itself, it should be a specimen plant and have a unique size, distinctive shape, or attractive seasonal features.

Rhythm

Rhythm addresses the factors of time and movement within a landscape. Rhythm is organized movement. Just as with music, landscape rhythm is an underlying sequence of elements that combines to give the whole landscape a particular mood. This is a principle that when implemented effectively may not be noticed, but when done ineffectively it is obvious that something is missing from the landscape. Landscapes are seldom experienced in an instant, but rather through time as a person explores the space. The journey may lead a person out a set of French doors onto a raised deck, down stairs to a large lawn, and around a corner on a stone pathway that ends at a small bench tucked beneath a flowering tree.

Rhythm in a landscape may be visual or physical, and it can be observed in both the horizontal and vertical planes of a design. In visual rhythm, the eye is directed through the landscape from location to location, and in back-and-forth or up-and-down sequences. One common example of vertical visual rhythm is a gradual progression of plants from short to tall. Another example of this type of rhythm might be reflected in finials atop fence posts (Figure 4-22). As a person looks down the length of the fence, there will be an obvious pattern created by the fence post finials.

Horizontal visual rhythm is often generated from bedlines or hardscapes (Figure 4-23). These ground-plane edges lead the eye through broad curves interrupted with sharper curves and establish a rhythmic sequence. Adjacent landscape beds parallel to each other strongly reinforce horizontal rhythm (Figure 4-24), and contrasting patterns and textures can develop spatial definition in the area.

FIGURE 4-23

Horizontal rhythm in this landscape is seen in the curvature of the bedlines and planting pattern of the annuals. There is also vertical rhythm in the natural progression of plant heights from the foreground to the background adjacent to the stone columns. *(Thomas W. Cook)*

FIGURE 4-24

The parallel bedlines of this design, combined with how the plants are arranged in the beds, create strong visual rhythm in the landscape. *(Ross and Beth Brockshus)*

How a person moves through the landscape and physically interacts with it can be determined, in part, by design rhythm. Examples of elements that influence physical rhythm include pathway width and route, dimensions and spacing of stepping stones, and locations of patios and other seating or resting elements. A pathway that is relatively wide and has a firm, uniform surface encourages a person to walk quickly through the landscape. If the path is relatively straight or gently curving, the pace may also be quick, but if the path has abrupt turns, the pace will be slowed as a person decides how to proceed. Large stepping stones spaced far apart may encourage a long stride, whereas small stones spaced close together encourage a more measured walking pace (Figure 4-25). And finally, a patio or small seating area gives the person a brief reprieve and allows him or her to stop and observe the surrounding landscape.

Physical rhythm can also be incorporated by the landscape designer in more subtle ways. For example, if a path narrows and then opens onto a large space, a

FIGURE 4-25

These large stepping stones placed close together make it easy to walk through the garden. *(Ann Marie VanDerZanden)*

FIGURE 4-26

This pathway is constricted by the plants on either side before entering a large expanse of turf. *(Ann Marie VanDerZanden)*

person's pace usually quickens as he or she passes through the constricted area and slows once he or she reaches the large expanse (Figure 4-26). In a similar way, a path that varies in elevation will impact the rhythm by slowing pace on the uphill stretch and increasing speed on the downhill stretch. Regardless of how rhythm is incorporated in the design, it is an important part of creating a landscape that is both beautiful to look at and interesting to walk through and experience.

FIGURE 4-27

Unity is created in this landscape by: the parallel lines of the planting beds and hardscapes and how the paver color echoes colors in the house and surrounding landscape. *(Kathleen Wilkinson)*

Unity

Unity is the principle that creates a link between the plants, hardscapes, and house. It creates a sense of interconnectedness within the design composition. A successful landscape design ties the house and landscape together and creates an outdoor living space that is an extension of the house.

Using building materials and colors inspired by the featured house is one way to achieve unity. For example, if the house is tan with wrought-iron detailing, select a paver blend that has tan, buff, and gray for the patio and walkway, a light brown mulch for the beds, and a piece of garden art made from wrought iron. Relating bedlines, patio edges, decks, and walks to the actual lines of the house also unites the house and landscape (Figure 4-27). Unity can be reinforced by using consistent bedlines such as curvilinear or rectilinear throughout the design. Even incorporating a single groundcover can unify the landscape. Just as with repetition, massing plants together and repeating these masses throughout the landscape is an important part of this design principle.

Balance

The two common types of **balance** in landscapes are symmetrical and asymmetrical. A third, radial balance, is less common, particularly in residential landscapes. Symmetrical balance is common in formal landscapes. These landscapes have an obvious central axis, and everything on one side of the axis is mirrored on the other side (Figure 4-28). While rigid and clear rules make symmetrical landscapes easy to design, the formality of this design type does not blend well with all home styles. The structured informality layout, as illustrated earlier in Figure 4-19c, is a good alternative to a purely symmetrical design.

Asymmetrical balance combines different objects on each side of a discrete axis or balance point, resulting in a similar visual mass on both sides of the axis or point (Figure 4-29). The bedlines in asymmetrical designs tend to be curvilinear, and the overall feel tends to be more informal than a symmetrically balanced design.

FIGURE 4-28

The obvious central axis of a symmetrical layout makes this design principle easy to see in a landscape. *(Ann Marie VanDerZanden)*

FIGURE 4-29

Asymmetrical balance is more organic and natural and often is not as obvious in a design when it is done correctly. *(Ross and Beth Brockshus)*

This informality, combined with the flow of curvilinear bedlines, creates a balance that is well suited for most residential landscapes. Common mistakes with this type of layout are a lack of bold bedlines and an absence of a dominant space within the landscape. Completing a number of form composition studies (discussed fully in Chapter 5) can help a designer organize the space more effectively and create spaces within the landscape that are appropriately sized to provide an adequate amount of visual weight (Figure 4-30).

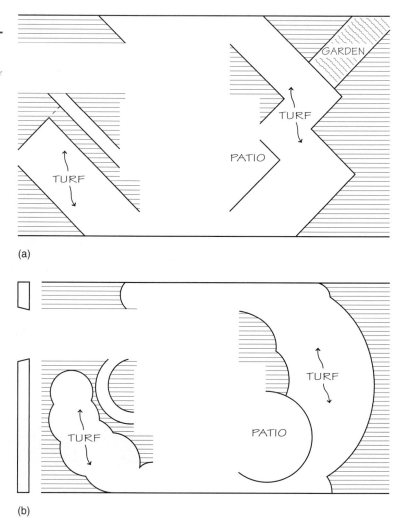

FIGURE 4-30

These two form composition studies show different bedlines and ways to arrange the space in this residential landscape. *(Courtesy of Julie Catlin. Steven Rodie)*

Radial balance is created by placing equally spaced elements around a center point (Figure 4-31), creating a strong focus in the middle of the pattern. Radial balance tends to feel formal, comparable to symmetrical balance, but varying the radial pattern or using just a portion of the pattern creates a more informal feel. Radial patterns can be used effectively in patios and paved areas, as well as in larger areas that integrate plantings and hardscapes.

Proportion and Scale

Proportion and scale refer to the size relationship between different elements within the landscape. Thomas Church (1955) noted that this principle is hard to define and no absolute rules govern it. He described two types of scale: relative and absolute. Relative scale is the relation of one part of the design to the others. Ordinarily, this is how the entire landscape relates to the house. The house is often the dominate element of a residential design, and it requires that at least some elements of the design be sufficiently sized. On this topic, Church (1955) wrote: "The best general

FIGURE 4-31

Radial balance is more formal but could be incorporated within either a symmetrical or an asymmetrical landscape. *(Steven Rodie)*

FIGURE 4-32

This patio is a comfortable landscape space in part because of the well-proportioned plants, hardscapes, table and chairs, and the overall size of the area in relation to the house and surrounding landscapes. *(Ross and Beth Brockshus)*

rule to follow is: When in doubt, make it larger. The eye detects a meager dimension more easily than it does a too-generous one."

Absolute scale is the relation of a particular design element to a human. Large elements such as a two-story house, a 50-foot-tall evergreen tree, or 500 square feet of patio may make a human feel dwarfed when standing next to these in a landscape. On the other hand, a two-story house surrounded by tall trees and enclosing a patio area may give the person a sense of enclosure and protection from outside elements. The opposite feelings might be true for a person if he or she is in a landscape in which the design elements are relatively small, such as plantings under 3 feet tall, a patio space that barely accommodates a table and four chairs, or a low fence that allows for an expansive view of the surrounding area. Understanding a client's preferences when it comes to this design principle is important when creating a space that meets his or her needs.

Plant materials help define size relationships within a landscape. The major relationships that involve plant materials are: plants to buildings, plants to other plants, and plants to people (Figure 4-32). Because plants are living and dynamic

Proportion Gone Awry

Everyone has seen it: the ugly, dwarfed line of junipers around the foundation of a house. All too often, foundation plantings are placed in a narrow border around the house. Since many shrubs have a similar height and width, there is not room for a very tall shrub in a narrow bed. The result is plants dwarfed by the house. This situation can be remedied by increasing the depth of the planting bed. A simple rule of thumb is for the depth of the beds to be equal to two thirds the height of the house to the eave. The correct dimension for bed depth will vary, depending upon the available bed space as well as the pitch and visibility of the roof (a steeply pitched tall roof may require a deeper bed to appear correctly proportioned). Small houses with low rooflines on city lots should be landscaped with small to medium trees and shrubs, while grander properties require larger plants and planting areas.

entities, these proportional relationships will change as the landscape matures and even season to season if the plants are deciduous or herbaceous.

To achieve correct proportion, the landscape design should be based on the mature height and spread of the plants. The scale may be off when the plants are young, but they will grow into the correct proportion. Since plants vary in growth rates, consider recommending that your client install larger specimens (e.g., 5 gallon or balled and burlapped) of slow-growing plants and smaller specimens (e.g., 1 or 3 gallon) of fast growing plants. This way, the entire design should come into proportion more quickly.

EVERYTHING IS LARGER OUTSIDE. When determining a sense of scale for an outdoor setting, specific measurements and dimensions that may be comfortable inside the home typically need to be enlarged for outdoor structures. For example, many outdoor overhead structures may only require 4 inch × 4 inch posts, but when an arbor or pergola is viewed by itself in the landscape, larger columns (6 inch × 6 inch) will seem to be a more appropriate proportion. These structures often create ceilings that are taller than indoor ceilings, which again seems more appropriate for outdoor settings. Another example is steps. Indoor steps are typically built with an 8-inch riser and 11-inch tread—a fairly steep angle, but necessary for indoor spaces when condensing stairway length from floor to floor. Outdoor stairs should be scaled for the larger space and longer strides people typically take outdoors. Stairs with lower risers, 4 to 7 inches, and wider treads, 12 to 13 inches and up to 18 to 19 inches, are well suited for an outdoor space. The wider treads not only accommodate long strides and complement the larger room size, but they also provide more space for planters and pots as well as an opportunity for use as an informal seating element. Thomas Church's comment about outdoor scale—"when in doubt, make it larger"—holds true in many situations.

FIGURE 4-33

Focal points, such as this bench, provide direction and invite guests into the garden. *(Ann Marie VanDerZanden)*

Emphasis

Focal points draw the eye to specific landscape locations when a person is viewing the landscape as a whole. They **emphasize** specific areas and create a contrast between memorable, vivid, or exciting landscape elements and the rest of the landscape. They stimulate the viewer's eye and visually draw the viewer from one location to another within the landscape. A focal point may be a specimen plant, garden accessory, or water feature. It captures attention by its unusual line, shape, texture, or color (Figure 4-33). Each major area in a landscape (e.g., frontyard and backyard) should have at least one focal point. Multiple focal points are acceptable if the area is large and divided into a number of smaller spaces. When placing focal points, keep in mind the vantage from which the landscape is most regularly viewed. For example, if the client will view the backyard most often from behind a sliding glass door in the breakfast nook, then consider locating a focal point so it can be seen from that area (Figure 4-34). An unseen focal point is useless.

 ## AESTHETIC PRINCIPLES APPLIED

Innovative and effective application of aesthetic design principles is central to creating a successful landscape. Once the design principles are understood individually, they can be combined and applied in virtually unlimited combinations. How these principles are applied has a significant impact on how the landscape is experienced, both physically and mentally, by people. Results of applying these principles include: connecting plants to landscape structures; creating well-defined spaces with and without plant masses; directing people through the landscape; and crafting outdoor living spaces that meet the physical, cognitive, and restorative needs of people.

FIGURE 4-34

Focal point location is critical to the value of the focal point. This focal point is enjoyed over breakfast every morning from the dining table in this homeowner's sunroom. *(Thomas W. Cook)*

Visually Connect Structures to the Landscape

Landscape structures (including houses, detached garages, sheds, etc.) typically contain straight edges, vertical lines, and relatively smooth textures. These characteristics create a high visual contrast in a landscape filled with irregular plant shapes and variable leaf textures. This contrast can be lessened (softened) by partially hiding corners and walls with plants (Figure 4-35). Appropriately scaled shrubs or small trees with rounded or oval forms are a good way to soften these straight lines. In addition, trees that are behind the roofline when viewed from the frontyard also break up the strong horizontal line of the roof edge, and they help to visually connect the structure to the surrounding landscape setting.

Landscape views are seldom panoramic or open in all directions. Instead, people tend to experience the landscape through windows or partially obstructed views. By strategically locating plants and structural elements, a designer can frame landscape views that will provide focus for the viewer and enhance his or her experience in the outdoor space (Figure 4-36).

Define Space: Mass and Void

Space should be a tangible, positive product of landscape design—it should not just be whatever remains after the design components (plants, paving, structures, etc.) are accounted for. In fact, one of the most difficult aspects of establishing effective empty space in landscapes is to establish it and then keep it open. The most obvious place to locate another element, tree, or planting bed is in a seemingly unused landscape space. Yet it is precisely because this space is void of elements that it has strong definition. A designer's ability to establish, define, and maintain empty space in a landscape design is paramount to the design's visual and functional success (Figure 4-37).

FIGURE 4-35

Plantings along building corners soften the shape of the building, blending it into the landscape. *(Ann Marie VanDerZanden)*

Direct People Physically and Visually

A landscape should direct people's views and movement, and the aesthetic principle of emphasis, which uses unique texture or a strong contrast between adjacent features, is a good way to achieve this. The front door of a house is the classic example of where people should be directed in a residential frontyard (Figure 4-38). When the front door is visible from the street, the landscape destination is obvious, so the designer works to further enhance it. In other cases, such as a front door that is hidden from view, visitors can be subconsciously directed to the feature by strategic location of emphasis elements in that area. Anyone who has looked for

FIGURE 4-36

These evergreen trees and shrubs combine with the containers on the steps to frame this door and direct people to this entrance. *(Ann Marie VanDerZanden)*

FIGURE 4-37

The empty space associated with the turf area in this backyard is equally as powerful as planting beds. *(Thomas W. Cook)*

the front entrance to a house only to end up facing an expanse of garage wall has experienced the lack of a strong visual pull to the correct destination (Figure 4-39).

Designing Outdoor Living Spaces to Meet the Physical, Cognitive, and Restorative Needs of People

Chapter 3 describes in detail the importance of human preferences in creating quality landscape designs and includes many examples of physical (water and shelter) and cognitive (mentally stimulating) preferences. Chapter 3 also includes a discussion of

FIGURE 4-38

When the front door is an obvious focal point for the frontyard, it is clear where visitors should go. *(Ross and Beth Brockshus)*

FIGURE 4-39

The colors, texture, and fountain in this garden direct visitors to the preferred house entrance, rather than to the front door located near the arbor in the distance. *(Steven Rodie)*

the restorative effects a landscape can have on people and the factors that lead to maximum restoration. Table 4-2 illustrates how aesthetic design principles can be used to meet the preference and restorative needs of people.

FUNCTIONAL DESIGN PRINCIPLES

In concert with the aesthetic design principles previously described, a designer should also consider basic issues that affect the functionality of the landscape. These functional principles are: working with existing topography, creating a useful outdoor space, accounting for maintenance, addressing irrigation needs, incorporating sustainability, and (depending on the client) providing wildlife habitat.

Several of these principles are covered in more depth elsewhere in the text, but they are highlighted here to emphasize the benefits of considering aesthetic and functional principles simultaneously during the design process.

TABLE 4-2

Examples of how aesthetic design principles can be used to meet the preference and restorative needs of people.

Application of Aesthetic Design Principles

Physical Preference

Water	Focal point: The unique characteristics (color, texture, shimmer) of water make it stand out from the surrounding landscape. Unity: A running stream through the middle of a garden represents a strong unifying element that ties the entire area together. Rhythm: The flowing edges of an irregularly shaped pond or the changing width of a stream creates visual rhythm. The sound of water gurgling, or moving in a stream, creates audible rhythm.
Prospect-Refuge	Proportion and scale: A space with a strong sense of refuge typically has a comfortable sense of scale and feels like an outdoor room with overhead structures, fences, and screening plants forming spatial walls and a ceiling.

Cognitive Preference

Coherence	Coherence can be created by consistent bedline patterns (order or design framework); symmetrical or asymmetrical patterns (balance); definition of a large space into smaller, mentally manageable areas (proportion and scale); and repetition of a landscape plants or features.
Complexity	Emphasis: Complexity can be generated through variations in color, texture, shape, and size and through highly contrasting emphasis elements like plants, paving, and architectural details. Landscapes that are highly complex can also be highly coherent but only if enough patterns (repeated textures and colors) and structure (rhythmic bedlines) exist.
Legibility	Order, repetition, unity, and emphasis: A legible landscape is well-structured, appears unified, and contains an occasional unique element or focal point.
Mystery	Order and rhythm: What is around the next bend? Or, what is partially obscured behind a row of shrubs? The assumption that there could be a next bend or a hidden location requires the landscape to have order, screening elements that limit views, and bends that result from the flow of rhythmic paths or bedlines.

Restorative Effect

Fascination	Emphasis, focal point, and rhythm: Bright colors, intricate details, and wildlife viewing can all catch the viewer's eye and enhance the focused, energizing concentration that defines the fascination. Rhythmic patterns created by contrasts and flowing curves can also generate fascination.
Escape or being away	Order, repetition, emphasis, and focal point: This factor may be achieved physically or mentally. A landscape that is organized and that has repeated elements to provide continuity and unique features to draw a person's attention can create an oasis away from a stressful setting. Examples include an outdoor patio used for lunch at work or a beautiful landscape that can be mentally wandered through by looking out a window.
Extent	Order and unity: The sense that a person can wander for days without running out of new places to explore is evident in landscape settings that do not seem to have boundaries. Using order by defining landscape rooms, or hallways between rooms, encourages a person to move through the landscape and explore new spaces.
Compatibility	Proportion and scale: Meeting the physical and psychological needs of the person using the space can be achieved by properly proportioned and scaled spaces that provide a sense of protection and allow a person to personalize the space with overhead awnings, tables, chairs, and other creature comforts found inside a house.

Working with the Existing Topography

The topography of a site is an important design consideration. It includes complex factors, including slopes, grading (the changing of topography to meet specific design requirements), walls, and drainage. A thorough awareness of topographical issues provides designers with a practical framework from which to create successful designs.

WORKING WITH LICENSED PROFESSIONALS. The technical expertise required to successfully manipulate a site (grading, retaining walls, drainage plans, and structures) requires that many of these should be designed by licensed professionals (landscape architects, architects, engineers, etc.). This is especially true for situations in which human health and safety can be compromised with poor design. Well-trained landscape designers should be able to conceptually design and work with technical experts on topographical issues. At the same time, they should also know their technical limits and seek the expertise of a licensed professional when conceptual design moves into the details of design and construction of significant grading, retaining walls, drainage systems, and other structural landscape elements (such as overhead structures, complex decking, etc). Capitalizing on the expertise of these professionals will ensure the project's success.

SLOPES. Slopes create aesthetic interest, add spatial definition, and ensure that a site will drain properly. Lack of slopes can lead to flooding and drainage problems that can damage plants, structures, and paving. Slopes are measured as a mathematical ratio between a vertical distance in relation to a specified horizontal distance (Figure 4-40). Refer to Table 5-1 for basic rules of thumb for slopes in a variety of landscape settings.

In situations in which slopes are too steep for plant materials to grow or other erosion control methods to be effective, retaining walls should be considered. Wall materials and design considerations are discussed in Chapter 7.

GRADING. Grading is the process of manipulating landscape slopes to change the natural contour of the site. Grading can be done for both aesthetic and functional purposes, including to create elevation changes to create a screen, to improve drainage or plant visibility, to flatten sites for buildings or improved mowing safety of turf areas, to make walkway and driveway slopes safe and usable, or to direct water to a specific site location (away from buildings, for example).

LANDSCAPE BERMS. Landscape berms, which are shaped mounds of soil, are a popular method of defining spaces and creating landscape interest (Figure 4-41). Several caveats should be kept in mind when designing berms.

- Berms with gradual slopes are easier to maintain and appear more natural. A relatively short berm can provide subtle yet strong definition to a planting bed.
- Berms should not be located in narrow areas because the base will be too small and the sides to steep.
- Avoid locating berms in the middle of landscape spaces (e.g., large open turf areas) where they appear as islands with little connection to the rest of the landscape. Use berms along edges of beds or pathways to reinforce the sense of space.

FIGURE 4-40

Slopes are measured as the ratio between a vertical distance in relation to a horizontal distance. *(Ann Marie VanDerZanden)*

Ratio
- H:V (horizontal distance:vertical distance)
- 20:1

Percent slope: 20:1 = 1/20 = .05 = 5%

FIGURE 4-41

A landscape berm is used to physically and visually separate the property between these two driveways. *(Ross and Beth Brockshus)*

- When berms are constructed, mix the new soil with the existing soil. Failure to do so will create a soil interface likely to cause drainage and rooting problems.

Creating a Useful Outdoor Living Space

The outdoor living space needs to be appropriately sized to accommodate the way the client intends to use the area. The space should also offer good circulation and access throughout. Outdoor living spaces are often undersized for their intended use. For example, the typical concrete patio or deck attached to a new, moderately sized home is rarely larger than 100 square feet ($10' \times 10'$—approximately the size of a small bedroom). This space is easily overfilled with a table and chairs or other outdoor furniture. A designer should help a homeowner see the benefits of increasing the size of this hardscape area before other landscape design decisions are made (Figure 4-42).

CIRCULATION AND ACCESS. The connecting framework of driveways and walkways throughout the landscape should be appropriately sized and made of durable materials able to withstand foot or vehicular traffic. Refer to Chapter 7 for

FIGURE 4-42

Combining a larger paver patio with the concrete slab of this new construction home resulted in a more usable outdoor space. *(Joseph E. VanDerZanden)*

discussion on driveway and walkway dimensions and materials. Additionally, gates need to be sufficiently wide to accommodate things such as wheel barrows and lawn mowers. Access to air conditioners and electrical boxes must be allowed for early in the design process. Shrubs that screen equipment can easily block air flow or physical access if planted too closely, especially when mature size is not accounted for. Household maintenance chores such as painting and window washing are made easier when spaces under windows and near the foundation are properly designed.

Accounting for Maintenance

All home landscapes need some level of maintenance. As discussed in Chapter 5, a client questionnaire should be used to determine how much maintenance the client is willing to assume. Keeping this information in mind during the design phase can help a designer meet those wishes. Chapter 6 describes maintenance considerations as part of plant selection and elaborates on both short- and long-term maintenance requirements as well as the seasonality of maintenance associated with some plant types.

Plants are not the only part of a landscape that require maintenance. Hardscapes, irrigation systems, water features, and lighting components also need upkeep. Some hardscape materials require significantly less short- and long-term maintenance than others. Refer to Chapter 7 for detailed information.

Addressing Irrigation Needs

Landscape irrigation serves a variety of important roles in the landscape. It helps ensure successful establishment of new plant material; it supplements water needs in drought conditions for established plants; and it increases the variety of plants that can be grown, especially in drier climates. Landscape designers may design irrigation systems themselves, with a specialized irrigation designer, or they may be asked to adapt an old system to a newly renovated landscape. The bottom line is that irrigation should not be an afterthought once the landscape is completed.

FIGURE 4-43

Planting areas less than 5 feet wide are difficult to irrigate and maintain and should be avoided when possible. *(Steven Rodie)*

When designing a landscape that requires a new or renovated irrigation system, consider the following:

- Irrigation equipment (back flow preventers, valve boxes, etc.) must be accessible. Access to faucets and equipment can be achieved by creating simple paths and providing proper clearance.

- Lawns and landscape shrubs have different water requirements and should be watered at different rates. Zoning plant materials will enhance plant health because water requirements can be matched to specific plants. This is especially important in irrigated landscapes, where over- or under-watering can easily occur on an regular basis.

- Planting areas less than 5 feet wide are difficult to irrigate effectively with overhead spray heads (Figure 4-43). Although irrigation heads are available for narrow strips, their efficiency in water application varies dramatically with wind and slope conditions. Spray patterns should always be overlapped to maximize efficient application. While narrow strips of turf between curbs and sidewalks may be difficult to avoid, narrow strips in other areas of the landscape should be avoided.

- Bedlines provide a strong sense of order when they are bold in their layout. Excessive curves or kinks in a bedline weaken the contrast between turf and other textures in the landscape. Simple bed layout also promotes efficient irrigation coverage and normally requires fewer sprinkler heads, resulting in cost-effective irrigation.

Renovating existing landscapes and reusing existing irrigation systems can be a difficult proposition. A designer should thoroughly assess the existing system to determine its current operational status, limitations, and flexibility. A wide variety of retro-fit irrigation solutions are available, but when these equipment changes are considered along with water pressure, volume and precipitation rates, a retro-fit system may still not meet the irrigation requirements of the new landscape.

FIGURE 4-44

This sustainable landscape includes many native and locally adapted species. *(Steven Rodie)*

IRRIGATION MAINTENANCE. While automatic irrigation systems reduce the time a homeowner spends moving sprinklers, the systems do require a minimum of annual maintenance. Some systems can be quite sophisticated when all of the valves, heads, and computer programming are considered. Working with a client to determine his or her comfort level with an irrigation system and recommending one that matches his or her expectations are important to the success of the system.

Incorporating Sustainability

A sustainable landscape minimally affects the environment by reducing inputs, reusing materials, or using recycled materials. This type of landscape can be both aesthetically pleasing and environmentally sound. Although it may cost more to install initially, a sustainable landscape has the added potential benefit of costing less over time (less maintenance, fewer chemicals, and less water). Proper plant placement can also prevent soil erosion, influence a household's energy consumption, reduce stormwater run-off, and attract beneficial insects and wildlife.

In order to have a truly sustainable landscape, a client may need to alter his or her expectation of how a landscape should look. Although many people expect immaculate lawns, amazingly manicured plants, and large shiny fruit, a sustainable landscape may not reflect perfection if low-input turfgrasses, unpruned plants, and selective chemical applications are part of a maintenance schedule. Creating a sustainable landscape means working toward a thoughtful balance between resources used and results gained.

Sustainable landscapes require as much, if not more, planning as a traditional landscape. The ultimate goal is to create an established plant community that becomes easier to care for as it matures (Figure 4-44). In nature, plants are typically found growing in associated communities. Establishing plant communities adds both aesthetic and functional benefits to a design. Examples of aesthetic benefits include order created by the massing of plants, increased diversity of textures among the mixed species, and habitat for wildlife. Conceptual plant communities are discussed in more detail in Chapter 6.

In addition to careful plant selection, sustainability can also be achieved by thoughtful selection of hardscape materials. A number of new landscape products made from recycled plastics and wood by-products are available that require virtually no maintenance and have a longer life span than wood. They can be used for decks, fences, benches, and planters; and they come in a variety of textures and colors (some of these products are described in Chapter 7).

Providing Wildlife Habitat

As discussed earlier, well-designed landscapes provide comfortable outdoor spaces for human landscape users. In addition, designers should consider the aesthetic and functional benefits of incorporating habitat value for wildlife. A significant interest in residential landscape wildlife habitat now exists in many areas of the country. From an aesthetic perspective, wildlife provides significant visual interest throughout the year and enhances enjoyment of a landscape. Functionally, birds and beneficial insects help lessen harmful or nuisance insect populations, and butterflies help pollinate many garden plants. While some wildlife can be destructive in a garden setting, selective use of resistant plants and strategic design of barriers can minimize damage.

SUMMARY

The design principles described and illustrated in this chapter are a starting point for design. Good design centers on meeting the client's needs by creatively applying the basic design principles to achieve an aesthetic and functional landscape. Incorporating landscape preference concepts in the design process and product can further enhance the quality of a design. As Church (1955) pointed out: "There are no mysterious 'musts,' no set rules, no finger of shame pointed at the gardener [designer] who doesn't follow an accepted pattern." Providing the design stays within the vague bounds of good taste, a landscape design is often a reflection of a designer's interpretation of what the client wants, given the constraints of the site.

KNOWLEDGE APPLICATION

1. Describe the five elements of art.
2. Apply each of the elements of art to landscape design, and describe two examples of each application.
3. List and describe the four overarching design principles.
4. Define the seven aesthetic principles of design.
5. Illustrate and describe the seven aesthetic principles of design in a landscape context.
6. Describe three applications of aesthetic design principles and the way each affects a person's experience in the landscape.
7. Describe the concept of functional design principles.
8. List and describe five applications of functional design principles in a landscape context.
9. Describe four examples of how aesthetic design principles are integrated with landscape preference.

LITERATURE CITED

Church, T. D. (1955). *Gardens are for people: Plan for outdoor living.* New York: Reinhold.

Doyle, M. (1996). *Color drawing: Design drawing skills and techniques for architects, landscape architects, and interior designers* (3rd ed.). New York: John Wiley & Sons, Inc.

REFERENCES

Austin, S. (1998). *Color in garden design.* Newtown, Connecticut: Taunton.

Kaplan, R., Kaplan, S., & Ryan, R.L. (1998). *With people in mind.* Washington: Island Press.

Lin, M. (1993). *Drawing and design with confidence.* New York: John Wiley & Sons, Inc.

Luke, J.T. (1996). *The Munsell Color System: A language for color.* New York: Fairchild Publications.

Stout, J., & Nelson, D. (1983). *Design posters for leaders.* Ames, Iowa: North Central Regional Extension Publication 110.

Stout, J., & Ouverson, C. (2000). *Design: Exploring the elements and principles.* 4-H 634. Ames, Iowa: Iowa State University Extension.

Wagner, J. (2003). *Basic psychological and physiological hydrostrategies.* Colloque/Symposium C 10.2 Chalcolithic and Early Bronze Age Hydrostrategies, Dragos Gheorghiu (Ed.). British Archaeology Reports International Series 1123: Oxford.

CHAPTER 5

Landscape Design Process

The process of design combines gathering information, setting goals, making decisions, and solving problems. In Chapter 3, we described design as a creative problem-solving process. In this chapter, we apply the process specifically to landscape design and include the necessary components for successful landscape design. These components are: establishing effective client–designer communications; helping the client develop a wish-list; collecting necessary information, including site documentation and analysis; producing the necessary graphic and written information to clearly describe the design; and completing the steps required to develop and refine a landscape design.

OBJECTIVES

Upon completion of this chapter you should be able to

1. describe the factors necessary for effective client–designer communications.
2. describe the contents of a client wish-list.
3. describe how to measure and produce a project base map.
4. describe how to inventory and analyze a project site.
5. define the components of a comprehensive design program.
6. define the role and importance of a design concept.
7. describe the role and components of each design drawing step (functional design, form composition, preliminary design, and final design).

KEY TERMS

Design Planning Questionnaire	setback	site analysis	landscape use area
covenant	site survey	sustainable design	preliminary design
plot plan	slope	design concept	
right-of-way	direct measurement	design program	
easement	baseline measurement	design theme	
	triangulation	functional diagram	

THE LANDSCAPE DESIGN PROCESS

Figure 5-1 shows the entire landscape design process from initial contact with a potential client to the client follow-up after the process is complete. This effort could take less than a month, or it could take several years—it all depends on the scope of the project. Some of these steps may be truncated or even skipped entirely, depending upon the particular challenges that face the designer.

There are as many approaches to the landscape design process as there are designers, clients, and landscapes. Amid these many variables, the most critical component to ensure success is process. A design professional's schedule can be hectic at times, especially during the spring and fall seasons. It can be tempting (and sometimes necessary) to cut corners and significantly shorten the design process. An abbreviated site inventory, a brief over-the-phone client interview, or final designs without a preliminary design review are all possible occurrences. These shortcuts may occur as economic reality overlaps with the notion of well-reasoned design. Good design and shortened process can go together as long as the designer is aware of what was skipped, why it wasn't required, or how the information might otherwise be included.

DESIGN PROCESS CASE STUDY

In order to illustrate the written and graphic products associated with each step of the design process, a case study design project, The Jones Property, will be used to summarize each section in this chapter. Additional case studies are found in the appendix.

CLIENT–DESIGNER COMMUNICATIONS AND THE SALES PROCESS

Good communication between a potential client and the designer must occur if successful and profitable projects are to be designed and implemented. This communication usually begins with an inquiry brought about by effective marketing and continues throughout the design process until the landscape is installed. Effective communication early on means soliciting your client for information and listening to the responses. It is the charge of the landscape designer to divine the desires of the client and to package them into a design the client will enjoy. This section will address these first contacts with the potential client, including meetings and meeting strategies, establishing outcomes, budgeting, and data-collection opportunities.

FIGURE 5-1

The landscape design process can vary greatly in approach, number of steps, and timeframe.
(Joseph E. VanDerZanden)

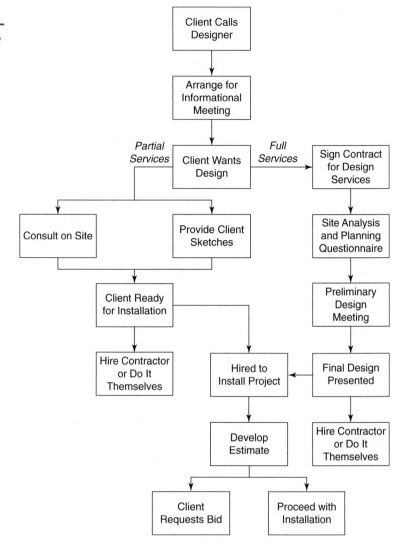

Making Initial Contact with the Client

The sales process starts when an interested party learns about the design services through some form of marketing, advertising, or referral and seeks more information by contacting the company. Generally, the potential client contacts the designer via telephone or e-mail. During this initial discussion, the designer may gather some preliminary information from the client about his or her landscape needs, but usually the details gathered during this brief exchange are few. From the designer's perspective, the most important outcome of this initial contact is to arrange for a formal, face-to-face first meeting.

Design Planning Questionnaire

Ideally, there will be ample time between the initial contact and first meeting for the client to complete and return a **Design Planning Questionnaire**. Some designers require this step before an interview so clients refine their focus. This will provide the

CASE STUDY

Introduction

The Joneses are a young couple who recently purchased their starter home in a Midwest suburban subdivision. The new house is located on a 70' × 125' lot, and is presently surrounded by a mix of built homes and open lots (Figure 5-2). The property reflects characteristics typical of many new suburban residences, including relatively narrow side yards, a double-car garage with a large driveway, and a walk-out basement with an overhead deck.

FIGURE 5-2

Frontyard (a) and backyard (b) views of the Jones residence. *(Steven Rodie)*

(a)

(b)

designer a chance to review a potential client's wants and needs and learn a little about how he or she plans to use the landscape. Responses to the questionnaire can be discussed in more detail and clarified during the first meeting.

The Design Planning Questionnaire also aids the designer in setting the meeting agenda for additional design discussions. Information about family members, pets, and personal plant preferences are important, as are determining whether there is a need for outdoor entertaining space, outdoor lighting, or an outdoor storage area. Questions pertaining to client knowledge of site conditions are also valuable, although the clients will likely know little about the site if they are new to the home. The broader the range of initial questions, the better the feedback. There are as many formats for questionnaires as there are designers. (A typical range of questions is found at the end of the chapter.) The client should be encouraged to approach portions of the questionnaire as a personal wish-list and to delve into the needs and desires that the landscape could address.

Company Web sites are a useful tool for delivering a design questionnaire. They also provide pictures of company projects that can stimulate potential client interests. While a premeeting questionnaire may not always be possible, it does help focus the client and designer.

The Business Side of Design

Every successful company typically has a plan that focuses business efforts on a particular group of potential clients. This group is referred to as a market segment. It is no different for successful design or design/build companies. Although a designer or design company just starting out may need to say "yes" to virtually every interested client who requests assistance, over time it pays to be more selective when choosing customers. Selections may be based on a combination of factors, including size of project, area of town or neighborhood, distance from business location, or type of design. By focusing on a particular market segment, you can increase profit returns on such things as efficiencies in travel or labor costs, specialized design training, and word-of-mouth advertising.

Meeting with the Client for the First Time

There is only one chance to make a first impression. Dressing professionally, setting a friendly tone, and getting to know the client are all important at this first meeting. Remember that you must sell yourself before you can sell your design.

DEVELOPING A FIRST MEETING STRATEGY. Consider developing a meeting kit to be used when meeting with a client for the first time. The kit might include a portfolio of work including hand-drawn and rendered landscape plans, computer-generated plans, section and elevation illustrations, photographs showing before and after pictures from past projects, locations of past projects, a list of

FIGURE 5-3

Company portfolio book or Web site image. *(Linda Grieve)*

Selling versus Marketing

There are fundamental differences between the terms *marketing* and *selling* that hold significant importance for landscape designers. To sell is to "give up, deliver or exchange (property, goods, services, etc.) for money or its equivalent" (*Webster's New World Dictionary*, 1994). Marketing, in comparison, is "the process of planning and executing the conception, pricing, promotion, and distribution of ideas, goods, and services to create exchanges that satisfy individual and organizational goals" (Pride and Ferrell, 1991). A designer who focuses entirely on selling design may be missing out on the potential benefits that marketing design provides—for the designer as well as the client. A marketed design will establish a designer's reputation and a framework for client references. It can also package good design as a lifestyle enhancer that leads to repeat business. When a designer goes beyond just selling and begins to market his or her services, he or she will likely realize greater rewards.

recent clients whom they could contact, and a Web site reference for additional project examples and company information (Figure 5-3).

In addition to work examples, develop a script or agenda that guides the discussion during the meeting. This will help ensure that all important points are covered and that specific opportunities to gather input from the client have been provided. It is easy to get distracted during a meeting, and following an agenda helps the designer leave the meeting with the minimum amount of information needed to proceed.

One of the most important first-meeting considerations is whether the design is for a new or renovated landscape. Each type requires a slightly different set of information.

FIGURE 5-4

A new home on a new site provides many design opportunities, but it can be difficult to know where to start. *(Jon Wilson)*

The New Landscape. New residential landscapes often present a blank canvas for design ideas. Homeowners facing an empty yard may have difficulty visualizing the design potential, especially if it is their first home landscape experience (Figure 5-4). When homeowners have lived on a property for a short time, site characteristics (such as a good view or a potentially poor drainage area) may not yet be evident. In such cases, the designer may need to explain site opportunities and constraints to the homeowner by using additional design drawings and descriptions. A barren new landscape will often compel homeowners to want immediate evidence of design implementation. In these cases, strategies such as planting proposed shade trees the first year (perhaps in relatively large sizes) or establishing some key hardscape areas will help start the project on the right note. In succeeding seasons, beds can be filled with plant material and design details can be added, and the homeowners will begin to personalize their landscape.

The Renovated Landscape. While similar in some ways to developing a new landscape, renovation presents distinct challenges. Renovation implies that a landscape needs to be corrected. In cases where a homeowner lives with a particular problem such as overgrown plants or fruit dropping on a patio, the needed corrections are obvious. For homeowners who have recently purchased a preowned landscape, however, careful thought should be given to proposed changes. Changes to address personal preferences are likely with the change in ownership, but other landscape problems may be less obvious. If certain plants are growing poorly, there may be a variety of reasons for the problem. Replacing the plants may correct the symptom but not address the underlying problem responsible for the poor growth, such as poor drainage, overwatering, or a change in sun and shade conditions. Identifying the source of a problem in an existing landscape can be a significant challenge when both the homeowner and designer are unfamiliar with the site (Figure 5-5). Renovation offers many challenges to the designer and prompts a different set of questions in the initial meeting discussions.

FIGURE 5-5

A wide variety of issues may be responsible for the poor turf quality in this landscape in need of renovation. *(Steven Rodie)*

Design Approach Case Studies. Four landscape design professionals were interviewed and asked to summarize how they would approach two different design scenarios. Scenario 1 is a landscape at the newly constructed Smith home, where the homeowners have a limited budget ($3000) and do not know where to start. Scenario 2 is a renovation of Mr. and Mrs. Newberry's existing landscape. They have lived in the home for 15 years, and recently the children have left for college. Now they want to update their landscape and include room for entertaining groups of 8 to 10 people. These detailed interviews are included at the end of the chapter and provide valuable insight for beginning designers as to how differently these experienced professionals approach these two common design scenarios.

First Meeting Location. The meeting location is also important. Some clients may be intimidated by commercial offices with large conference rooms and may feel they will be faced with an aggressive sales situation. They may prefer the comfort and safety of their own home, where they can meet at the kitchen table and look out at the potential landscape through a set of sliding glass doors. Meeting in the client's home also provides an opportunity to walk through the proposed landscape site and gather additional data. Other clients, however, may be impressed with an attractive commercial office. Such a space gives the landscape designer credibility and assures the client of his or her professional aptitude. Asking a client about his or her preference during the initial phone or e-mail contact can help you determine which location would be most appropriate. Regardless of the location, it is important to conduct this meeting where there are no distractions and in a comfortable setting where the client is at ease.

First Meeting Outcomes. The first meeting can have important outcomes for the designer and client. Using a consistent, even scripted, approach will ensure that the goals of the meeting are achieved.

Designer Outcomes. The designer has much to learn about the client during the first meeting. The designer will want to discuss the client and his or her lifestyle, how he or she plans to use the outdoor living space, maintenance expectations, what he or she expects from the designer, and the project budget. Examples of questions for the meeting include: Why do you want a change? What do you think is worth saving in the existing outdoor space? What do you think should be done about the parts that are not worth saving? If the meeting is held at the client's home, a walk through the landscape site usually yields even more information. Asking additional questions and taking preliminary field notes enables a comprehensive understanding of what the client wants and what the site has to offer.

Determining whether a client wants a sketch on a cocktail napkin or a three-dimensional computer-animated, fully rendered plan is another important outcome of that first client meeting. Some clients are quite capable of taking a germ of an idea and developing it into a functional and aesthetic landscape. Others have no interest in the hands-on approach to their design needs. They prefer that the landscape design be developed and installed with minimal input on their part. Develop questions for the first meeting agenda (script) that will help determine whether the client is a do-it-yourselfer or a complete hands-off sort of homeowner.

How Much Artistry Do They Want?

One of the unspoken expectations many clients hold about their designer is visible creativity in the drawings. Many residential design drawings are quickly sketched by hand or on a computer, especially in design/build situations where details may be developed in the field or the drawing fee is rebated. The drawings provide a clear rendition of the design but lack the aesthetic style or flair that some clients expect to see from an artist. As discussed in Chapter 2, the graphic style of a drawing can reflect as much about a designer as the clothes he or she wears and the attitude he or she brings to the initial meeting. Identifying the art expectation of the client in the initial meeting may be an important consideration in setting the framework for a successful design.

Client Outcomes. The outcomes from the first meeting for the potential client should also be clear. Ideally, the client will be informed about design philosophy, learn the designer's approach to design process, and determine if he or she wants to hire the designer. Although he or she will likely only see preliminary designs and a final design for the landscape, the designer's description of the other steps of the process (site analysis, functional diagrams, etc.) will demonstrate the work that goes into the final drawing. By explaining the multistepped approach and showing before and after photographs of successful projects, the designer can enhance clients' confidence in his or her skills and increase the likelihood of being retained.

The Budget Question

It is critical to discuss the client's budget during the first meeting. The designer needs to know the budget and what the client expects for his or her money. For example: How much has he or she budgeted just for the design? How much, if any, for the installation? What is he or she willing to spend for maintenance? It is also important to determine over what period of time the client is planning to implement the project. A large, multifaceted project can be completed in phases in order to better meet the client's budget.

In addition, homeowners sometimes enter design discussions with a limited sense of project potential, and they initially set their budget to match their expectations. Once they begin to appreciate how the outdoor living space can be enhanced or how problems with the existing site can be remedied by quality design, they may be willing to increase their budget to encompass the additional design opportunities.

THE BUDGET-SENSITIVE CLIENT.   Not all homeowners will have the requisite 8 to 10 percent of the value of their property to spend on their home landscape. Yet this does not mean they should be overlooked as potential clients. With a little creative marketing, budget-sensitive clients can enhance their property without depleting their savings. Designers should not be tempted to use low-quality hardscape or plant materials in the design or to skimp on bed preparation. This will only result in a substandard landscape that will need replacing in a few years. Instead, show the client how much each component of the design costs in relation to the total landscaping budget, and let him or her decide how he or she wants to allocate the landscaping funds. In addition, phasing the design should be promoted as an opportunity to maintain design quality while making the project more affordable. Given a structured yet flexible master plan, homeowners can proceed with a design installation over many years and still have a continually viable design.

Recommending less expensive alternative materials that still meet aesthetic and quality standards might be another method to save on project costs. For instance, use concrete pavers instead of natural stone for a patio or walkway. Another approach is to concentrate the more expensive materials near an entrance or other highly visible location. For example, if the client wants a wrought iron fence around the property, but is unable to budget for it, use the wrought iron for a gate area and substitute a black vinyl-coated pressed steel fence for the less-visible areas.

Another option is to combine two or three smaller projects in one area of the landscape to maximize the visual impact. Using bold color and textures will draw the eye to that area. Incorporating brightly colored annuals can be effective in highlighting the area, but this will also add to overall maintenance costs. A better alternative for this client would be to substitute flowering, woody ornamentals or low-maintenance perennials to achieve a colorful yet lower-maintenance effect.

A client can also stretch the landscape budget if he or she is willing to provide some sweat equity on the project. He or she might want to do some of the lighter work, such as installing plants, and focus the budget on more labor-intensive parts of the project that may require specialized equipment like a skid loader or plate compactor.

An additional option is to offer the client a discount, 25 percent for example, on design and installation if he or she is willing to wait until the slow season for the project to be completed. Although your profit margin may seem less because of the discount, you may actually be making a larger profit on the project. This larger profit could be realized because installation crews are not being paid an overtime wage, which is typical during peak demand times of spring and summer. In addition, it is likely that a design developed prior to the normal heavy spring and fall seasons will have a more flexible time schedule for completion, which can improve profitability as well as ultimate design quality.

CASE STUDY

Client Feedback

Mr. and Mrs. Jones are planning to live in the house for 4 to 5 years. They have a strong interest in gardening and outdoor entertaining, but typically do not entertain large groups. This will be their first experience landscaping a property, and although they have a lot of personal interest, they feel uninformed when it comes to landscape ideas. Their few specific requests for the design are as follows:

- favorite plants—Legacy sugar maple, blue spruce, red bud, burning bush, lilacs, morning glories (vines), shrub roses, hostas with variegated dark green and white leaves, hibiscus
- favorite colors—blue and purple
- low maintenance (a very high priority)
- privacy for patio and deck areas
- vegetable garden (150 square feet maximum)
- conveniently located but screened garden shed (10′ × 10′) for storage
- additional patio space (extra 100 square feet) (but cannot afford to expand deck)
- a design that accommodates children and pets (fence and turf play area), even though they do not currently have children and pets

SITE DOCUMENTATION

After the designer has been retained to provide landscape design services, the first step in the landscape design process is to document the site through mapping and photography. Some designers will begin their documentation during the first meeting by taking initial photographs and site measurements; others will arrange for a follow-up time soon after the meeting. Some designers may have staff assistance available for site documentation tasks. Regardless of how the site information is gathered, it needs to be completed in an efficient manner. The site should be documented thoroughly so information is not forgotten or misunderstood later. The data collected must be accurate so the designer can rely on these dimensions during the design process.

Photos and Videos

Photographic or video site documentation is a valuable, time-saving resource for the designer. In the time it takes to develop a design (especially if projects are being worked on simultaneously), it can be difficult to remember the locations or specific characteristics of important site features. Return visits to the site to verify measurements can be an expensive when compared to a quick reference to photos or a video. In addition, photography can be a powerful tool in communicating design ideas. Although most design documentation occurs on a site base map in plan view,

FIGURE 5-6

Quickly sketching landscape ideas over a site photograph can be an effective design tool. *(Courtesy of Karen Richards. Steven Rodie)*

Design Context and Creativity

A design sketched on the back of a napkin often congers up the romantic image of a designer's ideas that were quickly and elegantly fashioned at a moment's notice. But implementing those ideas without specific sizes and relationships can be nearly impossible, especially if someone besides the designer is implementing them. Without an accurate base map that provides a framework for scale and orientation, the napkin sketch is nothing more than, well, a napkin sketch.

elevation sketches on photographs can be quickly generated, and they clearly show ideas to the client (Figure 5-6). (See Chapters 1 and 2 for techniques.)

Developing a Base Map

An accurate, correctly formatted base map is required to develop a landscape design. In this section, we discuss base map development, including sources of existing site information and efficient methods to measure the site and gather additional information.

A complete base map will typically include the following information: property lines; utilities; house footprint with windows and room uses; roof overhangs; and existing features, including vegetation, setbacks, and easements. Formatted on a sheet with a title block, north arrow, scale (graphic and written), and room for notes and plant information, the base map provides the literal base for the entire design process (Figure 5-7).

EXISTING INFORMATION SOURCES. Existing base map information is derived from a variety of sources, including the lot drawing, a plot plan, and a plat or site survey. Each of these contains different types of information, and it is important for a designer to know where each can be located (county assessor's office, city planning office, etc.). In addition to information found in these sources,

FIGURE 5-7 A thorough base map provides an important foundation for every landscape design. *(Steven Rodie)*

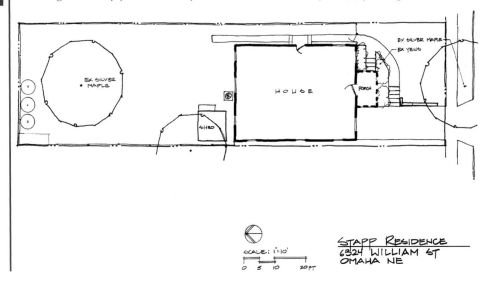

other property information that prescribes restrictions on landscaping or fence building should also be included on the base map. Land developers sometimes implement property agreements called **covenants** that limit individual land owner practices for the ultimate aesthetic or functional benefit of an entire development. Covenants are legally binding agreements and must be adhered to by homeowners, so they are an important component of background design information.

Lot Drawing. Lot refers to the client's property in the context of surrounding properties. Most residential lots are located in a subdivision of land that defines private property ownership and public land (road and street rights-of-way, parks, and school sites, etc.). Lots vary in size and shape, depending upon neighborhood, street layout, and site conditions (Figure 5-8). Lots are typically defined in the field by buried property or corner stakes or by other permanent site features (notch in a sidewalk, manhole cover, etc.). Property lines on maps and diagrams are defined by their length as well as their orientation to true north or south (bearing). When the boundaries of a property are surveyed, the surveyor begins at one property corner and proceeds to measure the length and bearing of each side of the property, completing the outline of the property at the beginning corner (Figure 5-9). Boundary bearings provide the designer with a precise indication of how the site is oriented for sun and shadow patterns, as well as **prevailing winds**—both critical factors in the analysis of site conditions.

prevailing wind: The direction from which winds most often and most typically blow at a certain location.

Plot Plan. A **plot plan** summarizes all of the survey information documented for a lot. In addition to property boundaries, a plot plan includes rights-of-way, setbacks, and easements.

Right-of-way is public property where streets, roads, and sometimes alleys are located. Right-of-way is normally wider than the street surface and typically includes the sidewalk, the parking strip or drainage swale along the street, and the trees

FIGURE 5-8

Lot sizes and shapes can vary dramatically depending on location and site conditions. *(Steven Rodie)*

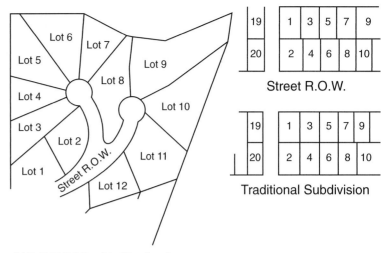

FIGURE 5-9

Lot dimensions are defined by the lengths and bearings of property lines. *(Steven Rodie)*

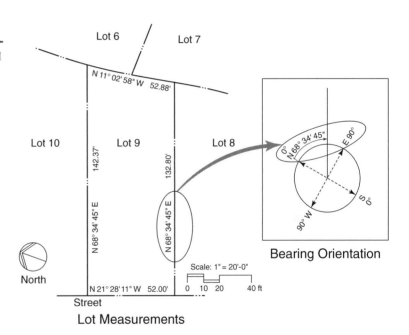

planted near the street. Since maintenance of the sidewalk and landscaping in this area is normally the responsibility of the homeowner, the ownership difference is sometimes overlooked. It is important to verify the relationship and width of adjacent right-of-way to a client's property so that design features are not inadvertently located in the area (Figure 5-10). Plants and other landscaping are typically allowed, but specific regulations (such as tree permits to ensure proper species selection) may apply to right-of-way improvements, and you should note them.

A property **easement** allows access to the property for utility maintenance, storm-water drainage flow, access to other properties, etc. It is important to know where easements are located and the restrictions that have been placed on their use, since these restrictions may influence design decisions.

FIGURE 5-10 The bold lines in this neighborhood aerial photograph illustrate how much residential frontyard can actually be public right-of-way. *(Steven Rodie)*

variance: An exception to land-use regulations (usually granted through a regulatory review process) for a property that may not otherwise meet the legal requirements of a zoning regulation.

A **setback** indicates the distance that structures must be located from property lines. Most lots have setback requirements for the front, back, and side yards to ensure minimum distances between structures on adjoining properties (Figure 5-11). **Variances** may be granted to allow a property owner to place a structure closer to a property line than development regulations typically allow. In order to maximize the functionality of outdoor space, some new developments are designed with a zero lot-line approach, which allows a structure sidewall to be placed adjacent to the property line.

Plat or Site Survey. A **site survey**, sometimes referred to as a plat, combines the plot plan information with accurately located site features, including the house footprint, driveway, sidewalks, etc. (Figure 5-12). The survey is usually available from the property owner since a copy is normally included in the closing papers of a property purchase. If it is not available from the owner, then the house architect or contractor, mortgage company, or county assessor will likely have a copy. It is normally the responsibility of the homeowner to furnish a site survey. Any time allocated by the designer to finding the information should be accounted for in the design fee. Using a site survey whenever possible can save significant drawing time and enhance the accuracy of the map.

Once all of the site information is gathered from the plot plan and site survey, additional information must be measured and added to produce a project base map. The types of information, as well as the required tools and techniques, are discussed in the following section.

FIGURE 5-11

Property setbacks define the minimum allowable distances between structures and property lines. *(Steven Rodie)*

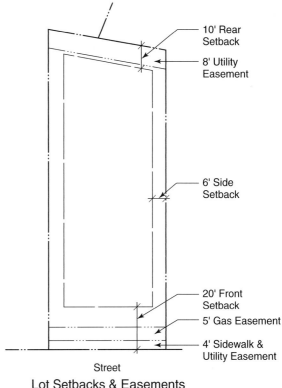

One type of information not typically included on a site survey is **slope** or contour information, which indicates the steepness of the site and indicates where water will drain on the site. This information can be relatively expensive to produce by a surveyor and may not be critical for sites that have reasonable slopes and drainage. For complex sites, however, where significant retaining walls will be required, a contour survey may be worth the additional cost. In some cases, the designer can use basic techniques to collect slope information; this will be discussed in a later section.

MEASURED INFORMATION. Site measurement is a part of virtually every landscape design project. As noted above, many site features are not documented in the property information typically provided to homeowners. Even if the information is available, it never hurts to verify a few critical measurements to ensure the accuracy of the information provided.

All measurements should be recorded legibly so they can easily be understood by the person producing the base map. In cases in which the map is not drawn immediately, information can easily be misinterpreted or forgotten if not saved in a legible, logical format. A clipboard with sheets of grid-lined paper allows the designer to quickly sketch the layout and measurements of basic elements at a set scale that keeps the drawing consistent (Figure 5-13).

FIGURE 5-12

A site survey (or plat) combines lot dimensions with structure, paving, utility, and other important information. *(Steven Rodie)*

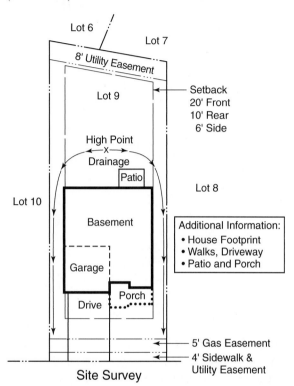

Site Survey

FIGURE 5-13

A clipboard is an essential tool for base map preparation; grid paper expedites note-taking and dimensional measurements. *(Steven Rodie)*

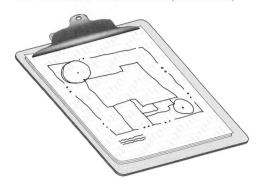

Important Additional Site Measurements. The following checklist identifies features not typically on a site plan that should be included in the measurement process (Figure 5-14).

- locations of known utility lines and boxes, including overhead and underground power, cable and telephone service; electric lines and water or sewer pipes are typically installed deeper than most landscape excavations; cable, telephone lines, and irrigation systems, however, are often near the surface
- locations of hose connections (hose bibs), irrigation valves, and system **backflow preventer**
- downspout locations
- the amount of clearance under windows
- the height and number of step risers in a set of stairs (so the amount of vertical drop can be calculated)
- the heights of retaining walls

backflow preventer:
A device on an irrigation system (normally near the water source) that prevents system water from flowing backward into the water source due to a water supply break or other loss of pressure. Water backflow can create a health hazard if the water source becomes polluted as a result of the backflow.

FIGURE 5-14

Adding additional measured and observed site information to the site survey enhances the value of the base map for design use. *(Steven Rodie)*

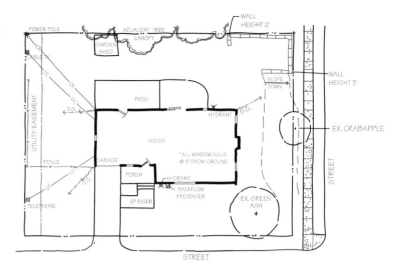

Locating Utility Lines....the Safe Way

To confirm utility locations, a resource called Digger's Hotline or One Call is available in many areas of the United States. Its purpose is to coordinate the notification of utility companies on proposed excavations so that utility locations can be verified and marked prior to starting the project. In some states, it is illegal to begin digging on a project if the proper contacts have not been made. Even with relatively small excavations such as fence post holes, digging near a telephone or cable television box along a property line can damage a shallow wire. With one phone call, all utilities can be located and avoided—quite a bargain when compared to the alternative of finding a utility line accidentally.

- the approximate top (crest) and bottom (toe) of major slopes
- the dimensions of structures or other features (including overlapping tree canopies or storage sheds) in adjoining lots that may have an impact on the design

MEASUREMENT TOOLS AND TECHNIQUES. Thorough and accurate site measurement is critical to accurate base map production. Appropriate tools are a significant factor in collecting accurate site information.

When used correctly, the following tools and procedures enhance accurate linear distance measurements (Figure 5-15):

- A steel measuring tape, preferably 100–200 feet in length, does not result in the distortion of a stretched cloth tape.
- A smaller 25-foot steel retractable tape is used for short horizontal and vertical measurements.

FIGURE 5-15

A variety of equipment is used to accurately and efficiently measure site conditions, including traditional levels (a), distance measuring equipment (b), and laser levels (c).

String Line Level

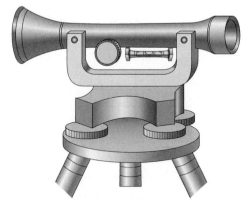
Dumpy Level

Carpenter's or Builder's Level

(a)

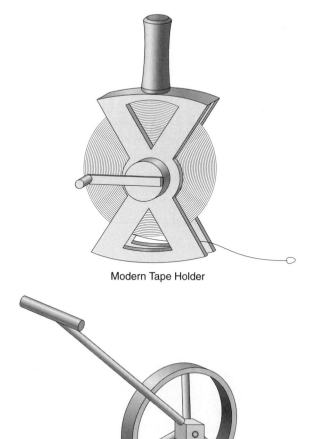
Modern Tape Holder

Odometer Wheel

(b)

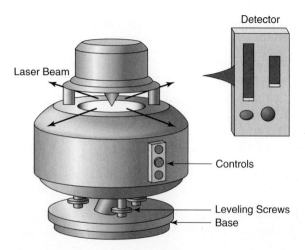

(c)

FIGURE 5-16

Direct measurement is the simplest measurement procedure. The measuring device must be repositioned for each measurement, however, and errors can accumulate when numerous individual measurements are added together. *(Steven Rodie)*

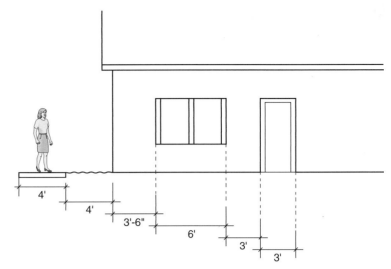

- Laser measuring devices provide an option to the traditional tape measure. They allow the designer to work independently and can be more efficient.
- A measuring wheel with a large wheel and strong frame is preferred for its consistency on steep slopes, uneven surfaces, and tall vegetation. When used correctly, a measuring wheel can save valuable time.
- Accurate pacing is an often-used procedure to determine lengths. In this case, the designer simply paces off the distance in question and multiplies the number of paces by a predetermined stride length. When pacing, keep in mind that anything affecting a normal stride will impact accuracy.
- A builder's level is a surveying device mounted on a tripod; when used with a measuring rod, it determines the rise or fall of a slope over a distance.
- Laser levels, available in a wide range of capabilities, can yield similar topographic site information as a builder's level. They project a perfectly level laser line across a great distance.
- A mason's line equipped with a line level is another useful tool to measure a slope over any distance.

COMMON SITE MEASUREMENT TECHNIQUES. Common site measurement techniques include: direct measurement, baseline measurement, triangulation, vertical or height measurement and estimation; and slope estimation. To accurately and efficiently measure a site, designers should be familiar with how each technique works and when to use it.

Direct Measurement. Measuring the distance between two points (e.g., width of a sidewalk) or two objects (between the fence and side of the house) is a **direct measurement**. This is the common technique for all measuring (Figure 5-16).

Baseline Measurement. **Baseline measurement** is a method that generates the linear dimensions of features, such as window and door locations along an exterior wall.

FIGURE 5-17

Baseline measurement generates distance increments along the entire length of a measured element; the measuring device is established once, and cumulative error is reduced. *(Steven Rodie)*

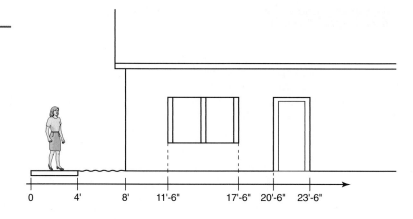

FIGURE 5-18

Incremental measurements along the baseline can be calculated by subtracting dimensions between two points. *(Steven Rodie)*

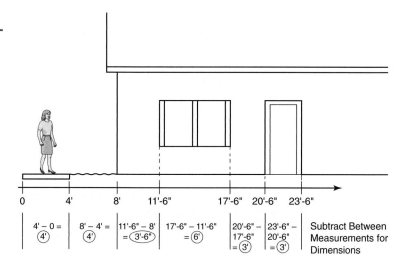

FIGURE 5-19

A baseline located along an irregularly shaped landscape feature provides an accurate measurement method. *(Steven Rodie)*

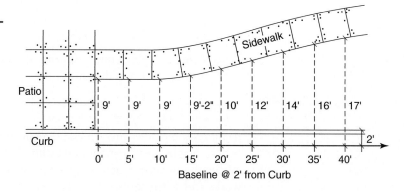

It is preferred because it is more efficient than continually moving the measuring tape (Figure 5-17). Multiple readings between the starting point and each of the points along the baseline are taken. The full measurement is required to draw major base map features, and subtracting adjacent dimensions provides the widths of individual features (Figure 5-18). In addition, a baseline can also be used to measure irregular shapes (Figure 5-19).

FIGURE 5-20 Triangulation is an efficient method for accurately measuring tree and other object locations. *(Steven Rodie)*

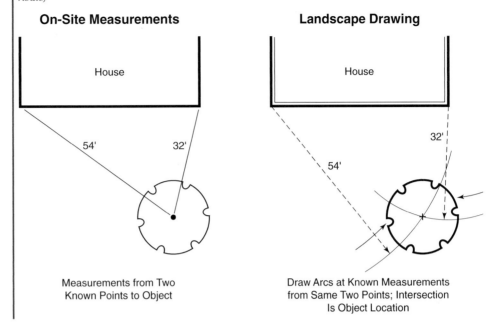

FIGURE 5-21 Multiple triangulation measurements can verify object endpoints and measure curving lines. *(Steven Rodie)*

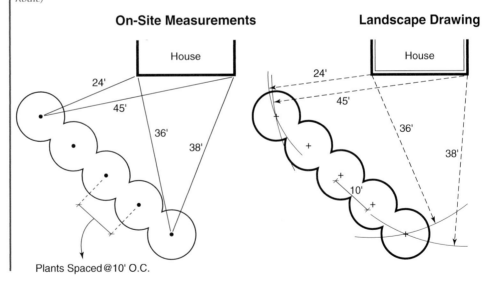

Triangulation. **Triangulation** is the process of using two known points in the landscape (such as two house corners) to locate a third point (such as a tree trunk) on the base map. As shown in Figure 5-20, distance measurements taken from the two known points are drawn so that their arcs intersect; this point locates the third element on the plan. Triangulation can also be used to locate the ends of an element (such as a hedge) or a series of points, such as those that form a curving bedline (Figure 5-21).

FIGURE 5-22

Visually scaling the height of an unknown landscape object with a known measurement can provide a reasonably accurate height estimate. *(Steven Rodie)*

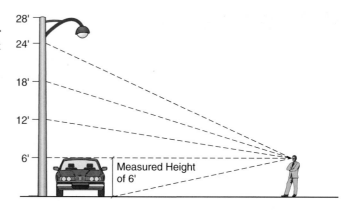

FIGURE 5-23

A mason's line and line level can be used to measure slope over relatively long distances. *(Steven Rodie)*

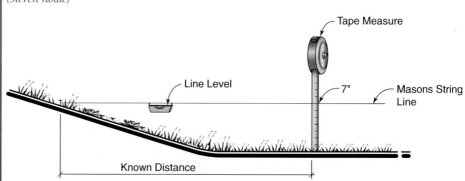

Vertical or Height Measurements and Estimation. It is valuable to surmise the height of landscape features so they can be referenced on the base map to document the scale of outdoor spaces. If the measurement is relatively easy to reach, such as the height of a single-story roof eave, it can be measured directly by using a steel tape measure. For taller measurements, heights can be estimated by using the height of a known object to gauge the length of the element in question. For example, a power pole might be estimated to be five Land Rovers tall. Multiplying the actual height of a nearby Land Rover by five will yield the estimated height of the power pole (Figure 5-22).

Slope Estimation. As previously discussed, slope information is not commonly included on a plot plan or site survey, yet it can be a critical feature of a site. While accurate topographical information can be determined using surveying tools, other field methods can also provide usable data. A mason's line and line level provide a reasonably accurate slope reference (Figure 5-23). Over shorter distances, such as a small patio or walkway, an 8-foot carpenter's level can also be used to gauge a slope (Figure 5-24). A house's foundation can be used as a reference to project the vertical drop of slope over a known distance (Figure 5-25). Sample calculations for determining slope are included in the Site Inventory and Analysis section.

FIGURE 5-24 A carpenter's level is used to measure slope over relatively short distances. *(Steven Rodie)*

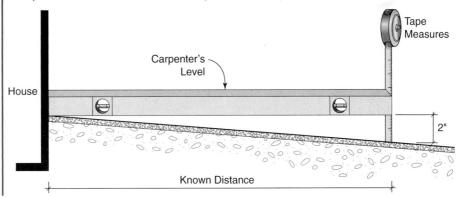

FIGURE 5-25 Existing level lines and vertical height measurements can be used to accurately estimate slopes and grade changes. *(Steven Rodie)*

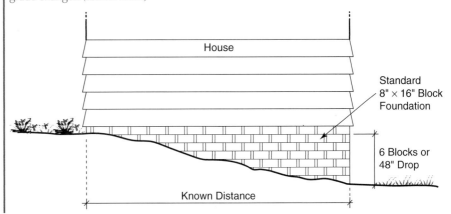

REMOVED BASE MAP INFORMATION. Having an accurate and complete base map is important, but it may be difficult to determine just what to include. For example, prior to the time the base map is completed, a designer or client may decide to remove an existing patio or move an existing retaining wall. Does it make sense to include this information on the base map, even though it will not be part of the final design? The answer is yes, and there are a number of reasons why.

- Elements planned for removal, but left on the base map, can aid a client's understanding of how the landscape will be changing.
- Leaving removed items provides a clearer indication of proposed changes for the landscape installer.
- In all cases, removed items should be shown by light, dashed lines (outline of a removed patio) or other unobtrusive symbol (a small 'x' at the center of a removed tree trunk, for example) so that the drawing does not become visually cluttered. The information should be drawn in pencil, so it can be easily removed later.

CASE STUDY

Base Map Information

A variety of site photos were taken to document site information. In addition to the photos of the front and back of the house (Figure 5-2, page 175), additional viewpoints included side yards, the view from under the deck, and pan views of the front and backyards to provide neighborhood context (Figure 5-26). All viewpoints were noted on a draft copy of the base map (Figure 5-27) for easy reference.

Mr. and Mrs. Jones provided a copy of their property site survey for base map development. The survey contained most of the important site features. Field measurements were required to locate the air conditioner, hose bibs, the specific layout of the patio under the deck, and the locations and dimensions of windows and doors. After adding and verifying measurements, the designer developed a final base map (Figure 5-28).

FIGURE 5-26

A variety of photo viewpoints should be selected to record existing site conditions, including side yard (a), views from interior rooms and exterior patios (b), neighborhood street context (c), and neighborhood backyard context (d). *(Steven Rodie)*

(a)

(b)

(continues)

CASE STUDY
(continued)

(c)

(d)

FIGURE 5-27

Photo locations should be recorded on a base map copy for future reference. *(Steven Rodie)*

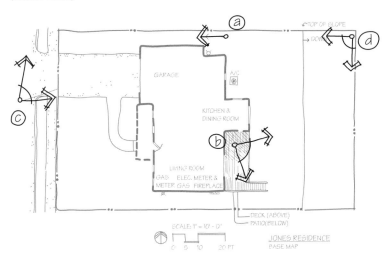

PHOTO LOCATIONS
FIGURE 5-26 a, b, c, d

(continues)

CASE STUDY
(continued)

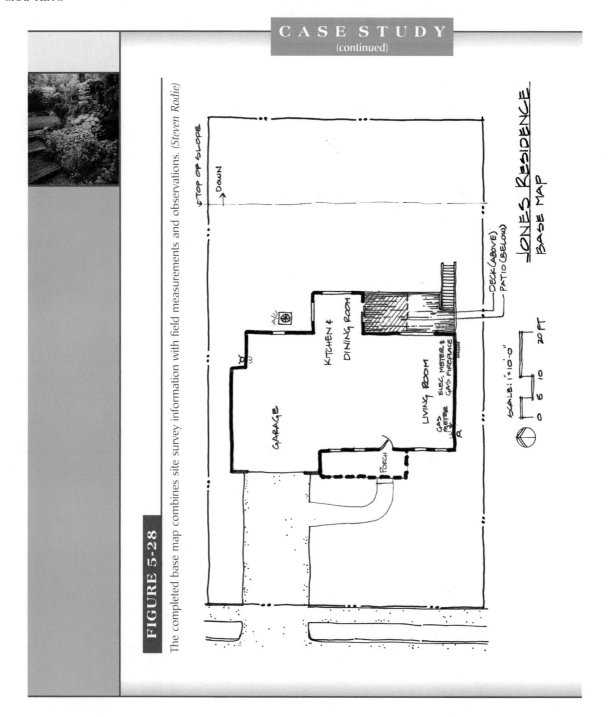

FIGURE 5-28 The completed base map combines site survey information with field measurements and observations. *(Steven Rodie)*

 ## SITE INVENTORY AND ANALYSIS

Evaluating a site provides an important opportunity to assess the site's opportunities and constraints. During this process, the designer and client discuss current site conditions and decide on design alternatives that will take advantage of opportunities and minimize problems. As part of this process, there are a number of things to

FIGURE 5-29 Site inventory records existing conditions. *(Steven Rodie)*

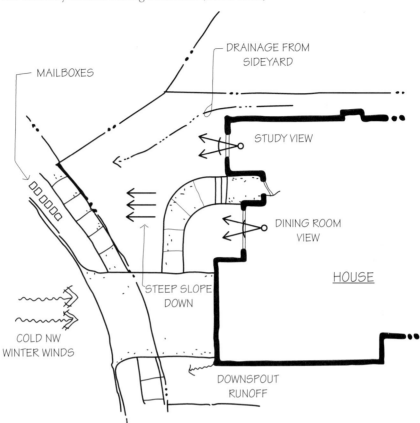

consider, including: making the most of the homeowner as an information resource, completing a site inventory followed by a site analysis, and thoroughly documenting the gathered information.

Capitalizing on the Homeowner as a Resource

The homeowner is an important resource in understanding site conditions. Regardless of how long the client has lived in the house, he or she will have spent more time on the property than the designer. Even though most homeowners are not trained in site analysis, they will still be able to identify poor drainage areas (or the corner of the basement that always floods during heavy rains), views they appreciate, or times of the day when shade would be a benefit. Portions of the Design Planning Questionnaire are typically devoted to site assessment and should provide useful feedback from the homeowner.

Two Distinct Steps: Inventory and Analysis

There are two distinct steps in assessing a site, and both are essential. Site inventory is the identification and description of issues. There are no judgments involved, just the documentation of an issue or set of circumstances. Identifying a kitchen window view, noting a poor drainage area, verifying the health of an existing plant, and identifying prevailing wind directions are all inventory steps (Figure 5-29).

FIGURE 5-30

Site analysis responds to site conditions with judgments, assessments, and proposed design ideas.
(Steven Rodie)

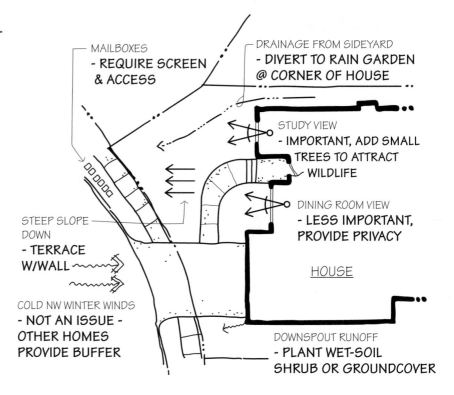

Analysis-Paralysis

Although a thorough site analysis for each design project should be on every designer's checklist, each analysis may include a slightly different issues focus. Steep slopes may be the primary concern on one project, while lack of privacy may be the major issue on another project. Because of this diversity, a designer must be flexible in determining which site factors to address. None should be ignored, but many may be quickly acknowledged as nonissues for a particular project. A designer's ability to prioritize these issues will alleviate the potential for analysis-paralysis, where unimportant factors complicate the process, but small critical design issues are not forgotten.

Site analysis goes a step further and judges the issue and, if warranted, proposes changes or enhancements. Noting that a kitchen window view is important and deserves attention because the kitchen is the most active room in the house represents analysis. The designer might take the analysis a step further and suggest that a strategically placed ornamental tree and birdfeeder outside the window would enhance the view. Analysis of poor drainage might identify it as a problem to be solved (through the construction of an underground drainage system) or an opportunity to maximize (such as creating a wetland with the proper selection of plants and soil amendments). In either case, based on client input and the creativity of the designer, a judgment will ultimately lead to a design response (Figure 5-30).

FIGURE 5-31 A typical site analysis for a suburban residential property. Courtesy of Tanner Draemel and Arik Solberg. *(Steven Rodie)*

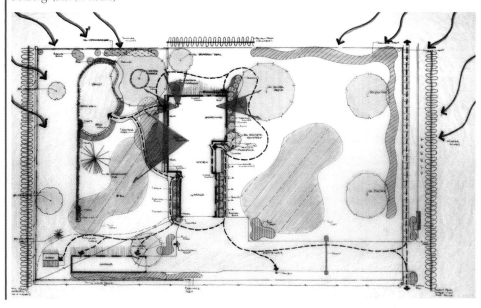

Documenting the Information

Site analysis information is best documented on a copy or overlay of the base map (see Chapter 2 for information on creating overlays and the graphic style used). The base map provides a logical context for the information and allows the designer to graphically show the inventoried and assessed landscape factors. The graphic style of a site analysis tends to be informal, including numerous, detailed, written notes capturing the judgments generated in the analysis process as well as the factors considered unimportant. This sheet not only provides the designer with a working summary of issues that need to be addressed, but it also reflects the initial graphic stage of design process. For some projects and clients, this documentation can give important clarification to the reasons behind the design decisions being considered (Figure 5-31).

FACTORS TO DOCUMENT. The following section provides an overview of the basic factors that should be considered in a site inventory and analysis, such as the site location, natural and physical site features, and constructed and client-related features. As previously discussed, some factors may not warrant documentation on every site, but the designer's checklist makes sure the assessment is comprehensive and nothing is overlooked. A sample site analysis checklist is included in the appendix.

Site Location. Site assessment addresses on-site information and inventories of off-site considerations that may affect the design. Adjacent property uses, especially if they vary from the site being assessed, can have a significant impact on the site design. Views to adjacent properties are typically the most important issue to consider. Screening may be needed for undesirable views, and good views may deserve framing. Noise, odors, and storm water drainage from adjacent properties

FIGURE 5-32

Neighborhood character should be considered as design ideas are developed. *(Steven Rodie)*

FIGURE 5-33

Headlight glare can be generated from off-site sources. *(Steven Rodie)*

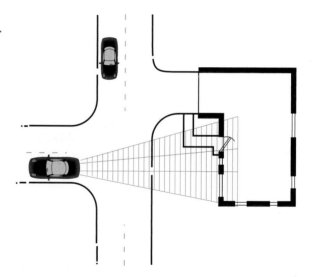

may also need to be addressed. In some cases, there may not be a reasonable way to solve off-site issues, but noting them in the inventory process ensures that they are not overlooked.

The architecture and landscape styles found in the neighborhood should also be considered. If an existing design style has already been established on adjacent properties through consistent color schemes, house styles, and landscape plants, then the aesthetic fit of proposed ideas should be considered. For example, if neighborhood parking strips are normally manicured turf grass, substituting tall prairie grasses may look weedy rather than naturalistic (Figure 5-32).

Adjacent streets, alleys, and other rights-of-way (a linear trail corridor, for example) should also be included. Is traffic heavy or light? Is curbside parking available? Is the alley used for garage access? Do any streets line up with the property perpendicularly (Figure 5-33), which could create headlight glare? Since the base map typically does not include off-site areas, notes for these issues must be made on the appropriate edges of the plan or in an additional overlay (Figure 5-34).

FIGURE 5-34 Notes for off-site issues should be documented in the site analysis. *(Steven Rodie)*

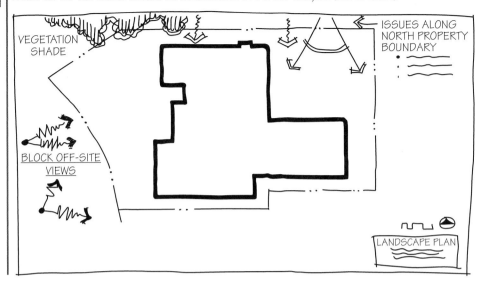

Natural and Physical Site Features. Many residential sites appear to have few natural features remaining. This can be especially true in new subdivisions where existing vegetation has been removed and the soil has been graded. Even in these situations, however, a variety of natural factors remain. Soil type and conditions, wind characteristics, topography, drainage, microclimates, and seasonal variations in sun and shade patterns all reflect natural site conditions that require assessment. In addition, on established sites, the health of existing vegetation should be included in the site analysis.

Soils. Soils are classified by their physical, chemical, and biological properties; they are described in detail in Chapter 6 in the context of plant selection. Ideally, site soils should be tested for every landscape project so that correct recommendations are made concerning plant selection, drainage, and compaction for paved surfaces. Soil test kits are available through County Extension offices throughout the United States. To conduct the tests, soil samples are collected and averaged from the site, then sent to a lab for analysis.

Local experience can help a designer understand what types of soil conditions to expect, what treatments and amendments will improve the soil, and what types of plants seem well adapted to localized conditions. County Master Gardener programs, as well as Extension horticulture publications available through most county offices, provide excellent local information to assist with the investigation of soil conditions and the preparation of recommendations.

Wind. Prevailing winds should be verified for every site. Relative exposure can be gauged on site. For instance: Is the site open and on top of a hill? Is the house nestled in a neighborhood valley with large trees? Average wind properties can vary by season and location, but historic data from the National Oceanic and Atmospheric Administration can confirm seasonal prevailing wind patterns (National Climate Data Center, 2006). Most residential properties do not have adequate space to plant an effective windbreak, so creatively finding ways to limit wind exposure may

FIGURE 5-35 | Wind exposure around a home can significantly affect microclimates and living conditions. *(Steven Rodie)*

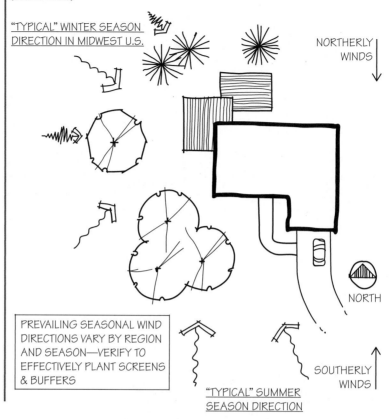

be necessary. Documenting wind information can make a significant difference in the design of an exposed site (Figure 5-35), while in other cases once it is inventoried it may be written off as having minor impact.

Topography and Drainage. Site topography and drainage assessment address slope locations and steepness, slope directions, and locations where water is expected to collect and flow from the site. Ideally, developed sites are graded so that excess storm water will flow away from buildings and toward the street, alley, or on-site storm water detention feature. The following factors should be considered in slopes and drainage assessment:

- Existing and designed slopes should meet the slope standards criteria (Table 5-1) that address safety issues and drainage issues. Gentle slopes are difficult to verify without the right equipment.
- The direction a slope faces (aspect) can significantly affect conditions on the slope. North-facing slopes receive little direct sunlight during certain times of the year, while steep south-facing slopes will typically contain the driest portions of the landscape, due the year-round solar exposure (Figure 5-36). Slopes may not be documented on the base map, but they should be noted on the site analysis.
- Site problems associated with poor drainage are best solved prior to plant and hardscape installation. It can be disheartening for a client to spend a

TABLE 5-1
Slope Guidelines

AREA	MAXIMUM	DESIRABLE	MINIMUM
Street	17%	1–8%	0.5%
Parking	6%	1–4%	0.5%
Service area	4%	1–3%	0.5%
Walks			
Building approach	4%	2–3%	1%
Major walk	5%	2–4%	0.5%
Ramp	8.33%	6–8%	0.5%
Paved pedestrian areas (seating, plazas)	2%	1.5%	1%
Lawn (mowed)	25%	4–10%	2%
Lawn (mowed, adjacent to building)	25%	4–10%	2% (absolute minimum)
Unmowed bank	50%	33%	—
Swales	5–6%	1–4%	0.5%
Playfield	4%	3%	2%

Handicapped Access Considerations

- Maximum slope of 8.33% (1-inch vertical climb per 12 inches of distance)
- 5-foot-long level landing provided for every 30 inches of vertical climb (30 horizontal feet of ramp at 8.33% slope)
- Maximum cross-slope of 2%
- Railing needed on both sides if over 5%
- Handicapped parking space access 2% or less in both directions

significant portion of his or her landscape budget on site drainage problems only to have poor drainage result in a wet basement and poor plant health.

- Downspout locations are sometimes overlooked in the site assessment. Depending on the size of the roof and the frequency of downspouts, large flushes of water during heavy storms can wash out landscape beds. Plants near downspouts may also have to endure continually saturated soil (Figure 5-37).
- Drainage solutions (including fill material, drainage swales, French drains, dry wells, and gravel trenches) may need to be considered in response to the site drainage problems (Figure 5-38). The landscape designer should keep in mind that above-ground runoff solutions are generally preferred to underground systems. Storm water runoff and water quality have become significant environmental management issues in many communities. Above-ground systems such as rain gardens, porous paving, storm water detention basins, and groundwater recharge areas are seeing broader acceptance and implementation in urbanized areas. Once water is collected underground and piped off site, water volumes and velocities increase, solutions become more complex, costs generally rise, and maintenance may become a complicating issue.

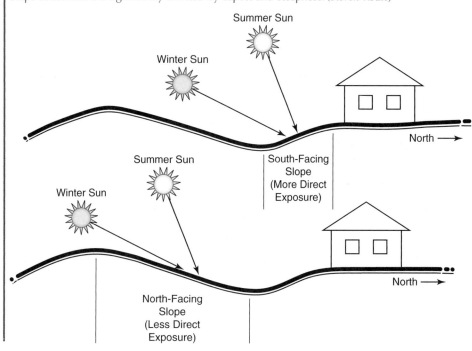

FIGURE 5-36 Slope conditions are significantly affected by aspect and steepness. *(Steven Rodie)*

FIGURE 5-37

Downspout locations should be considered in this frontyard as landscape beds near the house are planned. *(Steven Rodie)*

Slope Terminology and Calculations. Slopes are generally referenced as a ratio or percentage. In either case, slope represents a change in elevation over a specified distance. A slope ratio (4:1 or 1.5:1, for example) states the relative horizontal distance followed by the elevation change (Figure 5-39). A slope expressed as a percentage compares the elevation change divided by the horizontal measurement in increments of 100 ($\frac{4}{100}$ = 4% or $\frac{1}{50}$ = $\frac{2}{100}$ = 2%, for example) (Figure 5-40).

The terminologies and calculations are easily confused because horizontal distance is the top number in slope ratio and the bottom number in percentage ratio. Whenever slopes are measured, it does not hurt to verify the calculations.

FIGURE 5-38 A variety of drainage solutions are available to address drainage problems. *(Steven Rodie)*

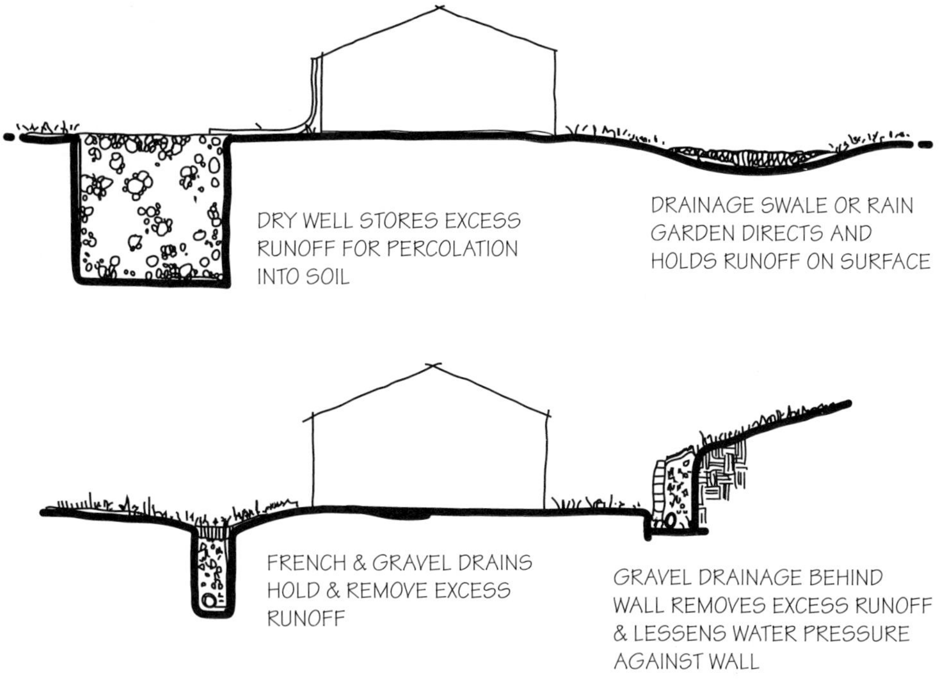

DRY WELL STORES EXCESS RUNOFF FOR PERCOLATION INTO SOIL
DRAINAGE SWALE OR RAIN GARDEN DIRECTS AND HOLDS RUNOFF ON SURFACE
FRENCH & GRAVEL DRAINS HOLD & REMOVE EXCESS RUNOFF
GRAVEL DRAINAGE BEHIND WALL REMOVES EXCESS RUNOFF & LESSENS WATER PRESSURE AGAINST WALL

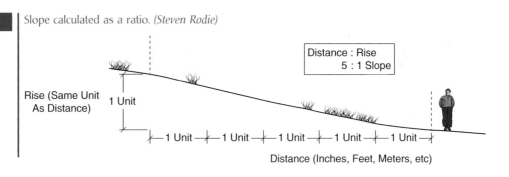

FIGURE 5-39 Slope calculated as a ratio. *(Steven Rodie)*

Distance : Rise
5 : 1 Slope
Rise (Same Unit As Distance)
1 Unit
1 Unit
1 Unit
1 Unit
1 Unit
1 Unit
Distance (Inches, Feet, Meters, etc)

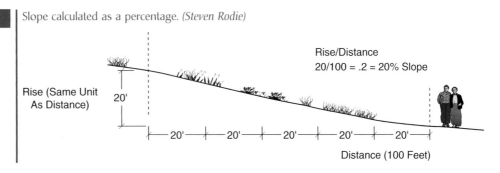

FIGURE 5-40 Slope calculated as a percentage. *(Steven Rodie)*

Rise/Distance
20/100 = .2 = 20% Slope
Rise (Same Unit As Distance)
20'
20'
20'
20'
20'
20'
Distance (100 Feet)

FIGURE 5-41

Microclimate conditions vary dramatically between the north (a) and south (b) sides of this home. *(Jon Wilson)*

(a)

(b)

Microclimates. Microclimates are site-specific conditions affected by wind, shade and sun exposure, temperature, and humidity in a relatively small area. Every landscape site, regardless of size, will have a variety of microclimates that can affect planting conditions and the physical comfort associated with outdoor rooms (Figure 5-41).

The following descriptions summarize the typical conditions that surround a two-story house on an exposed residential site in a temperate climate somewhere in the Northern Hemisphere (Figure 5-42).

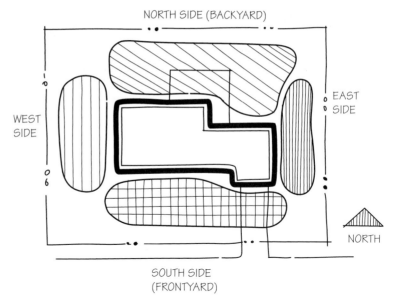

FIGURE 5-42

Typical conditions exist around the four sides of a residence. *(Steven Rodie)*

North side. This is the coolest, wettest, and shadiest location; it has direct exposure to cold, drying winter winds and some protection from midday sun and hot southerly winds in the summer months. This is a good side of the house for morning and midday summer patio space.

East side. This location receives morning sun and afternoon shade, which is beneficial to many plants not adapted to intense sun and heat. It is typically protected from winter winds and receives late afternoon and early evening shade and less intense morning sun.

South side. Full sun at midday exposure during the entire year (some early morning shade in summer) and hot southerly summer winds typically make this the hottest and driest microclimate. Because this location is the warmest area in winter, it may extend possibilities for less hardy plants.

West side. The effects from morning shade and afternoon sun in this location are stressful for many plants. It also has potential exposure to prevailing winds and full late-day sun exposure during summer months.

Tree canopy shade. Here, temperatures are decreased; humidity is increased, and soil moisture will vary, depending upon density of tree canopy, type of foliage, and root competition with the tree and plants growing under the canopy (Figure 5-43).

Fences or other structures. Conditions here will be like those in similar exposures along building walls but will not extend as far into the landscape. On an exposed site, a fence can moderate conditions and enhance plant growth (Figure 5-44).

Sun and Shade.   Sun exposure influences a variety of site issues beyond plant health. Accurately identifying sun patterns and exposure changes in the site analysis will

FIGURE 5-43

Significant microclimate differences exist between the full sun landscape and the dense canopy shade cast by a red oak. *(Steven Rodie)*

FIGURE 5-44

Microclimate conditions will vary between two sides of a fence in otherwise exposed conditions. *(Steven Rodie)*

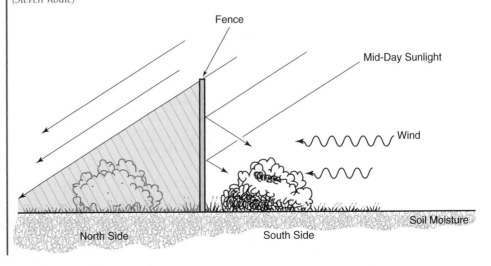

provide many options for the designer. For instance, a garden that requires full shade for comfort during the summer months may also benefit from full early morning sun in October, thus extending the utility of the space (Figure 5-45). Cost savings for home heating and cooling can be also realized when tree location is guided by sun-exposure patterns.

The sun's height in the sky and its position on the horizon when it rises and sets vary each day of the year. In the Northern Hemisphere, the sun rises in the east, travels across the southern sky, and sets in the west. The sun's path and exposure angle will vary during summer and winter, thus changing the intensity and duration of the solar rays. Calculated sun angles and seasonal changes for any latitude

FIGURE 5-45

Seasonal variations in sun/shade conditions can be used to the designer's advantage.
(Steven Rodie)

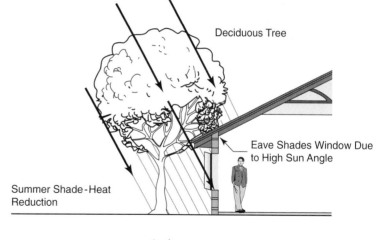

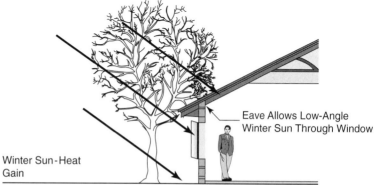

An Exception to Every Rule

From March through September, there is potential for sun exposure on the north sides of structures in areas that are typically assessed as shade only. This exposure is not usually a major factor but can generate unexpected and extreme conditions. For instance, during the early evening of a long summer day, an aluminum storm door handle on the front of a northwest-facing home can become too hot to touch (at least without a pot holder). In such an exposed site, the hot summer sun could potentially injure any shade-loving plants planted in this part of the yard. Conditions like this can teach designers that every site is different and there are exceptions to every rule.

can be found in landscape design reference books (Architectural Graphics Standards, 2000) or through Web site calculators (U.S. Naval Observatory, 2006).

Variations in the sun's angle and path must be considered in the design process for several reasons. First, design strategies that are fine tuned to seasonal sun and shade characteristics can extend the comfort range for outdoor rooms. Also, by knowing the sun angles during summer and winter months, the designer can specify plants and other structures that will take full advantage of sun and shade benefits

FIGURE 5-46

Seasonal variations in sun angles can complicate microclimate conditions on every site. (Steven Rodie)

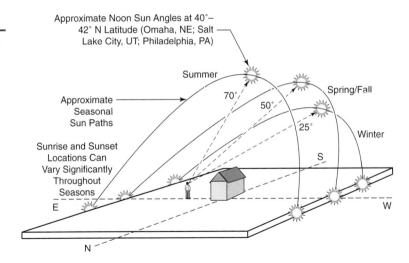

(Figure 5-46). Finally, understanding sun and shade variability strengthens the designer's ability choose the right plant for the right place.

Constructed and Client-Related Features. In addition to the location of a site and its natural and physical characteristics, the designer must consider any client-related features and legal requirements that pertain to the site. This information includes site restrictions, circulation and parking, the house and other structures, utilities and utility areas, and views.

Site Restrictions. Site covenant and easement information was discussed previously as part of the base map production process. The designer needs to carefully consider any restrictions that may apply and confirm that they have been adhered to.

Circulation and Parking. Circulation for automobiles is generally limited to an existing driveway and garage, but some designs may require additional off-street parking or a driveway redesign for better automobile maneuverability. On new sites, the designer may have the opportunity to work with the architect or builder to design the driveway configuration. Driveways take up a considerable amount of space and many times are the most visually prominent feature of the frontyard. Any opportunity to enhance new driveway design or improve an existing driveway (new paving materials or paving patterns, extending driveway width for better automobile access, etc.) should be seriously considered.

Pedestrian circulation should be assessed as a hierarchy. Primary paths, such as the front walk or a path to a major garden space, should be inventoried and analyzed for effectiveness, comfort, and safety. Secondary paths in garden areas must also function properly but are typically less formal and smaller than primary paths. Additionally, there may be indications of needed paths or paving that does not presently exist. Worn trails through a turf area or muddy sections of a well-used meandering path may indicate that additional changes are needed.

House and Structures. House colors should be noted in the assessment. Is there a theme of color use already reflected in the house that could be extended into

FIGURE 5-47

A shade of tan used for the porch lattice and awning trim was extended into the landscape through its use in the retaining walls. *(Steven Rodie)*

the landscape? If not, could the landscape implement a scheme that would complement the house? Is there an accent color used on the house (Figure 5-47) that would work well as a repeated landscape accent?

Emergency access to basement windows, as well as natural light access, should be noted so that plants and landscape features can be located appropriately. Site structures, such as detached garages and garden sheds, may complement the landscape setting, or they may require screening or relocation for better access and effective site use.

Utilities and Utility Areas. All utilities on the property should be previously located on the base map. Some may require minimum clearances for access (electrical and telephone boxes) or air circulation (air conditioners); some may require height restrictions on plantings or overhead structures, such as overhead power lines; and other utilities such as septic sewage systems and irrigation lines may need to be precisely located so that they can be avoided during site excavations.

Utility and storage areas, including clotheslines, garden areas, dog kennels, firewood storage, and trash containers, tend to be cluttered and unsightly. They should be assessed for screening requirements as well as for easy access and security.

Views. On-site and off-site views can be enhanced through sensitive landscape design, so it is important to accurately assess the viewpoints most valuable to the client. Outward views from the house should be verified by room, and the designer should experience each view if possible to fully understand the client's wishes for view enhancement, screening, or privacy. View adjustments are typically very strategic in their application. A plant may be desired to frame an outward window view but not block the entire view; a small tree, in just the right location, can provide privacy from a neighboring deck view but allow visual access to the remainder of the backyard (Figure 5-48).

Suburban development patterns tend to place houses close together, creating the need for privacy between neighboring backyards. Prospect-refuge, the universally preferred see-without-being-seen characteristic of outdoor spaces (see Chapters 3 and 4), can be a tricky proposition in wide open, small-lot backyards. When privacy

FIGURE 5-48

View screening should be carefully considered for efficiency as well as effectiveness. *(Steven Rodie)*

Well-placed trees and screens closer to the viewer (b) may be more efficient and provide quicker screening than trees placed farther from the viewer (a).

potential is assessed in such situations, designers often consider a property line screen of fencing or plants around the entire backyard. This strategy, due to deck heights and sloping lots, may or may not provide the anticipated privacy and will typically take many years to mature.

On the other hand, installing a fence or dense plant screen adjacent to a patio or deck provides privacy but cuts off views to the yard and can create a claustrophobic feel. As a compromise, a properly located see-through screen of lattice or multistemmed trees should be considered to effectively block views from adjacent yards but allow the client to see through the screen to the yard and beyond. This solution requires an intimate sense of the view characteristics in the landscape, which typically requires an additional investment of design time. But the payoff provides the type of design implementation that all designers strive for—a more effective solution, in a shorter timeframe, at a more affordable price.

Combining Site Inventory and Analysis with Sustainable Design

Sustainable design solutions are derived from the designer's ability to work with nature and think broadly about the natural systems that will exist in the landscape. Developing a true sense of place for a site requires a working knowledge of site conditions and the ability to reflect on the pre-existing natural systems. Adopting this paradigm helps the designer see problems as potential opportunities and be better equipped to custom fit design solutions to specific site issues. For instance, a heavily shaded garden would be seen as a chance to support a collection of unique plants rather than a problem area where typical sun-loving landscape plants will not grow.

A site analysis is easily ignored during the design process when time and resources limit direct designer involvement with the project. Sustainable design requires a commitment to understanding the site context for the design, and the enhanced quality of design is typically worth the investment.

Summary: Site Inventory and Analysis

Site inventory and analysis are important steps in the design process. It should be thorough but focused for each design project, and it should integrate personalized client feedback with designer awareness of potential site issues. Finally, the analysis should reflect judgments about the site and design potential, rather than simple inventory. In combination with a design concept and program, which are described in the following sections, the designer should have a strong framework to develop a successful design.

CASE STUDY

Site Inventory and Analysis

Site conditions were verified and summarized on a site analysis overlay of the base map (Figure 5-49). Key considerations that were identified included:

East–west yard orientation

Morning sun and afternoon shade near the house in the backyard will provide ideal growing conditions for many plants and shade for late-day summer entertaining. The afternoon sun exposure in the frontyard will generate relatively hot, dry conditions.

Privacy and views

Views from the east side of the house are important, but backyard privacy is also required from adjacent neighbors, so screening will need to be implemented.

FIGURE 5-49

The Jones residence site analysis provides a thorough assessment of site conditions. *(Courtesy of Julie Catlin, Steven Rodie)*

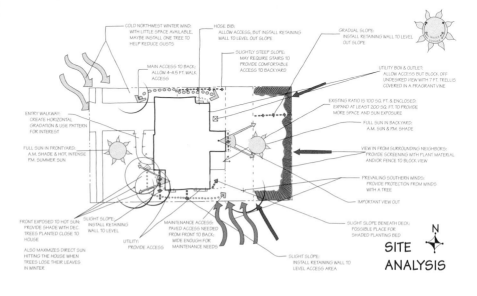

(continues)

CASE STUDY
(continued)

Narrow, sloped side yards
Side-yard areas have little functional value other than access to the backyard, and turf may not be the ideal surface in sloped shaded conditions; consider other plant materials, paved surfaces, or need for steps.

Deck and patio spaces
Combined outdoor entertaining area needs to be slightly expanded, and better use made of space under deck.

Backyard slope
The slope is not steep enough to require a retaining wall, but a wall may enhance use of the space and provide a better space for the requested pet and child play area.

Utilities access
Ensure access and clearance for all utilities, including hose bibs, air conditioner, electrical boxes, etc.

DEVELOPING A DESIGN CONCEPT AND A DESIGN PROGRAM

The final step in gathering information and setting the design and maintenance framework is developing a design concept and program. The **concept** is the theme for the project while the **program** includes the list of components that will make up the design. When combined, they integrate all of the previous design process information, forming a framework of specific design products and solutions that are desired in the completed landscape.

Design Concept

The design concept, also referred to as **design theme** or statement of design intent, is the core that holds the design program together. It represents the essence of what is expected from the design and what the design should or should not do. When the design concept is personalized, it becomes a powerful focus for the design and helps generate additional landscape value for the client. As an example:

The landscape will encourage intimate family gatherings while celebrating fall, the clients' favorite season of the year; it will allow them easy garden access to satisfy their need for fresh-cut flowers; and it will also allow them to play golf every Saturday morning during the growing season.

This statement reflects the characteristics of an effective design concept. It is personal and is written as a broad statement that does not stipulate how the design will be accomplished, but it does suggest how the design will feel. It states simply what the design must do to satisfy the clients. A clearly stated concept can be an effective test for the specific elements that are proposed in the design program. For example, cut flowers must be included, but their maintenance must not be a weekend

Design Concept Basics

Assisting homeowners with developing a design concept can be a difficult task when they are unclear as to personal needs or the landscape potential of their property. The best place to start in such a situation is with the basics. Low maintenance, curb appeal, and enhanced home value represent the core desires of virtually every homeowner engaging in a landscape design project. By stating these goals up front and then personalizing their implications for a specific project, a designer can develop a foundation for a meaningful design concept.

"But we are paying you to tell us what we need...."

Sometimes, homeowners just have not thought about the opportunities their yard presents. Or perhaps they have thought about it and quickly realized it can be a lot of work to identify and assess personal and family needs. Sometimes, they only want it to look like the landscape next door. Or maybe they want everything and expect the designer to cast the tie-breaking vote when needed. Regardless of client motivation, it is important for the designer to listen to the client and provide solutions to his or her problems. Delivering whatever worked for the people next door may work aesthetically, but the design will likely fall short in addressing the personal needs of the client.

chore. For a successful project, any program element that does not support the design concept needs to be re-evaluated or discarded.

Design Program

A design program is an outline of proposed design elements and solutions. It draws upon the Design Planning Questionnaire and the designer's site assessment, and it serves as a valuable checklist to verify that the design aligns with the design concept. For the client, it represents a tangible product of the design process, enhances the sense of value gained in paying for design services, and verifies the designer's focus. For both client and designer, it clarifies the communications that have occurred thus far. Changing or eliminating a program element at this stage of the design process is much easier (and less expensive) than having to make the change once the design drawings are complete or the landscape is built. Throughout the remainder of the design process, the program remains a fluid set of information that can be revised as the design evolves.

 ## FUNCTIONAL DIAGRAMS

Functional diagrams are a first attempt at conceptually organizing the landscape space. In many ways, the diagram is similar to a sketch for a home floor plan. Also referred to as concept plans or concept diagrams, functional diagrams are effective tools for initial arrangement, layout, and scaling of design ideas.

CASE STUDY

Design Concept and Design Program

The design concept for the Jones residence is as follows:

The landscape will provide an enjoyable, flexible, outdoor setting for our growing family, and include a maintainable balance of play space, gardening space, and favorite plants. It will provide an intimate, private setting for our family, but also allow us to comfortably entertain guests and host family reunions. When we consider selling the property in five years, the landscape will be a significant factor in the curb appeal and resale value of our home.

Based on additional client input and analysis of site conditions, the following design program for the Jones residence was developed.

- A garden space of 10′ × 10′ will be located in a sunny area of the backyard. Raised beds will help alleviate poor soil conditions.
- Approximately 100 square feet of garden equipment storage is required. A portion could be located under the existing deck for convenient access; garden shed location should allow easy access for mower and other equipment.
- Approximately 100 square feet of **dry-laid pavers** will be added to the lower patio in a configuration that provides overflow access to the yard for the occasional large family reunion party.
- The front walk will be redesigned for a more welcome feel, including a minimum width of 5 feet. The material is undecided.
- Approximately 1,000 square feet of backyard space will be allocated to open turf for pet or child play area.
- The backyard will be fenced with a 6-foot wood privacy fence.
- A 3-foot tall retaining wall will be constructed on the east property line to lessen the backyard slope.
- Visual screening consisting of a combination of plant materials and trellis or lattice will be strategically located for the deck and patio.
- The requested sugar maple will be located in the frontyard to provide afternoon shade; all other requested plants and colors will be located for optimal growing conditions and aesthetic impact.
- The south side yard will be planted in groundcover with stepping stones and steps for informal access; the north side yard turf will be replaced with mulch and stepping stones to provide equipment access between frontyard and backyard.
- A budget of $15,000 has been established to cover a 3-year phasing plan, which will prioritize major tree planting, garden storage, retaining wall and paving installation, fencing, and landscape bed planting.

FIGURE 5-50

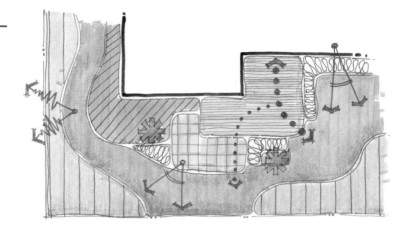

Functional diagram symbols include a variety of bubbles, arrows, lines, and asterisks. *(Steven Rodie)*

dry-laid pavers: Bricks or interlocking concrete pavers laid on a bed of tamped sand, usually over gravel; no concrete or mortar is required.

Functional diagrams define the ground plane elements of the landscape. As the name suggests, they organize the functional relationships between landscape areas and elements. Since they represent the first step in the design development process, they typically have a freehand, informal graphic style. They are derived directly from the information developed in the base map and site analysis, together with the specific information outlined in the design program. In essence, they become the initial framework for the drawn design. It is critical that a functional diagram reflects the previously gathered information; developing a diagram that does not correlate to client wishes, site judgments, and quantified needs for landscape spaces and uses will lead to a failed design.

As tempting as it may be to jot down names of specific plants for these sketched areas, remember that the purpose of functional diagrams is to organize general areas within the landscape. Deciding on specific plant material is a later step. The key consideration in developing a functional diagram is to assign every portion of the ground plane to a particular use.

Functional Diagram Elements and Information

Functional diagrams use graphic symbols to represent specific information and characteristics (Figure 5-50).

Bubble shapes. Bubble shapes delineate all ground-plane elements, including landscape beds, turf areas, hardscapes, activity areas or designated use areas, plants, and plant masses. These bubbles should be drawn freehand; they should have individual outlines, and they may slightly overlap to show relationships or closely parallel adjacent edges. Varying graphic patterns and value contrasts should be used to differentiate between represented elements. Sometimes, a pattern is selected that abstractly reflects the plant texture or hardscape material that it represents.

Dashed arrow and line symbols. Dashed arrow and line symbols are used to represent circulation patterns and access between use areas. A hierarchy of path importance can be established by using appropriate symbol line widths or pattern types.

Solid-line arrows. Solid-line arrows are typically used to represent views, with variations to show vistas, panoramas, and blocked views.

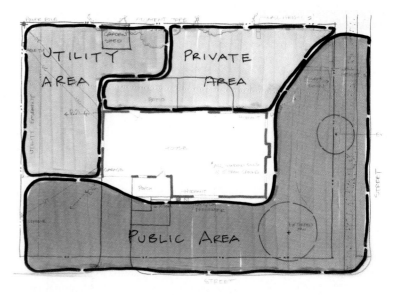

FIGURE 5-51

Public, private, and utility areas are initially identified in functional diagram development. *(Steven Rodie)*

Asterisks or star-shaped symbols. Focal points and emphasis are delineated with asterisks, star-shaped symbols, or other unique markings so the feature stands out on the drawing.

Developing a Functional Diagram

Designers must consider information from a number of sources in developing a functional diagram. As these sources are considered and space is organized, the landscape design will begin to develop. It will become clear how different spaces in the landscape will be used, how large they will be, and how people will move through the space.

IDENTIFY LANDSCAPE USE AREAS. Just as in a home, a landscape is composed of **areas** that are used for different purposes. The main areas are the public area, which is most often in the front of the house; the private area, often found in the back of the house; and the service area, which is generally in the back or side yard areas and contains utility features such as a garden shed or compost pile. Large bubbles are used to broadly define each of these areas (Figure 5-51).

The public area is most visible from the street. The main purpose is to frame the house and create a visually appealing and inviting landscape. An attractive entryway or walkway to the house is a primary feature and is often surrounded by an uncluttered area of grass or another type of groundcover. Planting beds containing shrubs and herbaceous plants around the front of the house also help to enhance the view from the street.

The private area, often equated with the backyard, is an important part of the American home. Providing easy access from the house to the outdoor area is a first step toward creating a space that will meet specific client needs.

Utility and service areas are functionally important to the success of the design but are often overlooked in the design process. How the client uses the landscape will affect how large of a utility area is needed and where it might be located. Avid gardeners may require a larger area to contain an extensive compost system or large shed. Regardless of the size of the area, it is important that it be screened from major views. Seldom does a compost pile make an attractive focal point! Screens such

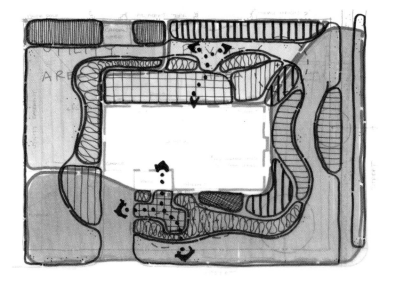

FIGURE 5-52

General areas are subdivided into more specific uses and designations. *(Steven Rodie)*

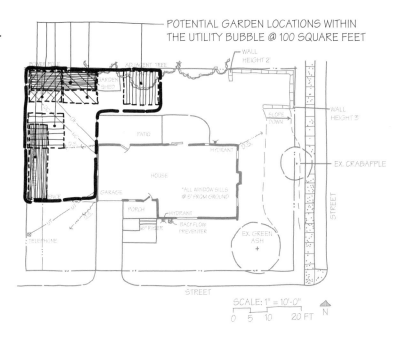

FIGURE 5-53

Within the utility area bubble, the designated garden space can be located in a variety of configurations. *(Steven Rodie)*

as a dense planting of shrubs, a vine on a trellis, or a fence can make it less noticeable. In addition to a compost area, the service area also might include garbage cans, tool and wood storage, or a dog run.

SIZING DIFFERENT AREAS APPROPRIATELY. Once the basic functional areas are defined, each bubbled area is divided and subdivided into more specific areas of plant materials, use areas, paved areas, garden areas, etc. (Figure 5-52). If an area has been defined in the design program with specified dimensions, then the appropriate-size bubble should be incorporated into the diagram to match the criteria. For example, if a garden area of 100 square feet is requested, then a scale 10′×10′ bubble can be designated on the diagram (Figure 5-53). Recommended

TABLE 5-2
Recommended Standard Dimensions for Outdoor Use Areas

USE	RECOMMENDED SIZE
Sitting, 2 chairs	5' × 5'
Sitting, 4 chairs	8' × 8'
Pair of lounge chairs	8' × 8'
Eating, 2 people	6' × 4'
Eating, 4 people	10' × 10'
Eating, 6 people at a picnic table	8' × 7'
Eating, 8 people at a picnic table	10' × 7'
Cooking/food preparation, grill	2' × 2'
Cooking/food preparation, counter top	2' × 4'
Cooking/food preparation, overall area	5' × 5'

dimensions exist for virtually every type of outdoor use area and element (picnic table, table and chairs, sitting area, etc.), and the dimensions correlate to the number of people that need to be accommodated in that area. Some common dimensions are shown in Table 5-2. Additional use area requirements are found in references at the end of the chapter.

Another issue to consider when sizing the bubbles is maintenance. Bubbles representing planting beds of perennials or annuals more than 4 feet wide are difficult to maintain without a work path. Bubbles representing turf strips narrower than 5 feet should be avoided since these areas can be difficult to mow, water, and fertilize. In locations where planting beds are located next to buildings, an accessible zone at least 3 feet wide should be provided next to the building to allow for maintenance (Figure 5-54).

Once he or she has determined the sizes, the designer then has the option of locating each bubble where it will fit best among other bubbles (a garden shed, flower garden, deck, etc). Both the location and the configuration of the landscape are flexible at this point because so many options exist (Figure 5-55). This process represents one of the essential sequences and challenges of design. It is a puzzle where most of the pieces are interchangeable, and many do not have a set size or shape. Approaching the process with the design program as a guide and remaining flexible to different ideas should result in a workable and creative functional diagram.

DOCUMENTING CIRCULATION AND VIEWS. As designers think about different ways to organize the space, they will want to consider circulation patterns within the landscape. How will people move from one area to another? How will people move between the house, garage, and landscape? The thorough designer will try different combinations to maximize views and enhance circulation.

FIGURE 5-54

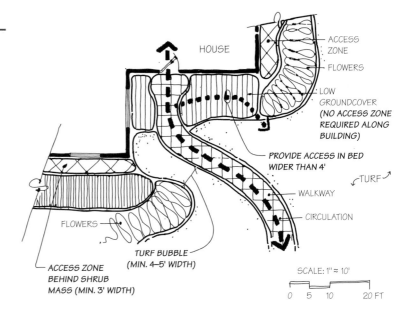

Functional diagrams should reflect appropriately sized bubbles and proper maintenance considerations. *(Steven Rodie)*

Views will vary in importance, based on where they are located and what is seen. Consider the location of utility areas, drainage patterns, and plant material when documenting circulation and views.

TREES AND OVERHEAD PLANE. The last items that need to be documented on the functional diagram are trees and other features in the overhead plane. Although most of the functional design process deals with organizing the ground plane, trees and other features located over the ground plane can have an impact on how the space should be used. These overhead features should be represented with shapes outlined with dashed lines. The symbols add an extra layer to the drawing, but they still allow the bubbles and other graphics on the diagram to clearly show their meaning (Figure 5-56).

Summary: Functional Diagram

Functional diagrams are effective tools for initial arrangement, layout, and scaling of design ideas. The designer derives a variety of benefits from using a functional diagram, including:

- It establishes a sound, functional framework for the design.
- The entire site is used in the design; no space is left over (Figure 5-57, page 226).
- The designer is able to adopt a big-picture view.
- It provides an organized technique to create a complete and complex design.
- Functional diagrams encourage exploration and creativity.
- Functional diagrams easily communicate design ideas, and because they typically lack the polish of a final design, clients are often more comfortable in requesting major changes to the design.

FIGURE 5-55

A simple walkway (a) or an elaborate entry patio (b) can be efficiently considered with functional diagrams. *(Steven Rodie)*

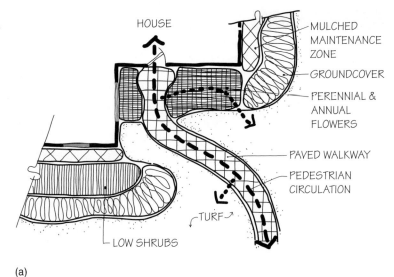

(a)

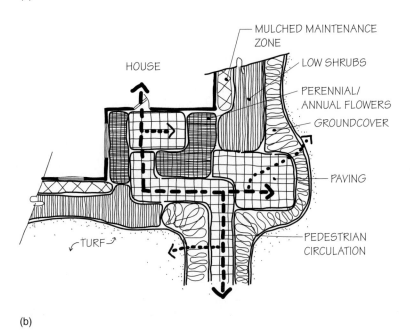

(b)

PRELIMINARY DESIGN

The **preliminary design** phase converts the loose, freehand bubble diagrams into a more refined, but still preliminary, design drawing. The landscape spaces in the design are clearly organized; the outdoor rooms become obvious, and the bubbles that previously represented masses of plants begin to illustrate specific plants. In addition, as specific design principles are implemented, a design theme begins to emerge.

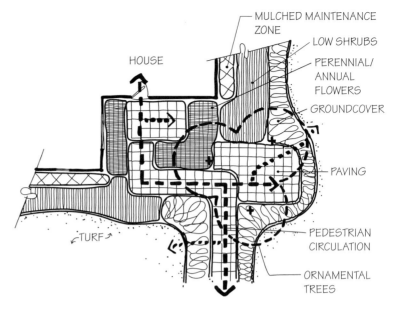

FIGURE 5-56

Outlines for canopy trees should be included in functional diagrams to reflect plant layering while clearly showing the ground plane bubble configuration. *(Steven Rodie)*

Purpose and Characteristics

Preliminary design focuses all of the design ideas previously proposed in the design process. The illustrative graphics used in this step are somewhat realistic and begin to identify surface textures and colors, plant types, distinct bedlines, and outdoor room edges. The drawing includes the refinement of applied aesthetic and functional principles. Material selections and the spatial layout of the design will also become clear during this phase of design. At a point when the design is becoming more understandable and well defined, the preliminary design also creates an opportunity for more valuable client feedback.

CLIENT FEEDBACK.   Client feedback is critical during the preliminary design. Changes can still be made, and every part of the landscape design can be reconsidered. Using realistic graphics in the preliminary design drawing may clearly show a design idea that previously was unclear. The designer and client will now be better able to discuss the relative merits and perceived shortcomings of the design.

AESTHETIC REFINEMENT.   Applications of the seven design principles discussed in Chapter 4 should be evident in a preliminary design. Emphasis areas of the landscape should be defined in key locations, such as the front door of the house or a view from the back patio. Specific plants, materials, or other elements should be repeated to unify the landscape. A strong framework of distinct bedlines creates a sense of order, and changes to vertical heights and horizontal widths create rhythm and interest. If specific requests for aesthetic characteristics were made in the design program, such as flower colors or seasonal flower display, this is the time to verify that the requests have been met.

FUNCTIONAL REFINEMENT.   The preliminary design should address all of the functional issues that were developed in the design program, such as confirming

FIGURE 5-57

Functional diagrams should establish purposes and uses for the entire ground plane (a); a diagram that leaves undesignated areas will lead to an ineffective design and "leftover" landscape spaces (b). *(Steven Rodie)*

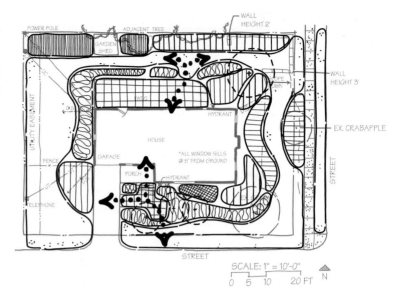

(a)

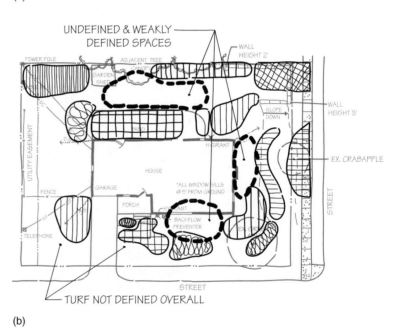

(b)

square footages for use areas; creating solutions for identified problems; and meeting requested allocations for public, private, and utility spaces. In addition, the designer should verify integration of elements and spaces that enhance the concepts of landscape preference and personal restoration.

SPATIAL REFINEMENT. Spatial composition focuses on the three-dimensional attributes of outdoor areas. At this stage of design, the relative proportions of spaces and the walls, ceilings, and floors that define them should be evident. How do

CASE STUDY

Functional Diagrams

The designer considered several alternatives for the layout of design elements for the Jones residence, including different arrangements of the additional patio space, alternative locations for garden storage and garden space, and possibilities to screen views and provide enclosure. The turf play area in the backyard is a major component of the design, so several sizes and layouts of that area were also considered. Figure 5-58 shows the two finalized functional diagrams that were developed.

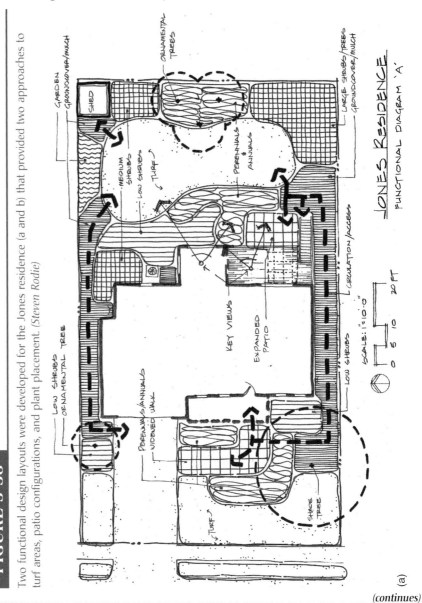

FIGURE 5-58 Two functional design layouts were developed for the Jones residence (a and b) that provided two approaches to turf areas, patio configurations, and plant placement. *(Steven Rodie)*

(continues)

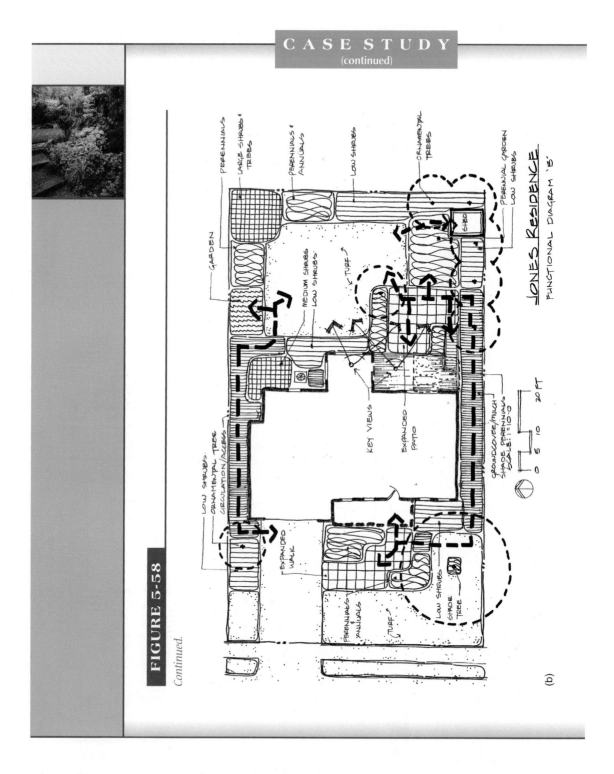

FIGURE 5-58
Continued.

the trees, shrubs, low-growing plants, and hardscape elements combine to make the ceiling, walls, and floor of outdoor rooms? All of these will influence how the landscape looks and feels to the user. For example, trees with a tall, loose canopy create a sense of openness; while a tree with a dense and relatively low-branching

FIGURE 5-59

Garden space can be effectively defined with lattice walls (a) and overhead structures (b). *(Steven Rodie)*

(a)

(b)

canopy creates a sense of enclosure. Privacy fences can be used to create visual barriers as well as to provide enclosure in locations with limited space. Instead of using trees to create an overhead canopy, the designer can use a pergola covered with vines (Figure 5-59).

ESTABLISH THE GROUND-PLANE FRAMEWORK. One of the steps in preliminary design is to refine the bubble shapes developed during functional design. At this stage of the design process, functional diagram bubbles should be sized properly, based upon the design program and plant sizing requirements. Oftentimes, however, the bubbles and corresponding edges lack the balance, flow, and structure that the preliminary design requires. Refining and adjusting bubble shapes and sizes during this step may be necessary.

FIGURE 5-60

Design themes include rectilinear (a), diagonal (b), angular (c), arc-tangent (d), circular (e), and curvilinear (f). *(Steven Rodie)*

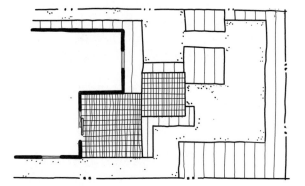

RECTILINEAR

(a)

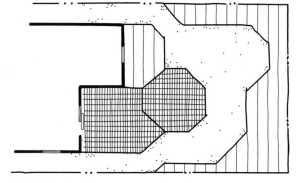

DIAGONAL

(b)

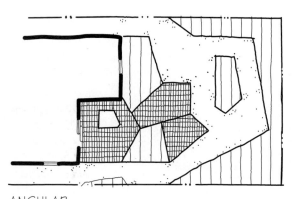

ANGULAR

(c)

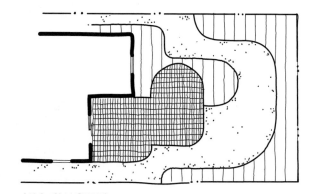

ARC-TANGENT

(d)

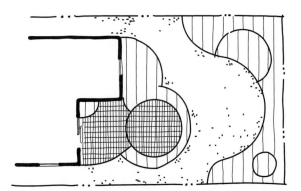

CIRCULAR

(e)

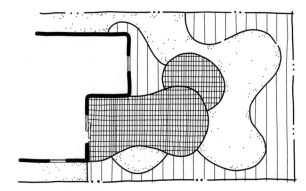

CURVILINEAR

(f)

ESTABLISH A DESIGN THEME. A design theme reinforces the landscape framework and helps unify adjoining areas. A theme is established through the consistent use of ground-plane patterns and shapes that are developed through visual and physical connections with existing landscape features. There are many approaches to developing a design theme. Booth and Hiss (2005) provide an elaborate discussion of design theme selection and a system for developing theme patterns. The information discussed here draws from their system.

The building blocks of themes are basic shapes and line patterns, including circles, partial circles (radii), straight lines segments (tangents), angles, and freeform curvilinear lines. Using these elements, the designer can develop the following themes (Figure 5-60):

- rectilinear: Straight lines and 90-degree angles
- diagonal: Straight lines, 45- and 90-degree angles
- angular: Straight lines and variable angles
- arc-tangent: Straight lines and varying-curve radii
- circular: Complete circles and variable radii
- curvilinear: Variable curves and no straight lines

Select a Design Theme. Selecting a design theme is based on a variety of issues, including architecture of the house or other structures in the landscape. For example, a landscape theme for a house with strong horizontal roof lines may use a circular theme with horizontal features to both complement and contrast the house architecture (Figure 5-61). An existing patio or deck with square edges might promote a rectilinear theme (Figure 5-62). A selected theme may also develop from the shapes, sizes, and orientation of functional diagram bubbles (Figure 5-63). The personal preferences of the client will also greatly influence the design theme selection. Most residential landscapes use a simple, curvilinear theme, which consists of a meandering bedline separating turf from landscape beds. Curvilinear lines are easy to mow along, and their freeform style complements the informality of most residential settings. Unfortunately, most curvilinear themes used in residential landscapes exhibit an excess of freeform style and are not as effective in defining space and creating interest as they could be if implemented correctly (Figure 5-64).

One of the most difficult steps in the design process for beginning designers to master is establishing a design theme that feels right. Correct feel comes from a balance of shapes, visual mass, and rhythm developed through curves or pattern sequences. Achieving correct feel requires effort, experience, and experimentation. It is one of those intuitive experiences that are an inevitable part of the design process. Regardless of selected design theme, it may take drawing lines over and over again, on multiple sheets of tracing paper, before the correct bedline or patio shape is found. Typically, the final forms will be an average of several lines or shapes, blended to give just the right feel. The time (and patience) spent in finding the right feel for a bedline or space relationship is usually paid back in design quality.

FIGURE 5-61

The horizontal, linear character of this house (a) is reflected in the horizontal line of the rock walls (b), while the curving bed patterns provide an interesting visual contrast. *(Steven Rodie)*

(a)

(b)

FIGURE 5-62

The strong circular design theme of this patio establishes the flow and sequence of curving bedlines in the adjacent landscape. *(Steven Rodie)*

Landscape Design Process 233

FIGURE 5-63 The long, narrow bubbles developed for the backyard area of this residence (a) provide a logical framework for a rectilinear design theme (b). *(Steven Rodie)*

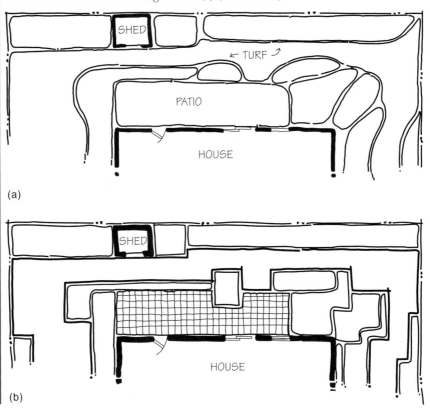

FIGURE 5-64 Curvilinear landscape installations can lack a strong sense of theme (a); simple flowing lines, obvious rhythm, and spatial definition result from effective curvilinear composition (b). *(Steven Rodie)*

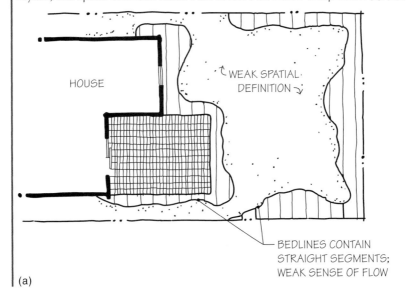

(continues)

FIGURE 5-64 *Continued.*

HOUSE

STRONG SPATIAL DEFINITION

BEDLINES REFLECT CONTINUAL VARIABLE CURVES

(b)

The Process of Developing a Design Theme. Once the designer selects a design theme or combination of themes, he or she draws a grid framework that simultaneously provides guidelines for possible line and shape configurations and aligns the theme edges and shapes with prominent building and site features. The framework is most helpful for the themes that include straight lines (rectilinear, diagonal, angular, and arc-tangent). For the circular and curvilinear themes, the grid will have limited value, but it can still help establish the relative sizes, shapes, and locations of the curving lines that define these two themes. Once the basics are established, circular and curvilinear themes are best developed with a repeated drawing of flowing lines until they appear correct and reflect the dimensions required in the functional diagram bubbles (Figure 5-65).

The grid is developed from a hierarchy of imaginary lines extended from the edges of structure walls, doors, and windows onto the site. Primary lines correlate to the outside walls and corners of the structure. Secondary lines extend primarily from doorway edges and changes in siding materials. Tertiary lines align with window edges (Figure 5-66). Intermediate lines can be proportionally included in the grid to provide additional guidelines for theme development (Figure 5-67). In addition, the lines can be projected at an angle from the structure (Figure 5-68).

Draw the grid on a sheet of tracing paper and place it between the completed functional diagram and a clean sheet of tracing paper (Figure 5-69). On the clean paper, you can now draw refined bubble shapes, using the grid as a general guide for aligning the landscape framework with the structure (Figure 5-70). Shaping the bubbles with the grid is not a straightforward process; one adjustment will likely lead to other adjustments. In addition, it may seem appropriate to use a rectilinear pattern in a formal frontyard setting and the more informal arc-tangent theme in the backyard (Figure 5-71, page 238).

FIGURE 5-65

Flowing bedlines typically require an averaging of many drawn lines to finalize a strong, balanced look and feel. *(Steven Rodie)*

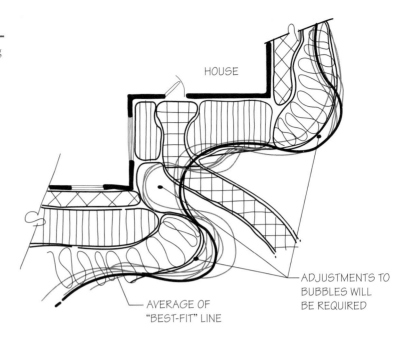

FIGURE 5-66

Initial guidelines for design theme development are extended from obvious house features such as corners, doors, windows, and changes in siding treatments. *(Steven Rodie)*

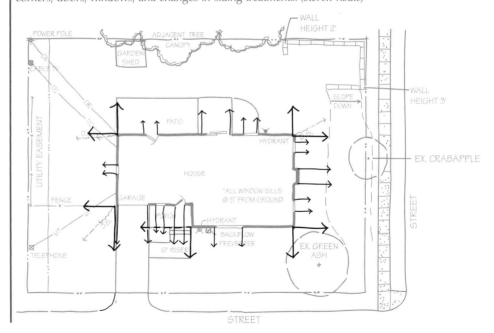

FIGURE 5-67 Intermediate lines can be added using initial line spacing and proportions as a guide. *(Steven Rodie)*

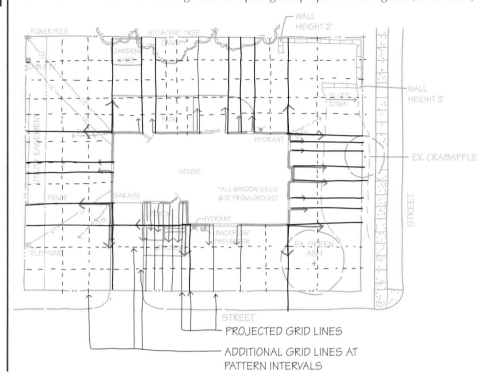

FIGURE 5-68 Guidelines can be projected at an angle from the house to add additional framework for angled landscape themes. *(Steven Rodie)*

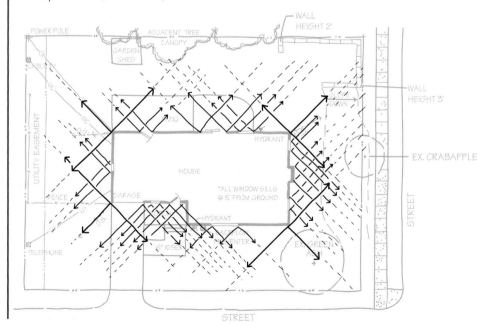

FIGURE 5-69

Combining the grid with the functional diagram creates a visual framework for shaping bubbles into a cohesive theme. *(Steven Rodie)*

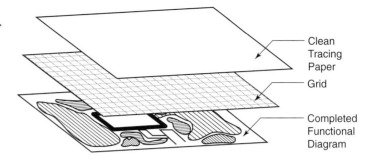

FIGURE 5-70

Establishing a design theme from the bubble layout requires various adjustments to landscape areas and elements. *(Steven Rodie)*

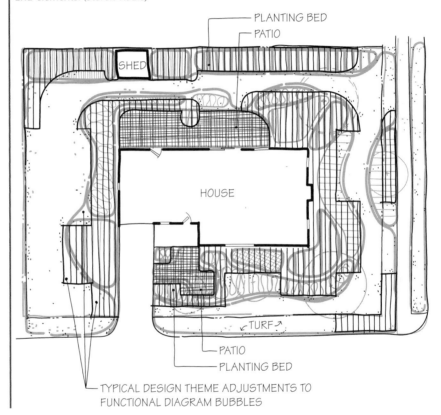

Several points should be emphasized when developing the design theme.

- Strong visual and physical relationships between landscape framework and other site features (buildings, gazebos) will make the design look and feel better.
- There are no absolute rules for framework development. A grid does not have to be followed exactly for every line. A functional diagram bubble may require location adjustment due to its importance. Different themes can be transitioned from one landscape area to another.

FIGURE 5-71 A formal rectilinear theme in the frontyard has been effectively combined with an arc-tangent theme in the backyard. *(Steven Rodie)*

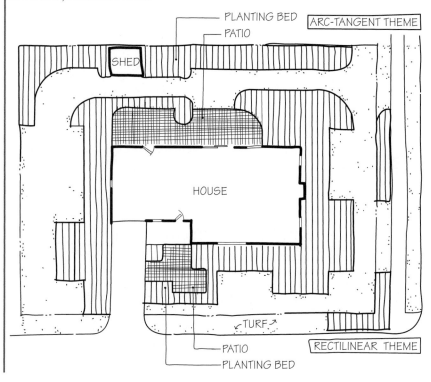

acute angle: an angle measuring between 0 degrees and 90 degrees.

- Although separate landscape areas can effectively vary in theme, such as a frontyard and backyard, one common area should not mix themes. For example, a rectilinear theme is diluted when angles are not precisely 90 degrees and straight lines are not perfectly straight. A curvilinear theme is weakened when straight lines, regardless of length, are randomly mixed with curving lines.

- Wherever patterns and edges join or intersect other edges, the designer should avoid narrow **acute angles** and strive to use broad angles (perpendicular angles at 90 degrees are ideal but not always possible). Acute angles may initially seem to be a better way to join edges. Visually, however, the connection is not as strong; and the paving surface, turf, or planting bed located in the angle will be difficult to maintain (Figure 5-72).

- While they are the most commonly used, curvilinear lines are often weakly implemented. Ideally, they should contain edges that flow back and forth while repeating the radius of each curve. They should reinforce or contrast other curving lines in the landscape. Kinked curvilinear are weak and should be avoided (Figure 5-73). Rectilinear or angled themes may be better suited for long narrow areas than curvilinear themes (Figure 5-74).

Landscape Design Process 239

FIGURE 5-72 In landscape areas where plant (a) and/or paving (b) materials join, acute angles should be avoided to minimize maintenance and structural considerations. *(Steven Rodie)*

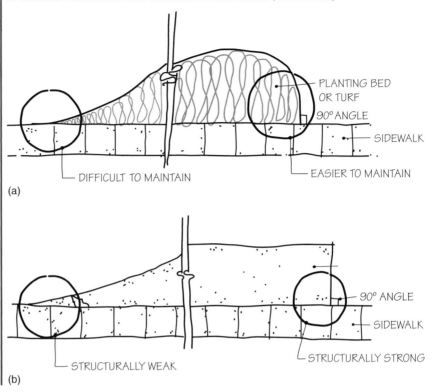

FIGURE 5-73 Effective bedlines (a), in comparison to weak bedlines (b), should repeat patterns, produce visual contrasts, and define landscape spaces. *(Steven Rodie)*

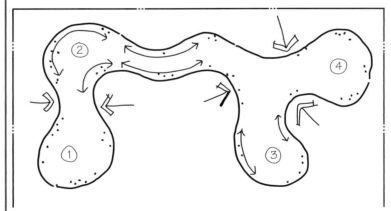

- DISTINCT SPACES ①
- CONTRASTING CURVES DEFINE SPACES, CREATE INTEREST ⟶
- PARALLEL CURVES DEFINE "HALLWAYS", CREATE STRUCTURE ⟵⟶

(a)

(continues)

FIGURE 5-73 Continued.

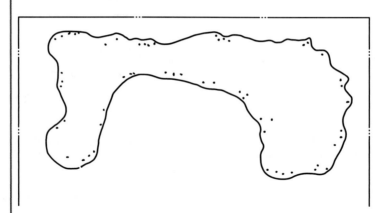

- WEAK SPACE DEFINITION
- NO COMPLEMENTARY/CONTRASTING EDGES

(b)

FIGURE 5-74 A long, narrow landscape space may not provide enough width to effectively use curvilinear patterns (a); rectilinear or angled themes likely offer more interesting patterns and efficient use of space (b). *(Steven Rodie)*

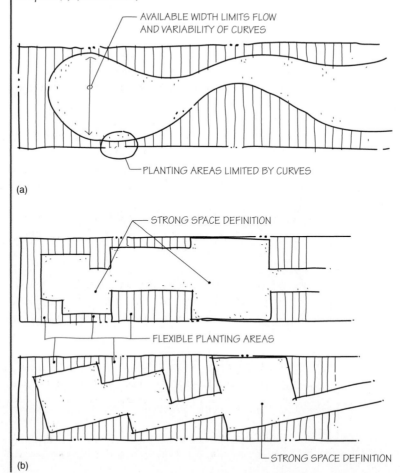

Graphic Style and Content

The preliminary design is typically drawn freehand, although the final product may be a combination of methods. Standard graphics should be used to reflect site information, plant types, and materials. Information that can be shown on the preliminary design includes:

- all of the base map information (property lines and house footprint, including window and door locations and room information adjacent to windows)
- use areas, such as patios, gardens, and utility areas (label or describe to document inclusion per design program)
- a design concept statement and other notes that help illustrate design
- estimated ground-plane elevations (deck or patio heights, wall heights, etc.)
- labeled plant materials identifying landscape role, relative size, and type; if massed, show as mass; groundcovers and herbaceous plants are shown as nondifferentiated bubbles
- a preliminary plant list, estimated plant prices, and quantities
- a rough cost estimate confirming that the preliminary design is within the budget
- elevation sketches of important landscape areas or areas that are particularly difficult to understand

Summary and Design Process Flexibility

Preliminary design synthesizes the site analysis, the design program, the functional diagram, and a design theme into a single plan. It clarifies design ideas that were previously unavailable in earlier drawings and allows for feedback from the client prior to finalization.

Depending on project scale and the design fees that the local market will bear, this stage of the design process may need to be combined with other stages. When shortcuts are required, designers may choose to combine this step with others or streamline the process in some fashion. In either case, full implementation of the design process enhances designer capability to provide long-term quality designs to satisfied clients.

FINAL DESIGN PACKAGE

Having accounted for any client feedback and further adjustments, the final design represents the complete landscape design package (see Jones case study at the end of this section). The final design elaborates upon design details, enhances the accuracy of the final cost estimate, and sets expectations for installation. Typically, a final design package includes a final design (or master plan) with the plant list, design concept, and additional notes; and detailed elevation drawings of key areas and constructed elements.

Landscape development is often a long-term process. Completing the landscape in phases over time has some advantages. It allows flexibility in the design and makes a large project more financially manageable for the client. A master plan is

CASE STUDY

Selecting a Design Theme

The functional diagram in Figure 5-58a, with several ideas borrowed from Figure 5-58b, was ultimately chosen by Mr. and Mrs. Jones. Two form compositions exploring different design themes were developed from the approved functional diagram (Figure 5-75). A rectilinear pattern was studied, based on the relatively small lot size and shapes of existing deck and patio spaces. A curvilinear composition was also developed, due to the clients' desire for an overall informal garden style. The rectilinear composition was ultimately selected.

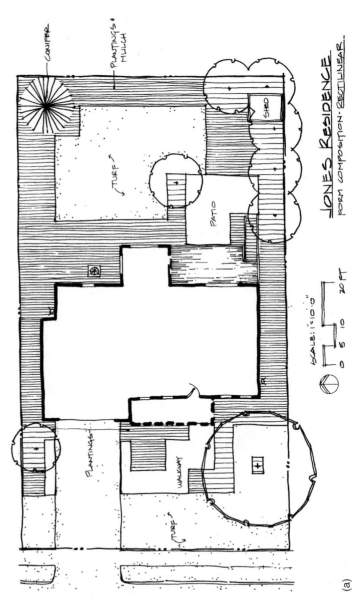

FIGURE 5-75 Rectilinear (a) and curvilinear (b) design themes were developed for the Jones residence. *(Steven Rodie)*

(continues)

CASE STUDY (continued)

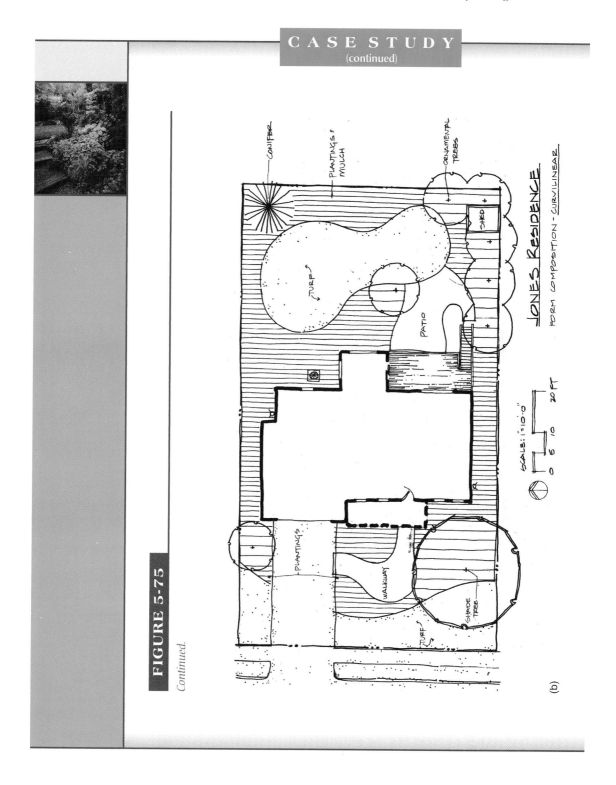

FIGURE 5-75

Continued.

CASE STUDY

Preliminary Design

The preliminary landscape design for the Jones residence is shown in Figure 5-76. The preliminary design incorporates all of the design discussions to date. Favorite plant locations have been identified, but most plants are still identified by function and purpose, and they will not be named until the final design is completed. Preliminary design drawings include a plan view of the site as well as a conceptual section elevation that shows the proposed spatial relationships for the outdoor living areas.

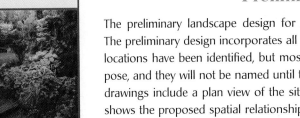

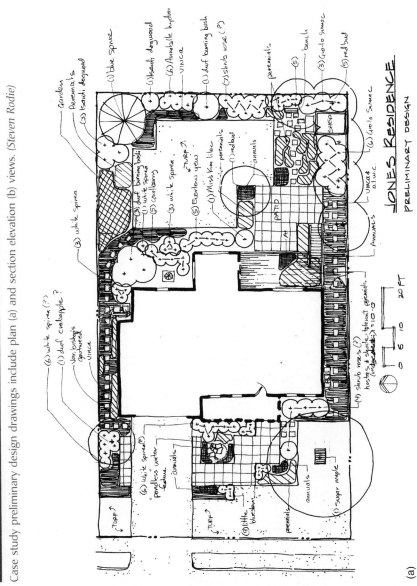

FIGURE 5-76
Case study preliminary design drawings include plan (a) and section elevation (b) views. *(Steven Rodie)*

(continues)

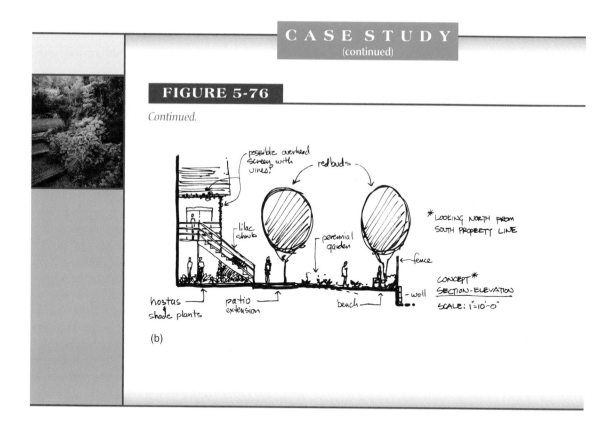

CASE STUDY (continued)

FIGURE 5-76

Continued.

important for any size project but is essential for large, phased projects. A solid, final design package will help the client set priorities, and it guides the project to a successful completion.

Final Design Checklist

Before developing the final design package, consider this checklist:

Does the design include everything the client wanted?

Does it reflect data from the site inventory and the Design Planning Questionnaire? If not, is there a good reason for the change?

Were utility lines, roof downspouts, hose connections, and other immovable features considered when placing planting beds and patios or walkways?

Are walkways convenient?

Are guests directed to the front door of the house?

Have any of the client's needs or desires changed since the design process began?

Has adequate privacy been incorporated into the private area?

Does the plan consider mature heights and widths of trees and shrubs?

Will the landscape be attractive when viewed from the living room? Picture window? Dining room? Kitchen? Porch?

Will the landscape be attractive all year long?

Do all parts of the landscape fit together into a unified design?

Does the design realistically reflect the maintenance budget and commitments the client requested?

Does the landscape reflect what was described in the design concept?

Does it integrate spaces and elements that have been shown to be universally preferred and restorative to humans?

The Final Design or Master Plan

In some cases, the final design is referred to as the master plan because the sheet often includes more than just a plan-view drawing of the final landscape design. Additional items included on the sheet are a detailed plant schedule, the design concept, and any additional notes necessary to implement the design. The final design should incorporate all of the elements outlined in the preliminary design, and the graphic style should reflect a realistic plan view of the landscape. Adding color is optional, but the inked graphics should stand alone, regardless of color application.

DETAILED PLANT SCHEDULE. The plant schedule becomes a shopping list for the client or installer; it includes botanical and common plant names, a number or letter abbreviation (if required), and plant quantities. The designer may also want to include a column with the recommended purchase size and a column listing the key reasons for selecting each plant (white flowers/red fall fruit, fragrant blossoms, disease resistance, etc.). Specified plants may not be available, due to seasonal variations in plant availability, or a phased project may require that plants be purchased several years into the future. By listing the desired characteristics that led to the initial selection, clients or designers can substitute alternative plants.

DESIGN CONCEPT. Any revisions to the design concept should be incorporated to the original concept. The final agreed-upon design concept should be included on the master plan sheet(s).

NOTES AND OTHER INFORMATION. This section on the master plan includes important information that may have been researched for the client during the design process, such as utility contacts, plant sources, or suggested soil amendments. Including this information may be a better alternative than using separate pages, which can become separated from the drawings, causing important information to be lost.

ELEVATION DRAWINGS, CONSTRUCTION DETAILS, AND OTHER GRAPHICS. Some designs will require section or elevation drawings to highlight important areas or those that are difficult to understand when seen in plan view. Section and elevation drawing may be hand-drawn illustrations or perhaps manipulated photo images. Other drawings, such as construction and installation details, may also need to be included in the final design package.

THE FINAL DRAWINGS: MULTIPURPOSE ROLES. Landscape drawings can significantly vary in their format, depending on the scale of the project,

Landscape Design Process 247

CASE STUDY

Final Design

The final design package for the Jones residence includes a sheet containing a plan drawing, plant list, concept statement, and notes; and a second sheet containing a section elevation and materials information (Figure 5-77). The designer developed an estimate and proposed schedule of work and delivered them to the client separately.

FIGURE 5-77

The final design package for the Jones residence is formatted on two sheets, including a master plan (a) and section elevation and frontyard views (b). *(Steven Rodie)*

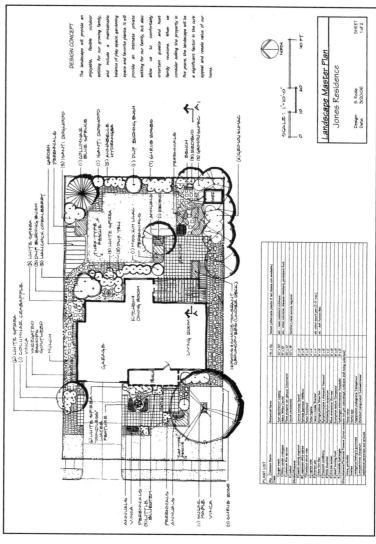

(continues)

CASE STUDY (continued)

FIGURE 5-77

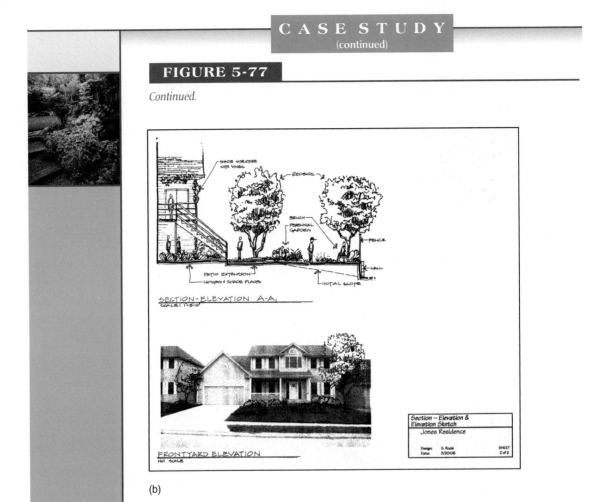

Continued.

(b)

the design fee, and the detail desired by the client. One factor that does remain constant, however, is the multiple roles that the drawings will play. On large design projects, several drawings are developed that separately show illustrative design ideas, plant locations, and dimensions for hardscapes. In these cases, each trade on the job site requires different types of information and, likewise, receives different types of drawings. On a small backyard project, the final drawing that closes the deal with a client is often the same drawing the landscape contractor will use to implement the design.

A single drawing, though, often is unable to meet the often divergent needs of both the client and the installer. Consider the illustrative, semirealistic graphics that get a homeowner excited about a project but do not clearly indicate the specified locations for every designated plant. Alternatively, a simplistic drawing that shows plant spacing dimensions and walkway widths does not contain the pizzazz that may be needed to sell the designer's ideas. Therefore, most final design drawings should be developed with a multitask role in mind. It potentially requires a more complex drawing but results in a drawing that will help sell and install the design.

SUMMARY

The landscape design process is a complex and multifaceted process. At its core, though, are the tools of quality landscape creation. Since no two clients or landscape sites are ever the same, it remains a constant challenge for the design professional to manipulate the process in order to generate quality designs, satisfy clients, and make company profits. With every project, designers will better understand how to integrate their design skills and experience with client personalities to produce creative and sustainable design solutions.

HOW FOUR DESIGN PROFESSIONALS WOULD APPROACH A NEW LANDSCAPE AND A LANDSCAPE RENOVATION

Four design professionals were asked to describe their design approaches based on two common landscape scenarios: landscape for a newly constructed home in a subdivision, and a renovation of an existing landscape. Each professional reviewed a plot plan of the sites and numerous photographs of each area. Their approaches are described in their own words.

Scenario #1—New Homeowners in a New Subdivision

Scenario 1 is a landscape at the newly constructed Smith home, where the homeowners have a limited budget ($3000) and do not know where to start.

LYNN KUHN, PERENNIAL GARDENS. Summary: This house needs distinctive landscaping that will set it apart from similar homes in the area. The question is how to accomplish that on a tight budget, satisfy the clients' needs, and make the client want to proceed with more landscaping in the future.

Step 1: Getting to Know the Client

During an on-site meeting, I would ask a series of questions that help me understand their lifestyle, their short- and long-term budgets, and what they envision in their landscape. As designers, we are tempted to blurt out a series of wonderful ideas; however, the client should do most of the talking. The information gathered in this meeting will guide the design process so the end result truly suits the clients' needs.

Step 2: Site Analysis

This process typically starts before I leave my car. Some initial observations I make note of include: the house's architectural style and color, awkward mowing areas, location of setbacks that could affect decisions concerning retaining walls and plant material selection, where water is coming from and going to, other drainage issues that will affect the landscape, and what neighboring landscapes look like.

The second part of the site analysis will come after I have gotten to know the clients. It is typical to have 5 to 10 pages of notes and 15 to 30 pictures as a result of the initial site analysis. Every time I visit the site throughout the design/build process, I am analyzing the site. If I miss a detail the first time, I hope to catch it the next.

Step 3: Budgetary Concerns

My recommendation would be to focus on the following:

- Master plan: They should invest in a comprehensive plan to implement over time. This will provide a cohesive look while preventing a hodge-podge lodge.
- Trees: They take longer to grow than shrubs, grasses, and perennials.
- Bed definition: Strip sod, amend soil, install edging, and lightly mulch with shredded bark. This step will ease mowing and provide areas for the clients to add plants to the landscape as their budget and schedule allow.

Step 4: Design

Finally, it is time to draw. Once a preliminary design is done, I ask myself a few questions to make sure the final design will meet the clients' needs. These include: Have I included elements that make this landscape distinctive and special to the client? Have I overdesigned for the client's short- and long-term budgets or for the neighborhood in general? Can this design be installed?

Conclusion

The success or failure of this project depends on whether or not the design meets the needs of the client and whether the initial $3000 was spent wisely. Both require good listening skills at the beginning of the process and throughout the project.

JIM MASON, COUNTRY LANDSCAPES. Country Landscapes, Inc. standard approach to a new house in a new subdivision with a limited landscape budget is ... PLANT SOME TREES! A $3000 landscape bed is not going to separate the Smith residence from any of the others on their block. Spend the $3000 on a collection of quality shade, ornamental, and evergreen trees that are strategically placed and will not affect any future landscape development.

For example, a really nice, large shade tree or two in the frontyard shading the many windows from the west sun would seem appropriate. An ornamental tree near the corner of the driveway and entry walk would highlight the entry and begin to soften the expanse of concrete. A collection of evergreen and shade trees along the north property line would begin to provide some privacy for the Smiths' outdoor living areas and also a backdrop for future landscape development. As houses go up behind them, the height of the trees may need to be scrutinized as there is a nice view to the northeast, unless the new houses ruin it. The tree selection would obviously take into account the soil and drainage make-up of the site, as well as exposure and microclimates.

As their new trees are growing, the Smiths can begin to think about what their landscape should include in the future. They can review their circulation patterns, window views, where the play structure goes, etc. After careful consideration of these patterns and with some disposable income, the Smiths and a landscape designer can make some informed decisions about the direction of their landscape. In the meantime, they will have a head start on the rest of their new neighborhood in the overall development of their landscape with the establishment of well-placed trees.

CHAD FRIESEN, MULHALL'S NURSERY, LANDSCAPING, AND GARDEN CENTER. Some of the first things I would discuss with the new homeowners with a limited budget of $3000 would be the phasing of a landscape plan. I would ask about family (kids), how long they plan to live here, their goals with the landscape (shade, screening for privacy, play area, and general landscaping interests).

Once I have a good idea of their interest level and what they want to do, I will address how it can happen. We would start by prioritizing their thoughts. I will recommend two approaches on how to do a plan, either draw an entire master plan or start by drawing up the phase they want to attack first. There are pros and cons to both approaches. An entire master plan is nice to give them a good idea of an overall approach to doing the project; the bad part with a master plan is most people's lives change before the whole plan can be implemented. The best approach for a client like this is to do a plan in pieces.

Based on their $3000 budget, I would recommend one of three current phases for this client: 1) front landscape; 2) tree layout for shade and ornamentation; or 3) screening for privacy in the backyard. Once they have identified their priority, then a plan would be drawn and (I hope) implemented and inspire them for further work throughout their yard.

BRYAN KINGHORN AND ANN HOUSER, KINGHORN GARDENS. We recommend a master plan to establish a design for the entire property and create for it a comprehensive vision. A master plan works well with a limited budget because the homeowner will be able to implement the design in phases as funding becomes available. From phase to phase, the initial established vision will be present and flow from area to area with ease.

To understand what the ideal master plan will consist of, we need to take clues from both the site and the homeowners themselves. While visiting the site, we take many things into consideration, including: What direction does each portion of the house face? What climatic conditions will the plants be exposed to? Where are the slopes, and how does the water drain in relationship with them? How do the other lots surrounding the site drain?

The clues we take from the discussions with the homeowner are key to the creation of the master plan. They may not be able to verbalize what they want in a garden, but they still hold a taste and a style that we need to find and understand. Design is subjective. There is no absolutely perfect answer, but there is an answer that will satisfy the homeowners beyond their wildest dreams. We need to take clues from their lifestyle, their interior décor, the architecture of their home, and their family. We discuss with the clients how they will use their yard. Do they have children? What ages are the children? What activities will they do outdoors? Do they have pets? Will they entertain outdoors? Do they like to garden? Do they have an irrigation system? Do they want privacy? These are among the many vital questions that will help us understand the homeowner.

After gathering our information, we analyze what the homeowners want, what they need, and what the site will allow. This will help us create our master plan. After presenting the master plan to the homeowners, we determine what he priorities are for installation. Typically, the front entry of the house is a priority. We would put establishing trees for this home on the top of the priority list.

After the priorities have been established for the garden, a long-term schedule can be drafted to direct installation. As the new homeowners live in their home and feel more comfortable in their surroundings, they may come up with modifications to the initial master plan. A garden is a dynamic thing and is always undergoing change, and we are open to having that happen. Offering the homeowners our maintenance and annual color service for their garden will give them assurance that the investment they are making will not only survive, but thrive.

Scenario #2—Renovating an Existing Landscape

Scenario 2 is a renovation of Mr. and Mrs. Newberry's existing landscape. They have lived in the home for 15 years, and recently their children have left for college. Now they want to update their landscape and include room for entertaining groups of 8 to 10 people.

LYNN KUHN, PERENNIAL GARDENS. The site has some interesting elements, such as several mature trees. The challenge in renovating an existing landscape is seeing the existing features as opportunities, rather than problems, as we move through the design process.

Step 1: Getting to Know the Clients

I would ask the clients a series of questions that tell me more about how they live and entertain. As empty-nesters, how long will they be in this house? I also would try to be a good listener, allowing them to expand on how they envision their new landscape. This profiling process is critical to a successful project.

Step 2: Site Analysis

This step includes initial observations about the style of the house, the neighborhood, drainage, and topography. Some additional considerations include:

- an inventory of existing plant material and deciding if it is to be removed or incorporated into the design. It is important to note caliper and species of trees and their general condition.
- photographing problems such as diseases or pests, poor branching, pruning issues, drainage, topography, etc.; can be used as a communication tool when describing ideas to the client later.
- a picture taken from across the street or as you approach in a car; this is how most people view the house. Look for balance and scale. Is it lacking?

Step 3: The Budget

After site analysis is a good time to revisit the budget. Ask the clients about specific priorities. Would they prefer to budget a set dollar amount each year? Or approach the installation area by area? As with client profiling and site analysis, this information will also help guide the design process.

Step 4: Design

Armed with lots of information about the client and the site, it is finally time to draw. Design suggestions I have for the site include:

- Add a front entrance stoop or porch large enough for a bench (or two chairs) with wide steps leading to a patio surface adjacent to the drive.

- Place trees and plant groupings in frontyard as needed to improve the view toward the street and to frame the view toward the house.
- I recommend a larger back deck rather than a patio. It will work better with the features on the house, keep users up away from the grass, and be less stressful to nearby locust trees.
- Add a focal point to be viewed from the deck and out the picture window between the locusts—a special object or a unique tree.
- Add a small, informal water feature to distract from the sound of the air conditioner and to add ambience, either on the deck or as a focal point to be viewed from the deck.
- Install trees and plant groupings as needed to add interest to the backyard, especially as viewed from the deck.

Conclusion

As with any project, the success or failure depends on how well the designer listens to the clients' needs. In this case, special attention is needed in two key areas: the front entrance and the back deck. The goal is to enhance the experience of the clients and their guests as they enjoy their outdoor environment.

JIM MASON, COUNTRY LANDSCAPES. Landscape renovations are potentially the most exciting and profitable jobs for a landscape company. In most renovation cases, Country Landscapes, Inc. will develop a master plan for clients, unless they have a specific project in mind. The designer must assume that the clients want a landscape that is unique and creative, or why would they have called you—they have lived with what they have for 15 years! Creative and unique usually are not inexpensive—so a budget discussion is imperative.

The goal of the master plan approach is to do a quality, creative design that may be made up of several smaller projects or one big project. If the designer did not get the full story on budget, there will be some options. If the clients choose just one of the projects and the designer and crew provide excellent customer service, they will likely be asked back to complete the installation of the master plan.

In established neighborhoods, it is important to get a feel for the existing landscape: the neighbor's landscape styles, the streetscape (what you feel coming down the street), and historical elements. Attempt to get a feel for the area. The Newberry residence, given its setting, does not need a general plan; it needs a detailed landscape development. Details such as paver pattern, plant selection, and path material will be critical, since the house is not that large and the landscape development will occur adjacent to it.

The front entry is cumbersome—steps next to the drive and no landing. The steps could be rounded and extended, and a small, paved sitting area could be set off to the west. A classy bench and maybe even a small bubbling fountain or some other detail could be added. The foundation plantings need to reflect the style/vintage of the house and could wrap around the new sitting area. The use of stone in these areas may be appropriate to repeat the stone in the chimney.

Giving the expanse of the backyard, it seems appropriate to maintain a minimal-sized deck to get out of the house; it could drop rapidly to a ground-level patio that acts as an entry to the backyard. Circulation will be critical here, to allow for entertainment and to get from the garage door through the patio to the deck.

With the large, existing honey locusts, clean-up will be continuous. So a water feature is probably out of the picture.

The patio will be large enough to entertain and be creative in design and material, possibly repeating the front sitting area. A nice shade garden would be developed at the base of the honey locusts. A fire pit or some other destination point would be remotely located in the large backyard to encourage the Newberrys to venture out into that space. Additional shade trees would be added over the years to perpetuate the feel of the neighborhood since the honey locusts will not last forever.

CHAD FRIESEN, MULHALL'S NURSERY, LANDSCAPING, AND GARDEN CENTER. For these clients, who are interested in updating their landscaping with no children at home, I would first start with some questions more focused on their overall intent for their landscape. Do they want a large deck or patio space for entertaining? Would they like to use some of the large backyard for more formal entertaining spaces? Do they want more privacy and screening? Do they want to include the front in renovation plans?

Once some of those questions have been answered, I would start to make suggestions and recommendations for them, starting in the front. I would like to see the front updated by possibly changing the entrance to the house. Possibly enlarging, adding pillars, whatever it might be because that is the entrance to the home, and it needs to be much more inviting than it is now. I would recommend removing all landscaping in front and updating. I would also recommend the possibility of a new driveway to aid in the parking situation that would be present if they have a group (8 to 10 people) over for a party. Current parking would be pushed into street, and they have enough property in front where some parking could be handled on their property.

In the back, I would start by recommending enlarging a deck or patio. It would be nice to have both, so that there could be two different levels for their entertaining. They already have nice shade with the honey locusts, so they should stay. I would also recommend that some screening be planted around the property for privacy.

Once they give some feedback on my recommendations, I would start with a plan focused on front, back, or the entire landscape, depending on their focus. Once I have a focus, I like to start with a concept plan and show that to the homeowners to make sure I am going in the right direction. This usually helps the homeowners if they have any doubt about how they want the space to work, before a lot of my time is wasted on a master plan. Once that concept plan has been reviewed, then a master plan can be drawn.

BRYAN KINGHORN AND ANN HOUSER, KINGHORN GARDENS. In order to develop a master plan for this site, a site analysis and inventory would need to be conducted. We would inventory the existing plant material and decide what plants were worth keeping, which plants could be renovated, and what trees need pruning. We would also decipher the flow of water on the property. Where does the water drain? Does water pool anywhere? Upon evaluation of the existing plant material, we would find out from the homeowners what plant material is sacred, if any.

We would ask the homeowners questions that will allow us to understand their needs and desires. What activities occur in the different parts of their yard? Are there certain views that they would like to accentuate or screen? Are there specific colors or plants they like (or do not like)? What type of entertaining will they do? Where do they entertain? Do they want a new deck or patio? Do they want a vegetable or herb garden? Do they enjoy gardening? What are the circulation patterns in their yard? Do they have an existing irrigation system? Do they do their own mowing? Is there a certain style of landscape that they enjoy? Would they like to explore the option of designing a new entryway to give them a more enjoyable entry experience?

Since these homeowners have existing rock mulch, we would discuss the benefits of using wood mulch with them. We would also discuss the opportunity to reduce their amount of maintenance by eliminating turf, especially around trees, shrubs, deck posts, and other structures. Because the homeowners' yard is so long and narrow, we would discuss with them the use of a destination point to help draw them out into the landscape to experience it.

After we gain an understanding of the site, a master plan tailored to the needs and desires of the homeowners would be developed. After presenting the design to the homeowners, we would determine the priorities for the installation of the landscape. Certain areas, such as a new patio or deck, may need to be finished first in order to avoid destruction of other parts of the landscape.

After we establish the order of priorities for the landscape, a schedule of installation can be produced. This schedule can take place over a matter of months or years, depending on the homeowners' budget and needs. We would offer the homeowners our continued services of annual color and maintenance to ensure they continue to enjoy the benefits of their investment.

KNOWLEDGE APPLICATION

1. Compare and contrast the client communication approaches described by the four professional designers in the appendix. What are the similarities and differences between their approaches and philosophies? Are there additional items you would include if you were asked to describe an approach?

2. Place yourself in the shoes of a client, and develop your own personal landscape wish-list using a landscape where you live or that you are familiar with. Then consider how you as a designer would address the list.

3. Working with another student, measure and document an actual site, produce base maps from your measurements, and compare your drawings for accuracy and completeness.

4. Select three diverse landscapes (residence, park, school site, urban city block, etc.), and thoroughly inventory/analyze the site conditions for each. As part of your assessment, determine which factors are unimportant and which are the most critical for each site.

5. Create a hypothetical design scenario (typical residential wish-list and site conditions), and develop three separate design concept statements that summarize the scenario from slightly different perspectives.

6. Assuming the role of a designer, describe to a hypothetical prospective client the full design process and drawing steps you would ideally implement to develop a complex design for his or her new residence.

LITERATURE CITED

Booth, N. K., & Hiss, J. E. (2005). *Residential landscape architecture.* Upper Saddle, NJ: Pearson Prentice Hall.

Newfeldt, V. (Ed.). (1994). *Webster's new world dictionary* (3rd College Edition). New York: Prentice Hall.

Pride, W. M., & Ferrell, O.C. (1991). *Marketing concepts and strategies* (7th ed.). Boston, MA: Houghton Mifflin Company.

Ramsey, C.G., Sleeper, H.R., & J.R. Hoke Jr. (2000). *Architectural graphic standards* (10th ed.). New York: John Wiley and Sons.

 ONLINE RESOURCES

U.S. Naval Observatory Astronomical Applications Department—Azimuth/Altitude Calculator: http://aa.usno.navy.mil/data/docs/AltAz.html. Accessed August 14, 2006.

National Climate Data Center—Climatic Wind Data for the United States (avg. 1930–1996). http://www5.ncdc.noaa.gov/documentlibrary/pdf/wind1996.pdf. Accessed August 24, 2006.

DESIGN PLANNING QUESTIONNAIRE. Use this questionnaire to help organize your thoughts when designing or renovating a landscape. It may bring to mind topics you have not considered, and it will give you a better idea of how to design a landscape to meet your client's needs.

Part I: Site Analysis
Color of house:
Architectural style:
Where are the most desirable views (inside as well as outside)?
Are there any undesirable views? Where?
Identify location of utilities—overhead: buried: gas/water meters:
Describe any unique features:
What is the soil type: Clay Decomposed granite Sandy Rocky Hardpan Rock shelf?
Does soil type vary in different areas of the property?
What is the prevailing wind direction in summer? _____ winter? _____
Are wind screens needed? Yes No Where?
Are sound buffers needed? Yes No Where?
Are there elevation differences? Minimal Moderate Severe slopes
Are retaining walls needed? Yes No Where?
Are there soggy areas (high water table)? Yes No Where?
Where will/does water drain?
Is a French drain required? Yes No Where?
Describe sun exposure:
Where is the yard too hot in the summer?
Describe location of existing trees, shrubs, and surface roots:
Are existing trees and shrubs desirable? Or expendable?
Describe location of existing site features and structures (fences, play equipment, deck/patio, garage, storage building, etc.):
Are existing features/structures in good repair?
Are existing features/structures in the best location?
Describe existing walks: Brick Cement Gravel Stone Bark
Are walkways in good repair?
Are walkways in the right places?
Is there a parking strip? Yes No Where?
What is your preferred level of maintenance: High Medium Low

Part II: Family Wants and Needs
Who will use your yard? Adults Children Elderly Pets
Is the yard used year-round? Mostly in spring/summer? Mostly in summer/fall? Mostly in fall/winter?
What is your preferred design style? Formal Semiformal Informal
Are there particular themes you want reflected in your outdoor space? (e.g., English, Oriental, Natural, Other)
What is your preferred shape (for lawns, walks, decks):
    Rectangular 45° angles Circles
    Straight lines Curving/free-form Combination Don't care
What is your preferred type of front entryway:
    Straight to the door Meandering Private courtyard Don't care
Is the backdoor or family entry suitable? If not, how would you like to change it?

(continues)

Continued.

What outdoor structures/features would you like to add?
Patio roof Raised planters Children's play area   Fenced vegetable garden
 Satellite dish Dog pen/run Storage shed BB area Gazebo Clothes line
 Deck Fence Swimming pool Spa/hot tub Sculpture Fire pit
 Boulders Dry creek Mounds/berms Pond Bench Fountain
 Waterfall and stream Greenhouse Putting green Rain barrel Other
What size patio/deck do you need? 2–4 people 4–8 people
 8–12 people 12+ people
Do you want walkways connecting parts of your yard? Yes No
Are additional walks needed? Where?
Do you want outdoor lighting? Landscape enhancement Security
What items need storage space? Garden equipment Garbage cans
   Bicycles Outdoor toys/sports equipment Lawn furniture cushions Other
Do you need off-street parking for guests? Cars RVs Other
How will plants be watered? Garden hose Sprinkler system Drip irrigation
How is your driveway used?
Do you have photographs of your yard? Yes No
Do you have photos of yards you like from books/magazines/garden tours?
(Photos can help you visualize what you want.)
Other comments:

Part III: Plants
What types of plants do you like?
Evergreens trees and shrubs:
Deciduous trees and shrubs: Flowering Nonflowering
 Conifer trees Fruit trees Shade trees Vines Roses
 Annual flowers Perennial flowers Vegetables Herbs
 Other
Do you like fragrant plants? Yes No
Favorite colors:
Least favorite colors:
How much lawn do you want? None Small Average Large
Where will the lawn be?
Are any family members allergic to specific plants? Yes No
   List plants:
Are deer a problem? Yes No
What special garden areas do you want? Vegetables Annuals Roses
 Perennials Herbs Wildlife/native Orchard Shade
 Rock garden Cut flowers Fragrance Wheelchair-accessible

Plant Selection: Blending Form and Function

CHAPTER 6

Many difficulties a plant might experience as a result of the growing environment can be avoided by carefully matching the right plant with a specific landscape location. Plant selection should be based on three broad categories: the environmental requirements of the plant; aesthetics or how the plant will look; and the functional qualities of the plant. Environmental requirements include a plant's low- and high-temperature tolerances as well as its light, moisture, and soil needs. These environmental requirements must be met before aesthetic and functional qualities are considered. The aesthetic qualities make the plant unique and attractive, and these ornamental features combine to create a distinctive outdoor living space. Functional qualities reflect the plant's role in the landscape (screen, focal point, etc.), and these qualities combine to make the space useful to the client. This chapter will describe guidelines for plant selection based on these categories as well as the need to consider plant maintenance, methods of developing conceptual plant communities, and the challenges of growing plants in urban conditions.

OBJECTIVES

Upon completion of this chapter you should be able to

1. describe important environmental considerations in plant selection.
2. describe important aesthetic and functional considerations in plant selection.
3. understand the impact of the urban environment on plant growth and development.
4. explain conceptual plant communities.
5. differentiate between native and exotic ornamental plants, and understand their design benefits and limitations.

KEY TERMS

hardiness zone	soil pH	woody plant	evergreen
heat zone	microclimate	deciduous	
soil texture	herbaceous plant		

Although choosing the right plant from so many options can be overwhelming, it is also one of the most enjoyable parts of creating a beautiful and functional landscape. A number of reference books (including those listed at the end of this chapter) are available to help designers make well-informed decisions and broaden their plant knowledge. A designer's ability to effectively select and incorporate adaptable plants for multiseason interest and function is one of the most useful skills a designer can develop.

SITE ANALYSIS AND PLANT SELECTION

As described in Chapter 5, the designer should conduct a detailed site analysis before starting the design process. Walking through the proposed landscape site and making extensive notes will lead to an understanding of the environmental features associated with the site. With these features in mind, a designer can determine what constraints and which desirable attributes need to be considered. Some basic questions related to the growing environment that need to be answered as part of the site analysis are: What are the soil conditions (fertile or poor, high or low pH, depth of soil, amount of drainage)? What elements (full sun, shade, wind, reflected heat, etc.) will the plants be exposed to?

ENVIRONMENTAL CONSIDERATIONS

Selecting a plant suited to the environmental conditions of the landscape site is central to its ultimate success in the landscape. Although matching a plant to the correct USDA plant hardiness zone is a good starting point, designers should also consider other factors, such as heat tolerance, first- and last-frost date, seasonal rainfall distribution, soil characteristics, light duration and intensity, and microclimates. Every plant tolerates a range of conditions for each of these factors, and it is the combined effects of these that determine true plant adaptability.

FIGURE 6-1

The United States Department of Agriculture hardiness zone map. *(United States Department of Agriculture)*

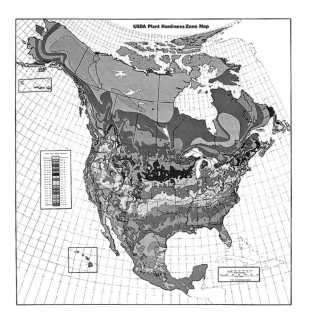

Plant Hardiness Zone

Plant **hardiness zones** (Figure 6-1) in the United States have been developed by the United States Department of Agriculture to depict average minimum winter temperatures based on the lowest temperatures recorded for each of the years 1974 to 1986 (USDA Miscellaneous Publication No. 1475, Issued January 1990; for additional information, refer to <http://www.usna.usda.gov/Hardzone>). The zones have been in use for more than 40 years as a rating system to match plant hardiness with a plant's coldest region of survival. However, no single winter is quite average; some are more severe than others in suddenness of freezing, severity of cold temperatures, or in daily or weekly temperature swings. Dramatic temperature changes in the fall and spring are common in regions such as the Midwest, where the climate is not moderated by oceans or other geographic factors like mountain ranges. Both the suddenness and severity of a winter can adversely affect plants that are otherwise hardy in a particular zone. For example, sometimes plants installed late in the season do not have adequate time to get established before cold weather arrives. Further, small, young plants tend to be less hardy than larger, more established plants. In both cases, these small or poorly established plants may not survive their first winter.

A plant's hardiness zone has become a critical piece of information included on plant tags in retail nurseries and garden centers, in plant catalogs and reference books, and even in landscape design magazines. Generally, a plant species that flourishes in one part of a given zone is likely adapted to other parts of the same zone, provided soil, light, and moisture conditions are similar between the different areas. The species is also capable of growing in warmer zones up to a certain point, when high temperatures then become the limiting factor. In addition to the stress from summer heat, some flowering shrubs and fruit trees require a minimum level of cold exposure in order to set flowers and fruit. Growing a plant in a hardiness zone that is too warm may not provide adequate chilling to trigger this event.

Just because a plant simply survives in a given zone does not necessarily mean it should be planted there. Japanese Pagodatree (*Styphnolobium japonicum*), for

FIGURE 6-2

Some plants, such as Japanese Pagodatree, survive into colder zones, but they really thrive in slightly warmer ones. Figure 6-2a shows trees struggling in zone 5, and Figure 6-2b shows a tree thriving in zone 7. *(William R. Graves)*

(a)

(b)

example, usually survives as a stunted, winter-damaged plant in the colder areas of zone 4. In zone 7, however, it grows into a robust tree with a height and spread of about 50 feet and provides a reliable profusion of large, creamy-white flowers in summer (Figure 6-2). Because of this discrepancy in growth, this species should be recommended for the warmer zones of its hardiness range. For a landscape to be lush and beautiful, it is important to choose plants that thrive, not merely survive, in the hardiness zone.

Heat Zone

In contrast to plant hardiness zones, **heat zones** are associated with the warm temperatures a plant will experience. Recently, plant scientists include a plant's performance in the summer heat as another means to evaluate adaptability.

FIGURE 6-3

The American Horticultural Society Plant-Heat Zone map. Used by permission AHS, 2006. (*American Horticultural Society*)

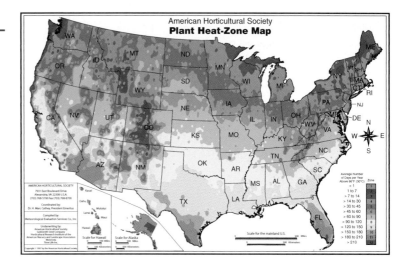

The American Horticultural Society Plant-Heat Zone Map (Figure 6-3) indicates the longest period of heat expected in all regions of the United States. Twelve different AHS Heat Zones are defined by their average number of days above 86°F (30°C), the level at which plants may experience damage on a cellular level.

Some plants may be able to handle the cold temperatures of a zone, but they do not thrive in the summer heat. A good example is yellow corydalis (*Corydalis lutea*). This herbaceous perennial can survive the cold winter temperatures of zone 5, but tends to scorch or go dormant during the high summer temperatures of zone 8.

Frost Occurrence

In addition to minimum cold temperatures, it is important to know when the first frost of autumn can be expected and how far into the spring a late frost is likely to occur. First and last frost dates have been calculated for each plant hardiness zone, providing the average date of the first fall frost and last spring frost for that area. Consult the weather service or Cooperative Extension office in your area to find information about your local conditions. Air temperature and movement have an impact on frost occurrence; the interaction of these two factors causes the formation of microclimates within a landscape. Because warm air rises and cold air sinks, cool air tends to accumulate in low spots and in areas with minimal air movement. This accumulation of cold, stagnant air often results in a frost pocket, which can form at the bottom of a hill or a low spot next to a building. The actual frost pocket area may be relatively small, only a few square feet, but it is still sufficiently colder compared to the surrounding area. Species that are marginally hardy in a given zone should not be planted in frost pockets.

Seasonal Rainfall

Total average rainfall has a significant effect on plant growth and development, and the distribution of that rainfall throughout the year is equally important. Some areas like Seattle, Washington receive substantial annual rainfall (37.07 inches), but most of the rainfall does not occur during the growing season. In contrast, Des Moines, Iowa and many other parts of the Midwest receive less total rain, but a

FIGURE 6-4

Seasonal rainfall distribution for Des Moines, Iowa; Seattle, Washington; and Portland, Oregon. *(Tigon Harmison)*

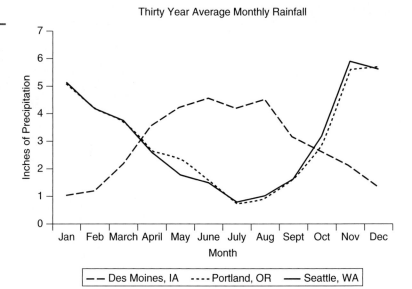

significant amount of rainfall occurs during the growing season (Figure 6-4). Another important aspect is the time between rainfall events. In Des Moines during June, July, and August, for example, there is a measurable rainfall event (≥ 0.01 inches) approximately every three days. This means that most home landscapes need minimal supplemental irrigation since the plants do not experience a prolonged period without water. In Seattle, on the other hand, there are usually five or six days between measurable events during the same three months.

Where summers are dry, plants may need supplemental water in order to survive. One way to minimize the amount of supplemental water required by a landscape is to incorporate plants native to that area, or to include plants adapted to a similar growing environment. For example, the dry summers and moderate winter temperatures of western Washington mean that many plants native to the western Mediterranean are suitable for that area of North America. Incorporating a mix of native plants and introduced species is one strategy a client can employ to reduce the amount of watering needed to maintain a landscape.

Soil Characteristics

Soil, and its characteristics, is a fundamental factor that influences a plant's long-term survivability in a landscape. Most, but certainly not all, plants can grow in a range of soil types from fairly heavy clay to quick-draining, sandy soil. For the vast majority of ornamental plants, the ideal garden soil is loam—a light, crumbly mixture of approximately equal parts of sand, silt, and clay particles, consisting of at least four percent organic matter. This mix of soil particles and organic matter creates a soil with the right balance of water-holding capacity and drainage, pore space, and adequate nutrient content. Organic matter influences plant growth and development because it holds water and nutrients, and it is loose enough for roots to penetrate.

SOIL BASICS. Terminology such as particle, texture, and structure are used to describe landscape soils. It is important to know what these terms mean and how they affect plant growth.

FIGURE 6-5

Soil pH is just as important as soil moisture and nutrients. The soil pH where this *Rhododendron mucronulatum* is growing is too high. *(William R. Graves)*

Soil is composed of clay, silt, and sand particles. Each of those particle types is a different size. Clay is the smallest; silt is an intermediate size, and sand particles are the largest of the three. In relative comparison in a scale we are familiar with, clay particles would be the size of apples, silt particles the size of a limousine, and sand particles the size of the White House. Having a mix of these three particle sizes creates gaps (pore space) between the individual particles and creates adequate pore space for water and oxygen needed by the roots.

Soil texture is determined by the relative proportion of sand, silt, and clay particles. The composition of the soil is a result of how the parent material has been broken down and weathered over thousands of years. It is difficult (if not impossible) to change a soil's texture. Adding sand to a clay soil, as is sometimes recommended, does not improve the soil but instead often results in a hard, cloddy product more like concrete.

Soil structure refers to how the individual soil particles are arranged. A soil with good structure has soil particles connected into large aggregates, allowing for air space between the aggregates. Adding organic amendments like compost, peat moss, and manure to the soil can improve structure. Greater pore space will enhance drainage, increase the amount of oxygen in the soil, and provide a good growing environment for roots. As the organic matter decomposes, it adds to the soil nutrients that are readily available to the plant.

In addition to soil type, soil pH can have a major impact on plant survival. Most woody and herbaceous ornamentals grow well across a range of soil pH, from slightly acidic to slightly alkaline (pH 4.5–7.0). Other plants, such as rhododendrons, azaleas, heathers, and heaths, prefer acidic soils (low pH) and may require periodic soil amendments to acidify the soil if it is normally alkaline or near neutral (Figure 6-5). Since extensive soil modifications can be quite expensive, a less costly alternative is to select plants adaptable to the soil environment of your client's site.

Soil pH. **Soil pH** is the measure of the acidity or alkalinity of soil. Soil pH can be measured by using soil testing kits or by sending a sample to a soil testing lab.

FIGURE 6-6

This large house creates a dense shadow next to it, and by late afternoon the shadow extends over most of the pond area. *(Ross and Beth Brockshus)*

Because some plants have specific pH needs, it might be helpful to complete a pH test before selecting plants for the landscape.

Soil pH is measured on a scale of 1.0 to 14.0, where 7.0 is considered neutral. A pH below 7.0 is acidic, and soil with a pH over 7.0 is more alkaline or basic. Soil pH can be raised or lowered, somewhat, by adding specific products. To raise the pH of an acidic soil (to make it more neutral), incorporate dolomitic limestone. The pH can be lowered by adding a sulfur product, such as aluminum sulfate or iron sulfate.

Although raising or lowering pH seems like a simple process, it can be a difficult proposition because pH is measured on a logarithmic scale rather than a linear scale. For example, lowering a pH of 7.0 to a pH of 6.0 sounds like a relatively small amount (a difference of 1), but it is actually a difference of 10^1, or 10. A pH of 5.0 is 10^2, or 100, times lower than 7.0. To create that much change in pH and keep the change from reverting back to natural conditions, ongoing soil amendments may be required, and this may be difficult to do in an established landscape. Designers should always consider selecting plants suited to the natural pH of the site.

Sun Exposure

How much direct sunlight a plant receives has a major impact on its growth. In residential landscapes, the buildings and varying sun angle between seasons (see Figure 5-46) influence the size and location of shadows in the landscape (Figure 6-6). The sun's angle not only affects shadows but also day length, which directly affects a plant's ability to create food for itself (photosynthesize). Plants that are in deep shade tend to grow more slowly than those that receive adequate light.

Some plants grow best in shade and have leaf adaptations that allow them to thrive in a low light environment (Figure 6-7). Others prefer part shade and perform poorly when grown in full shade or full sun. A good rule of thumb for plants requiring part shade is to situate them where they receive morning sun and afternoon shade. Often, the hot, intense afternoon sun is too much for these tender plants,

FIGURE 6-7

Most hostas thrive in shade conditions, and many have relatively large leaves to maximize their ability to capture sunlight. *(Ann Marie VanDerZanden)*

and it may cause them to bleach or scorch. Knowing a plant's light requirement and positioning it accordingly in the landscape are critical to its long-term success.

Another consideration is the slope of the site and its impact on light exposure. Slopes that face south or southwest receive more light and heat during the day than those facing north or northeast. Southern exposure slopes dry out more quickly, and plants growing in these conditions may require supplemental water. Capitalizing on different exposures in the landscape may extend the growing season, force some plants to bloom earlier, and create unique microclimates that allow plants to grow outside of their natural hardiness zone.

Microclimate

As described in Chapter 5, **microclimates** are landscape areas that exhibit variable growing conditions, including temperature, light, wind, humidity, and precipitation relative to a regional or local climate. One part of a designer's job is to match plants to a particular climate. However, utilizing either existing or newly created microclimates provides the opportunity to incorporate plants in an otherwise inhospitable zone.

A designer can create a microclimate by situating plants or hardscapes (or both) in such a way that light intensity is impacted, wind direction is modified, or temperature is changed (Figure 6-8). For example, if your client desperately desires a plant that is only hardy to zone 6, and he or she lives in zone 5, you may be able to increase the minimum winter temperature around the plant by locating it next to a south-facing brick wall. The wall will absorb light and heat during a sunny winter day and radiate heat during the night, when the temperatures drop. Another example is growing an annual vine over a free-standing pergola under which are planted shade-loving herbaceous perennials. The vine should reach maturity and cover the pergola by the middle of the summer, when the sun is at its hottest. The vine then provides shade for the plants through the rest of the growing season.

AESTHETIC QUALITIES

The aesthetic qualities of a plant make it unique and give each landscape its own distinctive look. These qualities can be categorized by such physical characteristics as form, texture, color, flowering and fruiting habit, and mature size. In addition to

FIGURE 6-8

This corner planting is protected on two sides and allows the homeowner to grow marginally hardy plants in this location. *(Ann Marie VanDerZanden)*

FIGURE 6-9

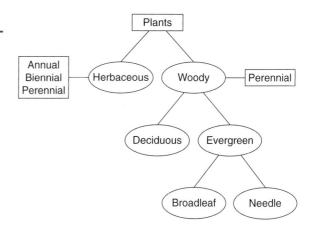

Basic plant classifications based on type and life cycle. *(Joseph VanDerZanden)*

these qualities, plants are classified by type (herbaceous or woody, deciduous or evergreen) and by life cycle (annual, perennial, or biennial) (Figure 6-9).

Plant Types

Herbaceous plants produce soft, succulent stems that die back to the ground each year in the winter. In contrast, a **woody plant** produces woody stems that continue to increase in diameter and length each year. Each spring, a herbaceous plant starts growing from the ground up and reaches its full size by the end of the growing season; a woody plant, however, continues to grow where it left off at the end of the previous growing season.

Woody plants are further classified as deciduous or evergreen. **Deciduous plants** drop their leaves each fall in response to the shortened day length and colder temperatures that accompany this seasonal change. Most deciduous plants have broad leaves, but there are also deciduous conifers, such as baldcypress (*Taxodium*

FIGURE 6-10

The needle-looking foliage of baldcypress turns a deep tan color in the fall and persists through most of the winter. *(Ann Marie VanDerZanden)*

FIGURE 6-11

Broad-leaved and needled evergreens provide different textures and colors in the landscape. *(Thomas W. Cook)*

distichum) (Figure 6-10) and larch (*Larix* spp.), that shed their needles every year. **Evergreen plants** keep their leaves or needles all year long. Evergreens are divided still further into broadleaf evergreens, such as rhododendrons, hollies, and boxwoods, or needled evergreens, such as pines, spruces, and firs (Figure 6-11).

Because evergreens keep their leaves through the winter, they can be susceptible to winter **desiccation**. This is a common problem in regions that have persistent, cold, dry winds throughout the winter. The continual wind blowing across the leaf surface draws moisture out through the leaves, in particular the broad-leafed types. Because the cold temperatures limit root growth and the availability of water in the soil, these plants are not able to replenish the water they lose through the leaves and stems. As odd as it may seem, these plants can experience drought stress during winter.

desiccation: The process of extracting moisture; typically from leaves.

Plant Lifecycles

Annual plants complete the entire growth cycle (germinate from seed, grow, flower, produce seed, then die) in one growing season (Figure 6-12). They must be replanted in the landscape each year. Some annuals, such as cosmos (*Cosmos* spp.) and moss rose (*Portulaca grandiflora*), produce a lot of seed, tend to reseed themselves, and come back each year without needing to be planted.

A perennial plant comes back each year. Herbaceous perennials include plants like hostas, daylilies, and coralbells; trees and shrubs make up the woody perennial group (Figure 6-13). In some cases, herbaceous perennials need winter protection of the crown (the point where the stems and roots meet at the soil surface; Figure 6-14) to prevent damage from cold temperatures. A layer of organic matter (leaves, mulch, pine boughs, etc.) 3 to 6 inches deep provides the necessary insulation (Figure 6-15). This layer also moderates the soil temperature and minimizes the amount of soil freezing and thawing that occurs during the winter. Extreme cycles of freezing and thawing can cause some perennials to be forced up

FIGURE 6-12

Annuals such as these petunias and snapdragons grow quickly during the growing season and are a great way to add a splash of color. *(Ann Marie VanDerZanden)*

FIGURE 6-13

Both herbaceous and woody plants should be used to create a diverse landscape and provide interest throughout the year. *(Ann Marie VanDerZanden)*

and out of the ground; this is called heaving. Once a perennial comes out of the ground and the roots are exposed to the cold, it generally dies.

Some perennials are termed tender perennials. This means that they have marginal survival rates through cold winters and cannot live without some form of winter protection. Many ornamental annuals are actually tender perennials that are native to warmer zones, where they grow year-round. Examples include petunias (South Africa), marigolds (Mexico and the Mediterranean), and snapdragons (southern Europe). These species thrive and grow quickly in the relatively short growing season in cooler climates.

The last lifecycle, a biennial, is the least common. Only a few ornamental plants, Sweet William (*Dianthus barbatus*) for example, have this type of lifecycle. Biennials require two growing seasons to complete their lifecycle. They will germinate from seed and grow only leaves (vegetatively) the first growing season. At the end of this first year, they form a **rosette** of leaves that will overwinter. The following year, they continue growing from the rosette, produce flowers and seeds, and then die.

rosette: A circular cluster of leaves that radiate from a center at or close to the ground.

FIGURE 6-14

The crown of a perennial (in particular tender perennials) is vulnerable to cold temperatures. *(Cynthia Haynes)*

FIGURE 6-15

Winter protection will ensure the crown of this perennial is insulated from cold winter temperatures. *(Ann Marie VanDerZanden)*

Four-Season Plants

When possible, select four-season plants, which provide year-round interest in the landscape. For example, a small tree such as a serviceberry (*Amelanchier arborea*) has attractive spring flowers; beautiful blue-green leaves in summer that change to a striking yellow-orange in the fall; small, edible, dark-blue fruit in summer (Juneberry is another common name for this plant); and pretty, smooth, gray bark all year long (Figure 6-16). When multifeatured plants are combined with others, the result can be a sequence of blooms providing continuous seasonal interest. One landscape scenario might include flowering trees with pastel blooms in spring; beds of perennials and annuals furnishing vivid hues in summer; trees and shrubs whose leaves turn yellow, orange, and crimson and brighten gray autumn days; and species with attractive bark and fruit throughout the winter.

Form

Form is the silhouette or three-dimensional outline of a plant. The many plant forms provide a designer with numerous options and applications for both aesthetic and functional uses in the landscape (Figure 6-17). When you are choosing plants

FIGURE 6-16

This Serviceberry has white flowers in spring (a), beautiful summer foliage (b), small edible fruit in June (c), and striking fall color (d)—a great example of a four-season plant. *(Ann Marie VanDerZanden, Steven Rodie)*

(a)

(b)

(c)

(d)

for a design, form is essential because it is visible all year long. Flowers fade and leaves die, but form persists. With all of this in mind, a plant's form is arguably its most important characteristic to consider. Incorporating a variety of forms in the landscape will create visual interest, but repetition of form gives order and continuity to the landscape. For example, using a number of different species all with a pyramidal form throughout the landscape creates continuity throughout the entire design.

Incorporating plants with a form similar to an architectural feature can also reinforce order and provide unity between the different elements (Figure 6-18). Columnar skyrocket juniper (*Juniperus scopulorum* 'Skyrocket') or the upright and

FIGURE 6-17

This planting of evergreens illustrates many different plant forms. *(Jeffery K. Iles)*

FIGURE 6-18

The upright forms of the evergreens and individual trunks of the palm trees echo the columns on the building and the stone pillars by the sign. *(Ann Marie VanDerZanden)*

FIGURE 6-19

Trees with upright and spreading shaped canopies are easy to walk under. *(Joseph VanDerZanden)*

narrow form of European hornbeam (*Carpinus betulus* 'Fastigiata') can be planted to echo columns on the front of a colonial-style home. These two species could even be used to accentuate the columns supporting an overhead pergola.

Plants with rounded or oval forms can soften the strong vertical lines of a house. Pyramidal forms tend to be rigid and stiff, and they provide a geometric look to the landscape. Plants with a weeping form are popular in many landscapes. Weeping trees and shrubs create graceful specimens and should be used individually to highlight the distinctive form. Many of these weeping plants are created by grafting a weeping form of the species onto a standard (trunk) of a compatible species.

Selecting a plant with the appropriate form based on its function in the landscape is yet another important consideration. Trees that are weeping or low branching, such as a weeping crabapple (*Malus* 'White Cascade') or cockspur hawthorn (*Crataegus crusgalli*), are difficult to walk under. Plants with these forms should not be located next to walkways, driveways, or other major circulation areas. Trees with an upright-spreading or vase-shaped canopy make good lawn trees because they allow adults, children, and pets to comfortably use the space underneath the canopy (Figure 6-19). Upright-spreading trees are also recommended adjacent to roadways to avoid low-hanging limbs that can be damaged by traffic and impede visibility. If the client wants a visual screen from a neighbor's property, a combination of trees and shrubs with rounded, oval, and pyramidal forms can be used to create a dense, overlapping barrier. A good plant selection strategy is to choose plants that naturally have the form you desire, rather than relying on severe or regular pruning to achieve it.

BRANCHING HABIT. Branching habit adds another aesthetic dimension to a plant's form, particularly if it is deciduous. When the leaves drop off these plants, the form remains unchanged, but the branching habit can give the plant quite a different appearance.

FIGURE 6-20

The unique branching habit of Pin oak makes these trees attractive with or without leaves. Give them plenty of room to grow so their form can develop naturally. *(Jeffery K. Iles)*

(a)

(b)

Some plants have unique branching habits that do contribute to the overall form. Pin oak (*Quercus palustris*) is a great example of a tree with multiple branching habits. The overall form of pin oak is pyramidal. This form is achieved because the youngest branches grow upright; as branches mature, they tend to grow more horizontally, and the most mature branches on the tree begin angling downward (Figure 6-20). Another example of distinctive branching is pagoda dogwood (*Cornus alternifolia*). The horizontal branches of this species give it a formal and rigid effect, either with or without leaves (Figure 6-21).

Branch density is an important component of branching habit. Some shrubs, such as spirea (*Spiraea* spp.) and honeysuckle (*Lonicera tatarica*), are densely branched. Others, such as rhododendrons (*Rhododendron* spp.) and staghorn sumac (*Rhus typhina*), have more space between branches (Figure 6-22). These more sparsely branched plants create an open and airy form in the winter when leaves have dropped, while densely branched species can function as a screen throughout the year.

FIGURE 6-21

Horizontal branching habits offer a unique aspect to a design. *(Steven Rodie)*

FIGURE 6-22

This spirea is an example of a densely branched shrub that forms a thicket. Branch density affects how a plant looks in winter and should be considered if the plant will be used as a screen. *(William R. Graves)*

Texture

A plant's textural characteristics may not be as obvious as some of its other features, and as a result, texture is occasionally overlooked as a design feature. Incorporating a mix of textures, or in some cases a mass of the same texture, can be an essential part of most every landscape composition.

A plant's texture is determined by the size of its leaves and stems and their three-dimensional arrangement (Figure 6-23). Plants such as big-leaf hydrangea (*Hydrangea macrophylla*) with large, widely spaced leaves and thick stems have coarse texture. Boxwood (*Buxus sempervirens*), with small, closely spaced leaves and thin stems, has fine texture. A majority of ornamentals are classified as medium textured. Coarse-textured plants tend to appear closer to the landscape viewer in comparison to plants with small leaves. The strategic use of textural contrast can influence the perceived size of a landscape space. For example, using coarse-textured plants in a

FIGURE 6-23

The combination of this boxwood hedge (fine) and hosta (coarse) is an example of strong contrast. *(Ann Marie VanDerZanden)*

small space can make the space seem even smaller. Contrasting textures add interest to a design, but too much textural contrast can result in a chaotic scene. Avoid too many combinations that include extreme differences in texture.

When evaluating the texture of deciduous plants, be sure to consider the stem size and arrangement, not just the leaf size. The silhouette of these stems will be present in the landscape for roughly the same amount of time as the plant will be in full leaf. For instance, Kentucky coffee tree (*Gymnocladus dioica*) appears to be medium textured when in leaf (Figure 6-24a). When the leaves begin to drop in the fall, the thick, dark-brown fruit pods become evident, giving the plant a coarse texture (Figure 6-24b). Once winter arrives and all of the leaves have dropped, the coarse, rugged texture is accentuated because of its stout branches and ridged bark (Figure 6-24c). The combination of these textural features creates a dramatic silhouette against the fall and winter sky.

LEAF CHARACTERISTICS. Glossy, waxy, hairy, crinkly, dissected, striped, splotched, and spotted—these are just a few terms that can be used to describe leaves. When these descriptions are combined with the vast range of colors found in leaves, the options can boggle the mind of the beginning or even journeyman designer (Figure 6-25). These leaf characteristics also influence the texture of a plant and should be considered as part of that aesthetic quality.

Interestingly, many of these leaf characteristics have a different impact, depending on the perspective of the viewer. Some are only apparent up close, while others are obvious from quite a distance. Leaves that are densely pubescent (covered with tiny hairs) may appear to be gray when viewed from a distance, but when viewed more intimately are actually light green. In other cases, leaves or needles with a waxy coating may reflect light and give the appearance of gray or blue foliage, but again when examined closely are actually a shade of green.

FIGURE 6-24

Seasonal changes of plants such as Kentucky coffee tree provide a mix of textures in the landscape. The summer leaves (a) provide a medium texture, while the fall fruit (b) and winter branching habit (c) give the plant a coarse texture. *(Ann Marie VanDerZanden, William R. Graves)*

(a)

(b)

(c)

FIGURE 6-25

Leaf characteristics alone provide numerous design options: hostas (a), Persian shield (b), Ornamental onion (coarse texture) and Agastache 'Golden Jubilee' (c), Chameleon plant (d), European ginger (e), Lungwort (f). *(Cynthia Haynes)*

FIGURE 6-26

This evergreen hedge provides a beautiful green backdrop for these showy perennials. *(Jeffery K. Iles)*

A final leaf characteristic to consider is whether the plant is deciduous or evergreen. Since deciduous plants drop their leaves in the fall, they take on a whole new look for part of the year. Needled evergreens and broadleaf evergreens do not change their appearance seasonally and can provide an attractive green backdrop in the landscape year-round (Figure 6-26). Broadleaf species like rhododendron and boxwood can be used in mass to create an attractive textural feature that can also set off more showy plants.

Color

The detailed discussion of color in Chapter 4 described the three dimensions of color (hue, value, and chroma), as well as warm and cool color distinctions, and complementary and analogous classifications. These multiple dimensions and classifications make color a complex design element. As a result, color is often overused or poorly used. Yet when it is used well, it creates a sense of unity throughout the entire landscape.

Color should be considered when selecting both plants and hardscapes. However, color associated with plants, unlike hardscapes, often changes through the course of a growing season. When compared to form or texture, color is the most fleeting plant characteristic, and a single plant can display a variety of colors throughout the year. For example, certain flowering crabapple varieties may appear burgundy or dark red when the new leaves emerge in the spring; bright pink when in flower; green in the summer; and rust, orange, or bronze in the fall (Figure 6-27).

The color wheel can be helpful in making decisions about color combinations (see Figure 4-5a). But it is important to realize that for each hue there are many combinations of value and chroma that will result in a whole range of color choices. For beginning designers, and sometimes even experienced designers, the choices and combinations can be overwhelming. One way to approach color selection is to start with a color theme or a limited color palette. From there, you can add additional colors if necessary, but at least with a predetermined theme or palette, a designer can begin to eliminate some color options.

FIGURE 6-27

The color of Prairie Maid crabapples changes throughout the season: dark pink blooms in spring (a), deep red fruit in summer (b), and yellowish-orange fall color (c). *(Jeffery K. Iles)*

(a)

(b)

(c)

One of the simplest approaches is to use the warm and cool color classifications. Choosing plants that fit into one of these broad classifications helps direct color selection without limiting it to a specific hue. Warm colors include reds, oranges, and yellows; cool colors include blues, greens, and violets. The warm colors tend to be bright and very visible in the landscape (Figure 6-28). Because these colors are so vivid, fewer plants can be used in a mass to achieve visual impact. The cooler colors tend to recede into the landscape and be more subtle (Figure 6-29). Larger plant masses of these colors are needed in order to achieve the desired visual effect.

Another approach to color selection takes into account where colors are located on the color wheel. Analogous colors are those located next to each other on the color wheel. Analogous combinations have subtle visual differences, and the colors tend to blend from one to the other. The analogous concept can be further combined with the warm and cool color classification. An example would be a planting composition that includes red, orange, yellow, and the array of shades for each of these hues. The result would be a continuous spectrum of warm-toned colors (Figure 6-30).

In contrast, complementary colors are located opposite each other on the wheel. Complementary color combinations create a great deal of visual contrast in a landscape

FIGURE 6-28

Warm colors are bright and showy. *(Ann Marie VanDerZanden)*

FIGURE 6-29

Cool colors are subdued and create a serene feeling in the landscape. *(Ann Marie VanDerZanden)*

(Figure 6-31) and seldom have a subtle effect in the overall planting. They can be used throughout the landscape to create a theme, or as a focal point to draw attention.

Designers who want to use a very limited color palette may choose a monochromatic approach. Monochromatic color schemes use one color with the full range of values from the darkest to the lightest shades. A blue scheme, for example, might include a range from a soft blue to a deep navy (Figure 6-32). Although the changes from one shade to the next may be subtle, including the whole range of shades provides enough visual interest to make the landscape appealing.

A good rule to follow when creating a monochromatic garden is to create plant masses of different sizes, depending on the lightness or darkness of the color. Increase the number of plants in the mass by one third the lighter the color. If the plant flowers range from pink to pinkish-red to red, for example, use nine pink-flowered plants, six pinkish-red flowered plants, and three red-flowered plants.

FIGURE 6-30

This annual planting creates a continuous spectrum of analogous warm colors. *(Ann Marie VanDerZanden)*

FIGURE 6-31

Complementary colors contrast with each other and have a bold look. *(Ann Marie VanDerZanden)*

In order for the lighter colors to provide adequate visual mass in the composition, more light plants are needed in comparison to the darker colors. This rule is just a starting point and may need to be modified, depending on the particular species being used and other characteristics of the specific plants.

In contrast to a monochromatic scheme with subtle color variations, a triadic planting scheme uses three different colors that are equidistant from each other on the color wheel. This results in lots of visual contrast and has a similar effect as using pairs of complementary colors. One example of a triadic scheme uses the secondary colors green, orange, and violet (Figure 6-33).

FIGURE 6-32

Monochromatic plantings include a range of shades of a single color, creating a subtle yet interesting effect. *(Ann Marie VanDerZanden)*

FIGURE 6-33

Triadic color combinations create a similar contrasting effect as complementary pairs. *(Ann Marie VanDerZanden)*

When a lot of color is used in a design, it is important to include resting spots for the viewer's eye. These lackluster optical oases allow the viewer a chance to pause and take a visual break from the stimulating color. Without these resting spots, the viewer may feel overpowered by the color. The subtleness of gray-foliaged plants, such as dusty miller (*Senecio cineraria*) or artemesia (*Artemesia schmidtiana*), makes these plants well suited for this role in the composition. Gray-foliaged plants can also be used to soften bright color combinations and to provide a graceful transition from one color to another (Figure 6-34).

Flowering and Fruiting Habit

Selecting plants with varied bloom times creates a landscape that is in bloom for the entire growing season. If season-long flowering and fragrance are important to the client, then a little extra time and attention to bloom times are warranted. There are a number of references that categorize herbaceous and woody plants by bloom time, including those listed at the end of this chapter. References like these are invaluable to designers.

Some clients may request flowering plants because they are also interested in the fruit produced by these plants. In particular, they may be interested in attracting

FIGURE 6-34

Gray foliaged plants such as lamb's ear provide a rest from the visual stimulation of bright colors. *(Cynthia Haynes)*

FIGURE 6-35

Some plants attract wildlife, which, depending on the situation, may be a desired trait. *(Nina E. VanDerZanden)*

certain birds to their landscape (Figure 6-35), and this should be clarified in the client questionnaire described in Chapter 5. Many species have persistent fruit that will remain on the plant through the fall and winter. It will ultimately dry up and fall from the tree or be eaten by birds, so it does not pose a maintenance problem. Other clients may find fruit (or the attracted wildlife) to be a nuisance, especially if the plant is near a patio or other living space. In this case, limit the number of fruiting species and, if possible, locate them in the landscape where they will not create a litter and maintenance issue (Figure 6-36).

Not all plant species are self-fruitful. Some species are referred to as **dioecious**, and they require both a male and female plant to produce fruit. In order for a client to have the bright red berries on a holly (*Ilex* spp.) shrub, there must be both male and female plants in the landscape. If there is a male plant nearby in the neighborhood, the client can plant just a female plant. In other cases, the fruit produced by the female plants can be quite a nuisance, and a homeowner may prefer having a male plant. Ginkgo (*Ginkgo biloba*) fruit are messy and have a rancid and putrid smell when ripe. The ginkgo is an excellent landscape tree; however, the reeking fruit produced by the female plants normally limits the planting of ginkgo trees to male selections.

dioecious: Having the male and female reproductive organs borne on separate plants of the same species.

Mature Size

The mature size of a plant must be considered during the design process. A common mistake is to select plants that sooner or later become too large for their location

FIGURE 6-36

When selecting plants with fruit, consider where they will be planted in relation to sidewalks, driveways, and parking areas. These crabapples were not the best choice for this landscape situation because of the mess they make when the fruit drops. *(Jeffery K. Iles)*

FIGURE 6-37

The plants in this design have outgrown their space. *(Ann Marie VanDerZanden)*

(Figure 6-37). A common residential landscape example of overgrown plants is large junipers (*Juniperus* spp.) planted under picture windows. Although the junipers may fit into their location for five or even ten years, eventually they will require regular and drastic pruning to preserve the view from the window. This not only increases maintenance costs but also creates excess green waste, neither of which are part of a sustainable landscape. Additionally, the drastic pruning often reduces the natural grace and beauty of the specimen; and it can lead to dieback, stress, and large bare spots where plants are incapable of generating new growth. A better alternative would be to select a compact juniper cultivar or small shrub with a mature height that reaches just below the bottom edge of the window. This plant will require minimal, if any, pruning; it will maintain its natural shape and will not block the view out of the window.

FIGURE 6-38

Dwarf plants have slower growth rates and may not reach their mature size in the lifetime of a particular landscape design. *(Paula Flynn)*

GROWTH RATE. The speed with which a plant reaches its mature size is equally as important as how large it will ultimately grow. Some plants, such as spirea (*Spiraea* spp.), honeysuckle (*Lonicera* spp.), and red twig dogwood (*Cornus stolonifera*), grow very quickly and can reach maturity in 3 to 5 years. Most trees and large shrubs, as well as evergreens, tend to grow more slowly and may take 10 or more years to reach maturity. These varied growth rates will influence proportion in the landscape and may make the overall composition appear out of scale until every plant has matured. Relative scale, as described in Chapter 4, is the relationship between the various landscape elements such as plants, hardscapes, and structures. To offset the effect on relative scale that varied growth rates have, a designer can recommend installing fast-growing plants in smaller sizes and starting slow-growing plants from larger specimens.

One last growth rate consideration is that of dwarf plants. Not only do dwarf plants have an overall small stature in the landscape, but their growth rates also tend to be extremely slow (Figure 6-38). For some dwarf conifer varieties, it is not uncommon for them to only have growth rates of an eighth to a quarter of an inch each year. Because of this, dwarf plants easily look out of place when planted in combination with larger, faster-growing plants.

 ## MAINTENANCE REQUIREMENTS

In addition to environmental requirements and aesthetic qualities, the plant's short- and long-term maintenance requirements must also be considered (Table 6-1). Having a client complete a Design Planning Questionnaire (as described in Chapter 5) will help a designer gauge how much maintenance the client is willing to shoulder. Discussing expected general maintenance requirements such as lawn care, light pruning, winter preparations, and specific landscape needs (fall leaf clean up, hedge shearing, care of annual flower plantings) with the client is an important part of the design process. Information gathered during this discussion and from the client questionnaire can significantly affect design decisions.

TABLE 6-1

Average maintenance requirements of common ornamental plants. *(Adapted from Hensley, 1994.)*

RANKING OF PLANT MAINTENANCE REQUIREMENTS: HIGHEST TO LOWEST
Turf
Espalier and topiary
Annuals/perennials
Groundcovers
Shrubs: deciduous and evergreen
Evergreen trees
Deciduous trees

Since a healthy landscape is growing and changing, it will always need some level of maintenance. (Neither author has yet to find a truly maintenance-free landscape. Even artificial turf and plastic plants need to be hosed down to remove the dust now and then.) In addition to the total amount of time required for plant maintenance, in some cases there is also seasonality associated with the maintenance. Some clients may enjoy spending time tending to their landscape. If this is the case, a very low-maintenance landscape likely will not fulfill their desires. The designer's goal is to create a beautiful landscape that meets the wants and needs of the client. Thoughtful plant selection can make this goal a reality.

Woody Plants

Woody plants generally have fewer maintenance requirements than herbaceous plants. Some woody plants even have different maintenance requirements, depending on age. For example, deciduous trees and shrubs require more pruning and fertilizing when they are young, less as they mature, and potentially more (dead limb or cane pruning, for example) as they begin to decline in old age. Fast-growing, weak-wooded, and relatively short-lived trees, such as poplars (*Populus* spp.), cottonwoods (*Populus deltoides*), and willows (*Salix* spp.), may need constant maintenance during their life spans. Needled evergreens as a group are relatively maintenance free, regardless of age. These tree and shrub maintenance requirements are true if the plants are situated correctly in the landscape. If a plant is too large for its location, too close to the house, or too close to the driveway or walkway, then additional maintenance will be needed.

Herbaceous Plants

Herbaceous perennials and annuals tend to require the most maintenance, particularly before and after the growing season. Many perennials need to be pruned to the ground in the spring to remove the previous year's growth that died over the winter (Figure 6-39). Other perennials, including columbine (*Aquilegia* spp.), New England aster (*Aster novae-angliae*), and coreopsis (*Coreopsis grandiflora*,

FIGURE 6-39

Perennials that die back to the ground each winter should have the previous year's growth removed in the spring. *(Ann Marie VanDerZanden)*

C. verticillata), can be cut back to the ground after frost has killed the foliage. As described earlier in the chapter, some perennials need to have a layer of mulch spread over the crown in the fall to protect the plant from cold winter temperatures. Providing a client with detailed perennial maintenance information will guide him or her in how to care for the plants. Knowing which perennials to cut back in the fall and which to cut back in the spring can determine whether or not they survive the winter.

Ornamental grasses have become popular landscape additions in the past few years because they provide an appealing texture not found in other herbaceous plants. Most ornamental grass species are easy to care for and just need to be cut back to the ground in late winter or early spring before new growth emerges. Maiden grass (*Miscanthus sinensis*), fountain grass (*Pennisetum alopecuroides*), and feather reed grass (*Calamagrostis* × *acutiflora*) are examples of clump-forming grasses that require minimal maintenance (Figure 6-40).

Although annual flowers are an eye-catching addition the landscape, they also require a significant amount of care and maintenance to keep them looking their best. Once planted in the spring, annuals benefit from regular and frequent fertilizing (every 2–3 weeks) with a water-soluble fertilizer throughout the summer. Water soluble fertilizer applied to the soil around the plants is quickly absorbed and

FIGURE 6-40

Clump-forming grasses are popular because they do not invade the entire garden space. *(Ann Marie VanDerZanden)*

FIGURE 6-41

Deadheading to remove spent flowers is an easy way to keep a plant blooming. *(Ann Marie VanDerZanden)*

provides needed nutrients. Quick absorption and rapid growth are important because the annual's lifecycle is so short.

Many herbaceous perennials benefit from deadheading (removing spent flower heads from the plant). The bloom period of purple coneflower (*Echinacea purpurea*), gooseneck loosestrife (*Lysimachia clethroides*), and perennial salvia (*Salvia × superba*), and annuals like marigold (*Tagetes erecta, T. patula*), zinnia (*Zinnia elegans*), and petunia (*Petunia × hybrida*) can be extended by regular deadheading (Figure 6-41). In other cases, deadheading is needed to keep the plants looking their best. Daylilies are a prolific-blooming perennial often used as a groundcover, but deadheading the flowers, each of which lasts one day, can be a major maintenance commitment. Including deadheading guidelines in a maintenance schedule for your client can help him or her extend the bloom time and keep the landscape looking beautiful.

FUNCTIONAL USES IN THE LANDSCAPE

Plants provide more than just ornamental value in a landscape composition. They define spaces and create the floors, ceilings, and walls of an outdoor room; screen or frame a view; determine circulation patterns; minimize erosion; and reduce reflective glare. These functional uses of the individual plants combine to create a successful landscape.

FIGURE 6-42

Privacy is created in this backyard retreat by the overhead tree canopies and dense plantings. *(Ann Marie VanDerZanden)*

Defining Spaces

Plant materials, either alone or in combination with hardscapes, define outdoor spaces. For instance, tree canopies can form the ceiling. The trunks of these same trees can be combined with shrubs to create the walls. Turf or another type of groundcover can create the floor.

How a space is defined is as unique as each individual landscape design. To be successful, a designer must understand how a client defines a space. If the client wants a private backyard, it is essential to understand what he or she means by private. One client may want a densely planted hedge so neighbors cannot see in, but so he or she can still see distant views. Another client may define a private backyard as a space with tall, heavily canopied trees, in addition to the hedge, to provide a more complete sense of enclosure (Figure 6-42). Yet another may equate the term private with just a fence. The type of plant, evergreen or deciduous, also influences how a space is defined. If a client wants backyard privacy all year long, it may be necessary to use a combination of evergreen and deciduous plants as well as hardscapes such as fences or overhead structures.

Because trees are the largest plants in the landscape, they have a major effect on how a space is defined. It is best to start by situating them in the design first and then filling in with the shrubs, groundcover, and turf. How the trees are arranged, how densely they are branched, and their canopy height all have a considerable impact on the outdoor space. Tall species with a loose and open canopy such as a thornless honeylocust (*Gleditsia triacanthos* var. *inermis*) provide some enclosure while retaining an open and airy feeling (Figure 6-43). In contrast, if littleleaf linden (*Tilia cordata*) is used in the same planting, its shorter size and denser canopy result in a more enclosed and intimate space.

Once the trees are in place, shrubs can be considered. Densely branched shrubs can create a nearly solid visual and physical barrier and in some cases can also mitigate noise. From an aesthetic standpoint, using shrubs with a mix of sizes, textures, and forms makes an attractive and functional barrier (Figure 6-44).

FIGURE 6-43

Honeylocust provides some overhead enclosure but still allows adequate light and does not block the street signs. *(William R. Graves)*

FIGURE 6-44

This water feature is situated along a visual and physical barrier using a mix of plants with different sizes, forms, and textures. It blocks the view of the neighbor and looks beautiful. *(Ross and Beth Brockshus)*

FIGURE 6-45

The size of a turf area can create a feeling of coziness, or in this case a sense vastness. *(Thomas W. Cook)*

FIGURE 6-46

The textural contrast between the turf and groundcover creates a clear definition of space. *(Ann Marie VanDerZanden)*

Groundcovers and turf are the most common plants used to define the floor of an outdoor room. Turf areas define a space and provide an opportunity to venture into the landscape so people can experience the space. Large turf areas can give the space a visual sense of expansiveness, even when the overall landscape is not particularly large (Figure 6-45). When groundcover and turf are used next to each other, the crisp edge created between the two types of plant materials makes an obvious transition between areas in the landscape (Figure 6-46).

Views

Often, a major design consideration of a residential landscape is the homeowner's desire to screen a view. Perhaps he or she does not want to see the neighbor's

FIGURE 6-47

The combination of a dense yews and plaster pillars creates an attractive and functional fence. *(Ann Marie VanDerZanden)*

FIGURE 6-48

The trees lining this circular driveway frame the view of the house from the street and are in the correct proportion for the size of the house. The front door is accented by the evergreens. *(Ann Marie VanDerZanden)*

property or vice versa. While evergreens will create the most consistent screen, combining evergreen and deciduous plants can be more interesting. Another option is to combine plants with fencing to maximize the effectiveness of the screen while breaking up the potential monotony of a long, simple fence design (Figure 6-47). A well-designed screen can be both beautiful and functional.

Framing a view is yet another function of landscape plants. The trunks of strategically placed trees can frame a piece of art, a specimen plant, and even the house. Placing trees at 45-degree angles from corners of the house, as well as tree canopy above the roofline, frames the view of the house from the street (Figure 6-48). Using smaller trees (20–25 feet) for single-story houses and taller trees (35–40 feet) for two-story houses creates the correct proportion to accentuate the house.

FIGURE 6-49

Visitors need directions through the landscape; these bright red shrub roses along the driveway help direct people around the driveway and to the house. *(Ross and Beth Brockshus)*

Directing Traffic

The next time you walk across a college campus or meander through a public landscape, notice the placement of low-growing shrubs that direct your movement. Designers often place these plants to block a perceived shortcut across a landscape. This prevents wear paths or damage to nearby plants or hardscapes. To be more persuasive, these plants are often densely branched and have thorns or prickly foliage. Compact yews (*Taxus* spp.) and junipers (*Juniperus* spp.) work well for this purpose. Particular favorites on many college campuses are the unfriendly looking barberries (*Berberis* spp.) and shrub roses (*Rosa* spp.).

Utility Purposes

Like certain hardscapes, plants can also be workhorses in the landscape by providing utilitarian functions such as influencing circulation patterns, controlling erosion, and deflecting light. A well-chosen plant can serve multiple purposes while still adding beauty and interest to the landscape.

CIRCULATION PATTERNS. Thoughtful combinations of plants and hardscapes create obvious circulation patterns in a residential landscape. Plants alone can play an important role in directing both foot and vehicular traffic. One example of this is the use of low-growing shrubs (1–2 feet tall) along a driveway or walkway. The plants create a physical barrier but not a visual one. This allows people to see where they are, where they want to go, and how to get there (Figure 6-49).

CONTROLLING EROSION. Planting groundcovers, turf grass, and suckering shrubs on sloped areas can reduce the amount of erosion that would occur if the area were left bare. Their dense, fibrous root systems help keep soil in place, and their foliage reduces the impact of water droplets hitting the soil surface. Turf should only

FIGURE 6-50

Ornamental grasses, in this case planted in parallel rows, can be used to control erosion while still providing an aesthetic element. *(Ann Marie VanDerZanden)*

FIGURE 6-51

Large shade trees on the west side of this two-story building provide summer shade and can reduce air-conditioning needs. *(Jeffery K. Iles)*

be used on slopes of less than 25 percent. Slopes steeper than this standard are difficult and dangerous to mow. If a client prefers the look of turf grass, consider using a low-growing ornamental grass like blue fescue (*Festuca glauca*) or dwarf mondo grass (*Ophiopogon japonica* 'Kyoto') (Figure 6-50). For steeper slopes, groundcovers can be used alone or in combination with a suckering shrub like fragrant sumac (*Rhus aromatica*) or red-twig dogwood (*Cornus stolonifera*).

DEFLECTING LIGHT. Light patterns experienced in a residential landscape can be modified by the deliberate location of plants. Modifying light patterns can be particularly important in areas with high light intensity and high year-round temperatures like the desert Southwest. Deciduous trees placed on the south and west side of a house can shade the house in the summer and reduce the amount of summer cooling required, yet still allow winter light inside the house (Figure 6-51). In contrast, evergreen trees and deciduous trees with persistent leaves (such as pin oaks)

FIGURE 6-52

The symmetrical architecture of this house is accentuated by the symmetrical and formal landscape. *(Thomas W. Cook)*

Combining Different Lines

Just because the house is rectilinear does not mean the planting beds need to be. Curvilinear beds used with a boxy, single-story ranch house create an attractive contrast between the landscape and house. The plants must be planted in curving masses or staggered patterns in order to maximize the curvilinear concept. Curvilinear beds with plants in straight rows create visual tension that is not appealing.

should be located carefully in climates where light access and solar heat gain is welcomed in the winter time. If the client has a large picture window or a swimming pool, careful placement of shrubs can reduce the glare from these reflective surfaces and make the outdoor living space more comfortable.

Aesthetic Purposes

In addition to the many functional roles just described, plants are also essential to the landscape's beauty. The house is generally the focal point of a residential landscape, so it is important to select plants that either complement or artfully contrast with this main feature. A planting that complements the house may mimic the house's architectural features such as line and color. In addition to the earlier example of upright junipers to mimic columns, perennials and annuals with reddish flowers can be used to echo a home's red front door.

A planting that contrasts with the home brings out differences between the two. Although a little contrast can be an effective design concept, too much contrast may result in a design that lacks unity and order. An example of contrast using a colonial-style home again would be using informal, cottage-style planting beds along the foundation of the home. The symmetrical and geometric lines of a colonial-style house beg for a planting composition with some sense of order and formality (Figure 6-52).

FIGURE 6-53

Accenting the main entrance gives visitors a visual clue about where they should go. *(Jeffery K. Iles)*

As described in other chapters, plants are generally used in masses to achieve order, repetition, and unity in the design. They can also be used individually to create accents and to draw attention to particular areas of the landscape. One of the most common locations for accent plants in residential designs is near the front entrance to the house (Figure 6-53). Depending on the size of the landscape and whether the area is divided into multiple outdoor rooms, other accent plant locations might be at the transition of one room to another, next to a water feature, near the edge of a patio or deck, or within a mixed-shrub border.

Small, ornamental trees work well as accents in residential designs, as do large shrubs. They come in a variety of forms, textures, and colors; and they easily stand out from a mass of shrubs. Selecting accent plants that have four-season interest will ensure that they are the center of attention all year long.

CONCEPTUAL PLANT COMMUNITIES

As previously discussed in this chapter, there are many aesthetic, functional, and environmental reasons for selecting plants. An additional consideration is how the plants will combine to create a conceptual plant community in the landscape. In nature, plant communities are created because plants grow in association with other plants with similar soil, moisture, and climatic requirements. As well, the shorter plants in the community are adapted to grow in the shade of the taller grasses, shrubs, and trees. This natural integration of different species enhances the overall landscape aesthetic, provides wildlife habitat, improves long-term ecosystem balance, and lowers overall maintenance costs.

Native plant communities have been identified throughout the United States and North America (Boon and Groe, 2004; Kartesz 1994; Kuchler 1985). References at the end of the chapter provide guidance in developing plant communities in designed landscapes. An example of a community listing is found in Table 6-2.

Grouping plants into identified communities adapted to a particular location does not necessarily guarantee success. Many times, environmental conditions (soil

TABLE 6-2

Example Plant Community Listing. *(From Boon and Groe, 1990.)*

BOTANICAL NAME	COMMON NAME
Trees—Canopy (over 48')	
Gleditsia triacanthos	common honeylocust
Juniperus virginiana	Eastern redcedar
Populus grandidentata	bigtooth aspen
Populus tremuloides	quaking aspen
Quercus ellipsoidalis	Northern pin oak
Quercus macrocarpa	bur oak
Trees—Understory (12'–48')	
Crataegus mollis	downy hawthorn
Malus ioensis	prairie crabapple
Prunus americana	American plum
Prunus virginiana	common chokecherry
Viburnum prunifolium	blackhaw viburnum
Shrubs (6'–12')	
Cornus racemosa	gray dogwood
Corylus americana	American filbert
Rhus glabra	smooth sumac
Shrubs (less than 6')	
Ribes missouriensis	Missouri gooseberry
Symphoricarpos occidentalis	Western snowberry
Symphoricarpos orbiculatus	Indiancurrant coralberry

structure, drainage patterns, etc.) have changed dramatically from preconstruction conditions, and this negatively affects plant growth and development. How people perceive the planting community can also affect its success. In some urban or suburban settings, designed plant communities may look too informal and may not aesthetically fit the setting. For example, while establishing a tall grass native prairie community in the backyard of an eastern Nebraska home landscape may work well, it would be out of scale and look weedy in a suburban setting.

FIGURE 6-54

Sandwich soils are created in urban settings when natural soil layers are modified. These topsoil mounds need to be incorporated into the existing soil to prevent a soil interface. *(Thomas W. Cook)*

An alternative for the suburban setting would be to include selected plants from a community rather than attempt to introduce an entire community.

Trying to Replicate Nature

Establishing plant communities or using native plants in a landscape setting should not be interpreted as a replication of nature. Designers may be able to approximate the aesthetics of a natural setting, but it is difficult to establish the integrated systems, cycles, and flow of energy and food associated with a natural system. Establishing this complexity from scratch, especially in the short term and especially in a suburban or urban landscape setting, is essentially unattainable.

THE URBAN GROWING ENVIRONMENT

The urban growing environment is often poorly suited for plants, making survival difficult. Although each situation is different, some of the most common challenges for plants growing in urban conditions include: poor and compacted soils, intense sun or shade patterns, and humans.

Soil

Many urban soils are called sandwich soils because they are composed of soil layers ordered differently than if the soil were in its native state. This sandwich effect is particularly prevalent on new home construction sites. Consider this: The topsoil is scraped off the building site; heavy equipment moves across the area during construction causing soil compaction; and sometimes before the new landscape is installed, a layer of topsoil is brought in and spread across the area. Generally, this layer of new topsoil is not tilled into the compacted layer below, so an interface between the two disparate soil types is created (Figure 6-54).

"To me, it is stupid to transplant trees into an environment they dislike and in which their length of life will be shortened and their beauty never revealed. Nature is not to be copied—man cannot copy God's out-of-doors. He can interpret its message in a composition of living tones. The real worth of the landscaper lies in his ability to give humanity the blessing of nature's spiritual values as they are interpreted in his art."

- Jens Jensen, Siftings

The effect of the heavy equipment on the soil can be exacerbated in wet conditions, causing even more severe compaction. A result of this compaction is that naturally occurring air pockets are compressed, and the soil tends to have inadequate oxygen for good root growth and limited water-holding capacity. This combination can be deadly to plants. Some plants are more tolerant of such conditions, but all plants will ultimately suffer from these conditions.

Because urban soils are not in their natural state, the 4- to 5-inch layer of topsoil often has a limited amount of nutrients, and the subsoil below it has even less. This lack of nutrients can be a real challenge to plants trying to get established. Additional nutrients can be added through fertilizer applications, but timing and frequency of applications are critical for the plant's long-term growth and development. In addition to physical limitations, the biological systems of urban soil also tend to be out of balance. Mycorrhizae, a type of soil fungi that symbiotically help plants uptake water and nutrients, are lacking in most disturbed soils. Reintroducing mycorrhizae back into urban soils has become a popular plant health enhancement, but mycorrhizae are living organisms that need reasonable soil conditions in order to live. Unless the complex properties of urban soil are improved, specific treatments may not enhance overall plant health. A designer's ability to persuade a client to invest in proper soil amendments at planting time, and continual improvements if necessary, will have a significant impact on the long-term success of the landscape.

Light Patterns

New construction can have a significant impact on existing plants. In addition to the roots being disturbed and drainage patterns being altered, sun and shade patterns can be drastically changed. Removing trees that provided an overhead canopy can turn a site that used to be part-shade into one that is full-sun. In contrast, a new building may cast a large shadow and transform a sunny area into a patch of deep shade. Buildings, patios, fences, and swimming pools can also drastically increase the amount of reflected light and heat.

Humans

Humans can be a plant's worst enemy. We often subject plants to excessive or insufficient water and fertilizer, we damage trunks with lawnmowers and string trimmers, and we get overzealous with electric hedge shears (Figure 6-55). Yet we still expect the plants to perform spectacularly. Minimizing the human impact on plants can be accomplished through thoughtful regard of the landscape.

Informing your client of water and fertilizer requirements can help him or her care for the plants appropriately. Installing tree rings (Figure 6-56) around trees or shrubs planted in a lawn area can ensure that trunks are not damaged by a mower or string trimmer. Making sure your client knows the maximum amount of growth that can be removed each time he or she shears the hedge (the hedge should be wider at the bottom than at the top to minimize low branch shading) will guarantee that the hedge performs its function in the landscape rather than turning into a disfigured, grotesque plant sculpture (Figure 6-57).

FIGURE 6-55

The trunk of this tree was damaged extensively by a lawn mower and will die. *(Ann Marie VanDerZanden)*

FIGURE 6-56

Installing trees rings around trees located in turf areas should help prevent damage from mowers. Apparently something else killed this tree. *(Ann Marie VanDerZanden)*

NATIVE PLANTS VS. NON-NATIVE PLANTS

Many horticulture professionals continue to debate the value of native plants compared to non-native plants (non-native plants are also sometimes called exotic plants). There is even debate as to what constitutes a native plant. Some experts disagree on how close a plant needs to be in relation to its provenance (origin) to be

FIGURE 6-57

Pruning can be done tastefully and in a manner that encourages the plant to grow, or to create unique living sculptures. *(Thomas W. Cook)*

FIGURE 6-58

Native plants are already adapted to the specific climate in an area and are well-suited to those growing conditions. *(Steven Rodie)*

considered native and even where that area of origination is located. Some argue further that plants are truly native only if grown from seed and planted in the same county where the seed originated.

Native plants are widely promoted as being the key to a low-input, low-maintenance landscape. Yet all plants, native or not, require some type of care. Using plants native to a particular region can certainly increase the likelihood that those plants will do well in the landscape (Figure 6-58). Other than hardiness zone, few residential landscapes provide the same environmental conditions that a native landscape provides. The urban or suburban planting environment will likely have a modified soil type, altered drainage patterns, different sun and shade patterns, and even microclimates that result in temperature differences. Sometimes native plants are not well suited to this urban environment. Non-native plants adapted to a particular growing area may be a better alternative. The use of native plants deserves consideration in addressing sustainable landscape practices, but their use should be placed in the appropriate context.

If a client is particularly interested in attracting native wildlife to the landscape, then it is important to consider native plants that will provide habitat for animals. Natives need not be used exclusively; a number of non-native plants also provide appropriate habitat.

It is true that some non-native plants (e.g., English ivy, *Hedera helix;* purple loosestrife, *Lythrum salicaria;* Japanese honeysuckle, *Lonicera japonica*) have become invasive and undesirable. Although a select group of ornamental invasive species have received plenty of attention, they are the exception rather than the rule. Climate, hardiness, and other factors impact a plant's tendency to invade landscape settings where it is not desired. To complicate matters, a plant that is invasive in one region may be a well-behaved member of the landscape plant community elsewhere. Most states have a list of ornamental plants that are considered noxious weeds or are on the invasive plants list. Consult your state Department of Agriculture for a list of invasive plants in your area.

As a result of plant collecting expeditions and mass-production practices, non-native species have been used extensively in landscapes around the world for hundreds of years. These plants have been important functional and aesthetic additions to landscapes and have not led to an environmental fiasco. As a design professional, you should not exclude non-natives solely on principle.

SUMMARY

A thorough site analysis to determine potential site limitations such as temperature extremes, soil, drainage, and light patterns is critical before the design process starts. Many difficulties a plant might experience as a result of the growing environment can be avoided by carefully matching the right plant with a specific landscape location. Thoughtful consideration of the functional and aesthetic roles the plant will play in the overall composition can also guide plant selection. Matching a client's maintenance preferences to the plants' needs will ensure that the client is willing and able to meet the maintenance requirements of the design. Ultimately, appropriate plant selection and continuing maintenance can make all the difference in the long-term success of a landscape.

KNOWLEDGE APPLICATION

1. Describe the following environmental factors that impact plant selection:
   - hardiness zone
   - heat zone
   - frost occurrence
   - seasonal rainfall
   - soil
   - sun exposure

2. Differentiate between the following terms:
   - annuals, perennials, and biennials
   - herbaceous and woody plants
   - deciduous, needled evergreen, and broadleaf evergreen
   - analogous and complementary colors
   - cool colors and warm colors
3. Describe the relationships between branching habit, plant form, and plant texture.
4. List and describe three functional roles plants play in the landscape.
5. Describe two examples of the urban environment's impact on plant growth and development.
6. Explain conceptual plant communities. Illustrate the layering concept found in these communities.
7. Differentiate between native and non-native (exotic) plants. Describe their design benefits and limitations.
8. Rank the following landscape plants in order of maintenance requirements: evergreen tree, deciduous tree, shrub, turf, espalier, annuals, perennials.

LITERATURE CITED

Boon, W., & Groe, H. (1990). *Nature's heartland: Native plant communities of the great plains illustrated in seasonal color.* Ames, IA: Iowa State University Press.

Hensley, D.A. (1994). *Professional landscape management.* Champaign, IL: Stipes Publishing.

Kartesz, J.T. (1994). *A synonymized checklist of the vascular flora of the United States, Canada, and Greenland* (2nd ed.). Portland, OR: Timber Press.

Kuchler, A.W. (1985). *Potential natural vegetation, national atlas of the United States of America.* Reston, VA: Department of the Interior, US Geological Survey.

REFERENCES

Duthie, P. (2000). *Continuous bloom.* Batavia, IL: Ball Publishing.

Free Dictionary by Farlex. (2006). <http://www.thefreedictionary.com>

Invasive Plants. (2006). <http://www.usna.usda.gov/Gardens/invasives.html>

Midwestern Regional Climate Center. (2006). <http:\\mcc.sws.uiuc.edu>

Miller, R.W., & Gardiner, D.T. (1998). *Soils in our environment* (8th ed.). Upper Saddle River, NJ: Prentice Hall.

The American Horticultural Society. Brickell, C., & Cole, T. (Eds.). (2002). *Encyclopedia of plants and flowers.* London: Dorling Kindersley.

Western Regional Climate Center. (2006). <http://www.wrcc.dri.edu>

CHAPTER 7

Hardscapes: Selection and Use

Occasionally the centerpiece, sometimes the backbone, but always essential—hardscapes are any nonplant part of the landscape. Examples include walkways, driveways, patios, decks, walls, ponds, fences, pergolas, and arbors. These features, individually and in combination, make up the ground, vertical, and overhead planes within a landscape and define outdoor living spaces. Hardscapes, including garden embellishments, are ultimately about the people who use the landscape, and they should be designed and selected with them in mind.

OBJECTIVES

Upon completion of this chapter you should be able to

1. define what hardscapes are and how they are used in ground, vertical, and overhead planes.
2. differentiate between hardscape material choices.
3. describe different types of landscape lighting.
4. describe different water feature characteristics.
5. compare forms of garden embellishment.

KEY TERMS

ground plane	dressed dimension	deadman	arbor
cut stone	composite wood	fieldstone	pondless water feature
flagstone	vertical plane	segmental wall unit	hyper-tufa
concrete pavers	freestanding wall	overhead plane	
nominal dimension	retaining wall	pergola	

Product diversity and creative integration of hardscape elements allow a designer to customize hardscapes to meet a client's needs. This chapter will describe different hardscapes common in residential landscapes, highlight appropriate uses, describe installation and maintenance costs, and provide an overview of landscape lighting and garden embellishments.

HARDSCAPES IN THE GROUND PLANE

The **ground plane**, or floor, is perhaps the most obvious part of an outdoor space. Patios, decks, walkways, driveways, and turfgrass areas are common ground plane elements in residential landscapes. Steps are also an important element of the ground plane since they help people navigate different levels of an outdoor space. Designing the ground plane to accommodate the way the homeowner plans to use the area is vital in making the landscape functional.

Patios

A patio is an ideal way to extend living space from inside the house to the outside, and it is the most common outdoor floor in residential landscapes. Patios are built from different materials and create a unique opportunity to customize the area. A patio should be designed with the house's architecture and landscape style in mind. For example, a linear or rectangular patio of milled flagstone creates a formal setting, while a patio with an irregular edge that blends into a surrounding planting bed creates a more naturalistic feel (Figure 7-1).

How the homeowner plans to use the patio determines the most appropriate material and most appropriate size. Questions to consider include: Will it be a space for entertaining or quiet reflection? Will kids use the space? What about pets? Also, keeping the patio in scale with the house is essential for it to fit into the overall landscape. A common practice to achieve this proportion is to keep the patio dimension that extends out from the house a minimum of two thirds the height of the house to the eave (Figure 7-2). This is just a starting point when designing a patio. Each landscape has a unique set of spatial relationships to consider, and the actual dimensions of the patio can be modified to make it best fit the space. A small patio can sometimes be lost; one that seems too large can visually dominate the area. Large patios can be made to look smaller by dividing them into different levels, by adding low walls used for seating, or by including planting areas for small trees that are completely surrounded by the hardscape (Figure 7-3).

Decks

As with patios, decks are a great way to create an outdoor living area. Decks can be particularly useful on lots where a patio would be difficult to build, such as one with a steep grade, poor drainage, or when excavation for a patio would damage roots of existing trees. Deck design concepts have improved, and decks are no longer just wooden rectangular slabs hung on the back of houses. Newer designs take into account the home's size and architecture so the deck will fit into the overall landscape. Although decks have traditionally been made of wood, new materials

308 CHAPTER 7

FIGURE 7-1

A patio made of natural limestone blends into the adjacent planting area. *(Joseph E. VanDerZanden)*

FIGURE 7-2

Appropriate patio proportion can be achieved when depth of patio equals two thirds the height of the eave. *(Thomson Delmar Learning)*

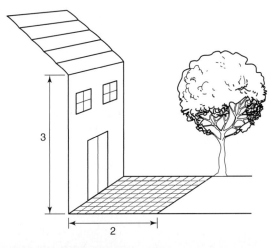

FIGURE 7-3

Built-in seating and planters make large hardscape areas seem smaller. *(Thomas W. Cook)*

FIGURE 7-4

Composite wood in complementary colors and a diagonal pattern create an attractive and long-lasting deck. *(Steven Rodie)*

such as composite wood products, metal, and coated wire can be used to customize decks (Figure 7-4). In addition to these new materials, aligning the decking on a diagonal or some other pattern can make the deck more appealing. Just as with patios, decks with multiple levels can make a large deck seem more intimate (Figure 7-5). Built-in seating and containers overflowing with plants can soften the area and provide the feel of a comfortable indoor space.

Walkways

Arguably, a well-planned walkway has the single greatest influence on a person's experience of an outdoor space. The walk may be the shortest path between two points, or it may meander through the landscape, bringing new vistas around each corner. Strategically placed walkways can link major landscape features such as patios, planting beds, and service areas together, creating a network of routes that is both nice looking and useful without resembling a highway interchange.

Most large, residential landscapes are designed with walkways that can be classified as primary, secondary, or tertiary. The walkway's function will determine which hardscape material is most appropriate. Primary walkways direct foot traffic from the driveway or street to the front door and should be a reasonably direct route between two points (Figure 7-6). These walks should be a minimum of 4.5 feet wide (the minimum width required for two people walking side by side), be made of a solid surface that is easy to walk on, be able to handle wear, and be maintainable and accessible in all weather conditions. Good choices for such walkways include natural stone, either dry laid or mortared; unit pavers; brick; or wood. Secondary walks tend to wander through a landscape and are not necessarily major thoroughfares (Figure 7-7). Because they have less traffic, they can be narrower (2–3 feet) and made of gravel or a thick layer of mulch. Tertiary paths are often hidden from view and used by the homeowner to access planting beds for maintenance (Figure 7-8). These walks should go unnoticed by the landscape visitor.

FIGURE 7-5

Dividing a large deck into multiple levels can make it appear smaller. *(Ann Marie VanDerZanden)*

FIGURE 7-6

Primary walks need to be adequately wide and made of a solid surface. *(Ann Marie VanDerZanden)*

Driveways

Large expanses of gray concrete are a common feature in the front of many homes. Although a necessity for most homeowners, driveways can be functional and more attractive if designed with a little creativity. The number and size of garages usually determine the driveway width. To be useful, the driveway needs adequate size to allow for vehicular and foot traffic. Where existing driveways are not wide enough to accommodate people getting in and out of a vehicle, additional hardscape should be added (Figure 7-9). The driveway also needs to be durable and able to

Hardscapes: Selection and Use   311

FIGURE 7-7

This secondary walk has stepping stones to provide a shortcut through the planting bed. *(Ann Marie VanDerZanden)*

FIGURE 7-8

A discrete mulched path allows access to the back of this bed. *(Ann Marie VanDerZanden)*

handle the weight of vehicles. While these constraints do somewhat limit options, there are still alternatives.

Materials such as concrete pavers; stamped, colored, or patterned poured concrete; or natural stone that complements the house and surrounding landscape can help soften the visual effect of the drive (Figure 7-10). Using the same material on both the driveway and an associated primary walkway will connect the driveway with another landscape element. Planting beds along the driveway edge or vines growing over a trellis next to the garage can mitigate the hard lines of a drive. Adding

FIGURE 7-9

Natural stone, similar to the adjacent walls, was used next to this driveway to provide a larger area for people to get in and out of a vehicle. *(Ross and Beth Brockshus)*

FIGURE 7-10

This red patterned concrete drive is an attractive alternative to traditional concrete used in driveways. *(Ann Marie VanDerZanden)*

other ornamental features such as containers, garden embellishments, overhead structures, and gates adds visual interest to a driveway (Figure 7-11).

Steps

Steps are essential in many landscapes for moving through the space, yet their usefulness should not overshadow their ability to be a striking landscape feature. Steps can be constructed from a range of attractive materials, and the amount of traffic they receive will influence which is most appropriate. High-traffic steps and those associated with primary walkways should be made from wear-resistant

FIGURE 7-11

A metal gate and distinctive lamppost accentuate the start of this driveway. *(Thomas W. Cook)*

FIGURE 7-12

These high-traffic steps are made of durable stone and are wide enough for two people to walk side-by-side. *(Thomas W. Cook)*

materials and create a firm, solid walking surface (Figure 7-12). Nonprimary steps can be made from a variety of materials but should still be relatively easy to negotiate. Regardless of the role, it is critical to make uniform steps and to use the correct ratio of risers to treads (Figure 7-13). Poorly designed steps are a serious trip hazard, regardless of landscape location or use.

FIGURE 7-13

Standard equation used to calculate dimensions for outdoor steps. *(Thomson Delmar Learning)*

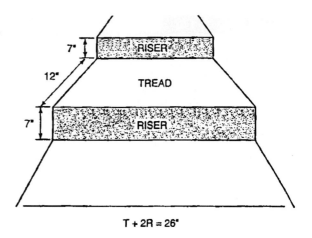

FIGURE 7-14

Example of how the standard equation can be modified to steps with a longer tread. *(Thomson Delmar Learning)*

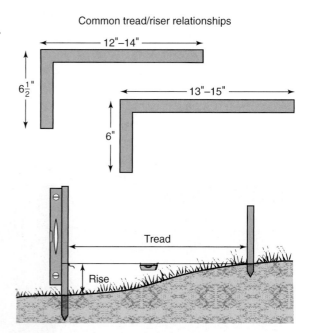

The standard formula to calculate riser height (R) and tread depth (T) for steps is: 2R + T = 26. For example, if you are designing steps with a 14″ depth (T = 14), the formula would determine a correct riser height (R) of 6″ (2[6] + 14 = 26).

To make deeper steps with a lower rise, the formula can be modified. For these steps, each $\frac{1}{2}$-inch reduction in riser height is accompanied by a 2-inch increase in the tread depth (Figure 7-14).

Outdoor steps that have a slightly lower riser height and a longer tread depth compared to indoor steps are easier to negotiate; this design results in steps more in scale with the larger outdoor space (Figure 7-15). These larger steps also equate with the typically longer strides a person takes in an outdoor setting. An additional benefit of deeper steps is that they can be used for seating while a person pauses along a landscape path, or for additional informal seating near a patio or deck.

FIGURE 7-15

Larger steps are in scale with the outdoor space.
(Thomas W. Cook)

(a)

(b)

Just as with walkways, steps that are part of a primary circulation pattern should be a minimum of 4.5 feet wide to allow for adequate circulation. When possible, a minimum of three steps should be grouped together, since a single step can pose a tripping hazard. Also, using a different material or changing the pattern on the nose of a step makes the edge more obvious and alerts the user to the location of the step.

FIGURE 7-16

These steps are functional and blend in with the landscape. The deep treads require a couple of paces for each step. *(Steven Rodie)*

Steps do not always need to traverse the shortest distance between two places. Steps that curve or zigzag across a slope result in an intriguing design feature as well as an easy way to move through the landscape. Breaking up a series of more than 10 steps with landing areas provides resting spots for users and can also add interest to the landscape, particularly if the landing has a bench, a piece of garden art, or containers displaying colorful plants. On long, gentle slopes, incorporating steps with long treads (2–3 feet) causes a person to take two paces on the tread and alters the tempo at which he or she moves through a landscape (Figure 7-16). This leisurely pace enables him or her to absorb the garden's beauty.

MATERIALS SYNOPSIS. Hardscapes used in the ground, vertical, and overhead planes serve many functions in the landscape and can be made from a variety of materials. How the client plans to use the hardscape will influence which material should be used to construct it. What follows is a synopsis of materials commonly used for ground plane hardscapes. Some materials such as stone, brick, concrete units, wood, and composite wood are also suitable for vertical and overhead uses. Additional materials appropriate for the vertical and overhead planes will be described in later sections.

Gravel. Gravel or crushed stone consists of angular pieces of stone in a range of sizes. It is available as a single size (e.g., $\frac{1}{2}''$ diameter pieces) or in a mix of sizes

FIGURE 7-17

The angular-shaped pieces make gravel distinctive, and the different-sized pieces will lock together over time. (Ann Marie VanDerZanden)

FIGURE 7-18

The rounded shape of pea gravel makes it suitable as a decorative feature. (Ann Marie VanDerZanden)

(e.g., $\frac{3}{4}''$ diameter pieces and smaller, also called $\frac{3}{4}$ minus) (Figure 7-17). In the latter, the various-sized pieces lock together to form a packed, firm surface. Pea gravel, on the other hand, consists of roundish marble-sized stones that do not lock together (Figure 7-18). Both types of gravel are available in a range of colors.

Angular and pea gravel are particularly suitable for secondary and tertiary walkways. Because they do not form a solid, uniform surface, snow removal can be difficult, but water seldom puddles on gravel surfaces. Both of these types of gravel should be laid over a compacted base layer to make the surface more stable, and they need some type of edging to keep the material contained (Figure 7-19).

Edging Materials. In addition to containing gravel walkways, landscape edging is used to hold pavers in place and to physically and visually define areas within a landscape. When used to separate turf and planting beds, edging should be installed flush to the ground surface to allow close mowing and to minimize the need for string trimming. Regardless of edging configuration, a vertical edge of 2″ to 3″ is required to allow for adequate mulch depth and to hold mulch in place. Examples of edging include the following:

Black Plastic: One of the most common choices for homeowners is black plastic edging because it is flexible and easy to install. It generally has a rounded piece on top with a 3-inch flat section underneath, which is buried below grade. Large spikes are used to secure it in the ground. Black plastic edging often works up and out of the soil due to **frost heaving**, and it occasionally needs to be resecured. It is prone to fading, cracking, splitting, and mower damage.

Steel and Aluminum: While similar to black plastic in how they are installed, steel and aluminum materials are less flexible, more expensive, and more permanent. If not installed properly, these products can also work out of the soil (Figure 7-20). Steel edging will rust over time, creating a natural **patina** that blends with the landscape.

frost heave: Upward thrust of the ground caused by the freezing of moist soil.

patina: A change in appearance produced by age and use.

FIGURE 7-19

Metal edging is used to separate gravel and turf areas and prevent the turf from encroaching. *(Ann Marie VanDerZanden)*

FIGURE 7-20

Aluminum edging that is not installed properly can work out of the soil due to frost heaving and become a hazard. *(Steven Rodie)*

Wood: Available in a range of sizes, lumber is relatively inexpensive and low maintenance once installed. Wood edging is most suitable for flat surfaces, and only decay-resistant species or treated lumber should be used. Continuous contact with soil will cause untreated lumber to quickly rot. The lumber (typically in 1"×4" or 2"×4" sizes) requires metal or wood stakes to hold it secure. Less expensive treated wood has a high tendency to warp and may need to be replaced after only a few years in the landscape.

Natural Stone: Although stone is one of the most expensive edging materials, it is long lasting and relatively maintenance free. It is available in a range of colors, sizes, and shapes; and it naturally blends well with the surrounding landscape. The flat surface provides a solid surface for a lawn mower wheel to travel on.

Bricks: Whether stood on end, laid end to end, or placed edge to edge, bricks create an attractive and formal edge. When laid flat, they provide a solid mowing edge similar to stone.

Precast Blocks: These manufactured concrete blocks come in a range of colors, forms, and sizes; and when installed properly they are low maintenance and long lasting. Smaller sizes create an edge with a tight radius when necessary.

Poured-in-Place Concrete Curbing: This product is custom mixed for color and poured on site to the height and width specifications of the designer or homeowner. Although it can be expensive, it is durable and long lasting, and it can be customized to meet the distinctive needs for each landscape. The curb can be shaped to provide a flat mower wheel surface on the outside edge, reducing the need for string trimming along the face of the curb. To create a lighting effect, rope lights can be imbedded into the curbing when it is poured.

Furrow Edging: This type of edging is done to create a distinct boundary between turf and a planting bed. A sharp shovel is used to cut into and remove the turf along the bedline to a vertical depth of 3" to 4". This process usually needs to be repeated a couple of times during the growing season to keep the turf from encroaching into the bed. Spot spraying of herbicide along the furrow may also be used to keep the edge defined.

Natural Stone. Stone has been a mainstay in landscapes for thousands of years and was a major component of gardens built by the ancient Egyptians, Romans, and Chinese. It is suitable for a wide range of applications; typically weathers well; and provides a rustic, casual feel to a landscape. The type of stone used in a landscape tends to vary by region. For example, **limestone** is common in the Midwest, **granite** and **bluestone** in the Northeast, **sandstone** in the Southwest, and **basalt** in the Northwest. Each of these stone types varies in color and properties, which influences the absorptive capacity, strength, and how it will weather and age (Table 7-1). Density of the stone has a major impact on shipping and freight costs.

Natural stone is available in a variety of shapes and forms. Some are suitable for use in the ground plane while others are more suitable for use as walls or features like outcroppings or dry riverbeds. The terminology used to describe different stone types can vary by region. Below are general definitions for two of the most common terms.

Cut stone: Also called dimension stone or wall stone, includes stone cut to size at the quarry, **split faced** if required, and polished or finished to different specifications. Most natural stone can be purchased in standard square and rectangular sizes such as: 6" × 6"; 4" × 12"; 6" × 12", 12" × 18"; etc. The stone is also milled to a uniform thickness, making it easy to use (Figure 7-21). Cut stone is commonly used to build walls.

Flagstone: Flagstone is a generic term used for flat stone slabs that range in thickness from $\frac{1}{2}$ inch to 3 inches and are used in ground plane hardscapes (walks,

---

limestone: A common sedimentary rock consisting mostly of calcium carbonate; relatively soft and porous.

granite: A common, coarse-grained, light-colored, hard igneous rock consisting chiefly of quartz, orthoclase or microcline, and mica; very dense and nonporous.

bluestone: A bluish-gray form of sandstone.

sandstone: A sedimentary rock formed by the consolidation and compaction of sand and held together by a natural cement, such as silica; relatively soft and porous.

basalt: A hard, dense, dark volcanic rock composed chiefly of plagioclase, pyroxene, and olivine; nonporous.

split faced: Natural stone that is split, either using a chisel for a more natural look or guillotined for a straight, relatively smooth surface.

TABLE 7-1
Natural stone characteristics.

STONE TYPE COMMON NAME	WHERE QUARRIED	COLOR	STRENGTH (psi)	ABSORPTION (%)	DENSITY (lb/cu ft)
Dolomitic limestone					
• Iowa buff	Eastern Iowa	Buff	10,550	7.50	133.8
• Eden	Wisconsin	Gray/white	25,698	0.63	165.6
• Cottonwood	Kansas	White, buff	9,613	6.30	142.7
Sandstone					
• Colorado red	Colorado	Red, brown	24,265	1.26	155.4
• Colorado buff	Colorado	Buff/tan	24,265	1.26	155.4
• Apache	Southwestern United States	Varied shades, buff, light rose	18,400	2.39	149.2
• Bluestone	New York/Pennsylvania	Blue, gray, green	18,255	1.21	162.5
Argilite sandstone					
• Echo mountain	Northwestern United States	Purple, rust silver, brown	18,000	0.64	155.0
Quartzite sandstone					
• Sandy creek	Wisconsin	Light brown, buff, cream	17,700	0.50	158.6

Source: Rhino Materials, Des Moines, Iowa

FIGURE 7-21

Cut stone was used to build this wall. *(Ross and Beth Brockshus)*

FIGURE 7-22

Milled flagstone ready for installation. *(Ann Marie VanDerZanden)*

drives, patios) (Figure 7-22). Common flagstone materials include limestone, bluestone, and sandstone.

The range of thicknesses available makes this material appropriate for many different hardscape applications. Stones that are 2 to 3 inches thick can be set onto a crushed stone base for patios and walkways. Stones that are 4 to 5 inches thick can bear the weight load necessary for a driveway, provided the base has been prepared properly. Thicker stones (6 inches or more) can be used to create naturalistic steps (Figure 7-23). In most cases, flagstone is milled to uniform sizes as described in the cut stone section above. Other times, it will be milled to a uniform thickness, but the shape of the pieces is irregular and random, which results in a more natural-looking hardscape.

Proper base preparation is essential to the integrity of stone walks, driveways, and patios (Figure 7-24). Base preparation will vary, depending on the use. Foot or vehicular traffic, weight load, and regional weather extremes all require different treatment (Table 7-2). Appropriate references are available, and the designer should consult them during the design phase to make sure the proper stone is selected and installed correctly.

FIGURE 7-23

Large slabs of stone make sturdy steps and blend in well with this natural stone path. *(Ross and Beth Brockshus)*

FIGURE 7-24

Typical base preparation for a stone, brick, or concrete paver walkway. *(Thomson Delmar Learning)*

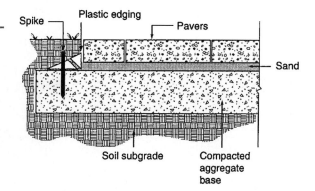

Brick. Brick is a traditional material that has long been popular in landscapes. The richly patinated surface and chips caused by time and weather of old brick lend a rustic look to the landscape (Figure 7-25). In formal gardens, new brick used in linear or geometric paving patterns creates a sense of order and neatness. Common paving patterns include herringbone, basket weave, and running bond (Figure 7-26).

Bricks are mainly available in a reddish-brown color, although there will likely be regional variations. Brick sizes and nomenclature have varied over the centuries, but historically a brick is small enough to be held in the hand and rectangular in shape. In 1992, a committee of professionals associated with brick manufacturing and distribution developed standard nomenclature for bricks (Table 7-3). Also, the American Society for Testing and Materials (ASTM) publishes the most widely accepted standards on brick. These standards predict the durability of the product in actual use and include classification of the bricks based on exposure, appearance, physical properties, and other factors. Before selecting brick for a project, consult the

TABLE 7-2

General base thickness chart for various paving materials.

PAVING TYPE	PEDESTRIAN TRAFFIC	OCCASIONAL VEHICLES	AUTO TRAFFIC	HEAVY TRAFFIC
4-inch concrete slab	3 inches	4 inches	6 inches	8 inches
Interlocking paving block with 1-inch sand setting bed	4 inches	8 inches	8 inches	16 inches
Clay paving brick with 1-inch sand setting bed	4 inches	8 inches	8 inches	Not recommended
2-inch unmortared stone with 1-inch sand setting bed	4 inches	Not recommended for this traffic type		
2-inch mortared stone with 1-inch mortar base	4 inches	Not recommended for this traffic type		
3-inch granular paving	3 inches	4 inches	Not recommended	

The depths are general recommendations from manufacturers. Verify actual base depths required for a project with an engineer.

industry standards to make sure you have selected the appropriate material. The manufacturing and installation costs of brick are high compared to other unit pavers. High-quality bricks need to be installed and cared for properly for the hardscape to be durable over time.

Concrete Pavers. Since the early 1970s, there has been an upsurge of **concrete paver** options available in the landscape industry. These products are available in dozens of colors and finishes, and the installation patterns are only limited by the designer's imagination. Some manufacturers will do custom blending to create pavers to specifically match a customer's color request. In addition to a range of color and finish options, pavers are available in a variety of shapes, including square, rectangle, circle, keystone, octagon, and trapezoid (Figure 7-27). Packaged kits are also available to create a round paving area, with all of the pavers precut to account for the

FIGURE 7-25

The rustic look of old bricks adds charm to a landscape. *(Thomas W. Cook)*

FIGURE 7-26

Common brick paving patterns. *(Thomson Delmar Learning)*

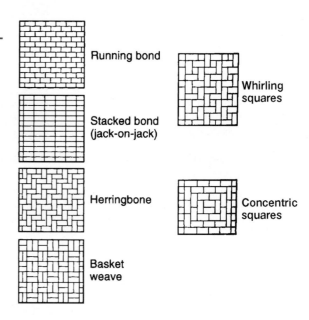

curvature of the design (Figure 7-28). Using pavers with a granite-like finish in a traditional herringbone or basket-weave pattern creates a formal look, while using tumbled pavers in a random pattern (Figure 7-29) creates an informal patio, walk, or driveway.

Concrete pavers are versatile and durable. The manufacturing process ensures uniformity in size, color, and finish. Pavers have a high compressive strength that allows them to handle extreme weight loads, provided the base is prepared correctly. Installation is relatively fast with low labor costs, and if a paver is damaged it can be replaced. Most pavers, except for those that that have been tumbled, are

TABLE 7-3
Standard Brick Sizes

	Modular Brick Sizes					
	Nominal Dimensions (Inches)			Specified Dimensions* (Inches)		
UNIT DESIGNATION	WIDTH	HEIGHT	LENGTH	WIDTH	HEIGHT	LENGTH
Standard	4	$2\frac{2}{3}$	8	$3\frac{5}{8}$ $3\frac{1}{2}$	$2\frac{1}{4}$ $2\frac{1}{4}$	$7\frac{5}{8}$ $7\frac{1}{2}$
Engineer standard	4	$3\frac{1}{5}$	8	$3\frac{5}{8}$ $3\frac{1}{2}$	$2\frac{3}{4}$ $2\frac{13}{16}$	$7\frac{5}{8}$ $7\frac{1}{2}$
Closure modular	4	4	8	$3\frac{5}{8}$ $3\frac{1}{2}$	$3\frac{5}{8}$ $3\frac{1}{2}$	$7\frac{5}{8}$ $7\frac{1}{2}$

*The specified dimension may vary within this range from manufacturer to manufacturer. The Brick Industry Association, Technical Notes 10B-Brick Sizes and Related Information.

FIGURE 7-27

Concrete pavers are available in a variety of colors and shapes and allow designers to customize any landscape. (Ann Marie VanDerZanden)

FIGURE 7-28

Prepackaged kits with appropriately sized and shaped pavers are available for circle patios, making them easy to install. *(Ann Marie VanDerZanden)*

FIGURE 7-29

The tumbling process randomly broke the edges of these pavers to create an aged effect. *(Thomas W. Cook)*

beveled edge: Results when two surfaces meet each other at an angle other than 90 degrees; typically at 45 degrees.

manufactured with a **beveled edge** (Figure 7-30), which makes snow removal easier and prevents the edges from chipping or cracking.

Poured Concrete. Concrete is a strong, long-lasting, and versatile landscape material. It is made from different proportions of a binding agent (Portland cement), aggregate (coarse and fine sand, and gravel), and water. Different ratios of aggregate materials are used to create various textures and concrete strengths. Concrete can be formed into a variety of shapes to create a unique landscape element.

Hardscapes: Selection and Use 327

FIGURE 7-30

The beveled or chamfered edge enhances the durability of concrete pavers. *(Ann Marie VanDerZanden)*

FIGURE 7-31

Colored concrete can be an eye-catching addition to the landscape whether used on the ground plane or in a sculpture. *(Darcy Loy)*

hand trowel: A finish on concrete created when using a flat-bladed hand tool for leveling, spreading, or shaping.

broom finish: A finish on concrete created when using a broom or other stiff-bristled tool.

rebar: A rod or bar used for reinforcement in concrete or asphalt.

Coloring agents can be added during the mixing phase to produce concrete in a range of colors (Figure 7-31). An alternative is to add a color hardener to the surface while the concrete is still in the plastic (not yet set) state. This method is more labor intensive, but the colors are brighter. In addition, various finishing techniques (such as **hand troweling**, a **broom finish**, stamping, staining, or etching different patterns) can result in a truly distinctive hardscape.

Poured concrete requires a significant amount of site preparation, including excavation, fabrication of forms, compacted base and sub-base materials, and laying wire reinforcing mesh or **rebar** in a $2' \times 2'$ grid in the base of the excavated area.

FIGURE 7-32

Applying a stamping technique to this concrete patio makes it a striking hardscape. *(Thomas W. Cook)*

Pouring the concrete requires skill in order to pour, spread, and smooth the material into a high-quality surface that will drain properly. Large, uninterrupted areas need to be installed in sections with expansion joints which allow for shrinking and swelling of the slab. Control joints are integrated into the surface to help control the locations of the inevitable cracks that occur due to slab settling and shrink-swell forces. Slabs that are 4 inches thick should have joints every 8′ × 8′; slabs that are 5 inches thick should have joints every 10′ × 10′. Joint placement creates patterns on the surface, and if done thoughtfully, these joints can meet the technical requirements of the concrete and can provide an interesting addition to the space. Concrete should only be installed when temperatures are above freezing, and it must be kept moist during the curing period (3–4 weeks) so it does not dry too quickly and crack. Even the best-poured concrete is susceptible to cracking.

Techniques to Enhance Plain Concrete. One popular method to enhance drab concrete is color. Adding earthy-toned coloring agents during the mixing phase results in a concrete slab that blends into the landscape. Many colors are available to meet a designer's needs, including exotic-sounding colors like willow, Omaha tan, Kaluha, palomino, and mesquite. Combining color with unique finishing techniques (such as stamping) increases visual interest and eliminates an otherwise plain, gray slab from the landscape.

Stamped concrete is a technique in which a pattern is pressed into newly poured concrete. There are a number of patterns to choose from, including brick, flagstone, tile, European fan, octagons, and even weathered wood (Figure 7-32). After the concrete is poured and floated (leveled) and excess water has evaporated, specially formulated hardening and releasing agents (so the stamps do not stick to the concrete) are applied to the surface. Large stamps imprinted with the pattern (brick, flagstone, etc.) are placed on the concrete surface. A hand tamper is used to firmly imprint the texture and pattern of the stamp into the concrete. It is important to keep the stamps fitted tightly together to create faux grout lines (these lines provide a good location for expansion joints). After the concrete has cured for two or

FIGURE 7-33

Figure 7–33a shows the edge of the slab after the stencil technique was applied. The end result is a faux stone patio (b). *(Ann Marie VanDerZanden)*

(a)

(b)

three days (depending on the weather), it is power washed to remove excess releasing agent on the surface. Three to four days after the surface cleaning, a muriatic acid wash is applied to remove any remaining releasing agent and to etch the concrete surface so the sealer can penetrate more deeply. The final step is to apply two to three coats of sealant 24 hours after the muriatic acid wash. Stamped and colored concrete should be cleaned and sealed on a regular basis (every 2–3 years) to protect the surface and prevent fading. It is important to let your client know about this added cost if he or she chooses this product.

Another decorative concrete technique is stenciling. In this case, patterned stencils are placed over an existing concrete slab or on a newly poured slab while still in the plastic state. Then a thin decorative coating in the color of your choice is sprayed onto the surface. The areas under the stencil remain uncoated, and after the stencil is removed they will resemble mortar joints (Figure 7-33). Decorative finishes such as troweling or a broom finish can be applied to the coating before it dries. Once the coating has dried (which takes between 2–8 hours), the stencils are removed, and a sealer is applied.

Acid-etching or acid-staining is another decorative technique applied to existing concrete. Although this is called an acid-stain, acid is not the ingredient that colors the concrete. Rather, the metallic salts in an acidic, water-based solution react with hydrated lime (calcium hydroxide) in hardened concrete to yield insoluble, colored compounds that become a permanent part of the concrete (Figure 7-34). The acid in chemical stains opens the top surface of the concrete, allowing metallic salts to reach the free lime deposits. Water from the stain solution then fuels the reaction, which takes a minimum of 8 hours and can continue for up to a month on concrete that is not completely cured. To get more uniform color development, it is best to apply the product to concrete that has cured for at least 3 weeks. Several companies manufacture chemical stains that are variations of three basic color groups: black, brown, and blue-green.

FIGURE 7-34

The acid-etching technique creates one-of-a-kind coloring on a concrete slab. (Dan Wirtanen)

Before the acid-stain is applied, the concrete must be cleaned and any sealants removed to allow for the stain to penetrate the surface. Once the stain has been absorbed, the chemical reaction will occur, and color will start to develop on the surface. Decorative cuts and sandblasted patterns achieved by using stencils affixed with an adhesive back can enhance the appearance of stained surfaces. If your client wants the overall stained effect to be uniform, the cuts and sandblasting should be done after the stain is applied since stain is absorbed differently on cut surfaces. As with the other decorative techniques, the final step is to seal the concrete surface.

These three decorative techniques offer endless possibilities for colors, patterns, and images. Stenciling and acid-etching are great ways to renovate an existing concrete slab and help it blend in with a new landscape. Working with a reputable installer will result in a truly unique concrete hardscape for the client.

Wood. Wood serves both structural and decorative functions in the landscape. The right type of wood and appropriate fasteners are fundamental to a project's look and lasting success. From an aesthetic standpoint, the grain pattern and surface texture of the wood are essential design factors. A smooth, fine grain species is a good choice for a deck surface, while a coarse-textured, rough-hewn wood is suitable for rustic fencing.

Many species, grades, and finishes of wood are available for landscape uses. Redwood and cedar are naturally rot and insect resistant and are commonly used for decks, fences, arbors, and pergolas. For landscaping purposes, wood is available as rough cut or as milled lumber. When wood is milled, it is cut into certain sizes (e.g., $1'' \times 4''$; $2'' \times 4''$; $2'' \times 10''$; $6'' \times 8''$) referred to as its **nominal dimensions**. After the wood dries and is planed, the actual dimensions are reduced, and that is referred to as the **dressed dimensions**. Sometimes, larger dimensional lumber is referred to as landscape timbers. For example, although a timber is called a $6'' \times 6''$ (the nominal dimension) its actual size is closer to $5\frac{1}{2}'' \times 5\frac{1}{2}''$ (the dressed dimension). These seemingly small differences can have a significant effect on how the landscape structure will be built (Table 7-4). It is important for a designer to differentiate between nominal and dressed (or actual) dimensions.

TABLE 7-4
Nominal and Actual Dimensions for Stick Lumber

NOMINAL DIMENSION	ACTUAL DIMENSION
1″	$\frac{3}{4}″$
2″	$1\frac{1}{2}″$
4″	$3\frac{1}{2}″$
6″	$5\frac{1}{2}″$
8″	$7\frac{1}{4}″$
10″	$9\frac{1}{4}″$
12″	$11\frac{1}{4}″$
Example comparison: 2″ × 12″	$1\frac{1}{2}″ \times 11\frac{1}{4}″$

An important consideration to make when selecting wood is whether it will be in contact with the ground. If this is the case, pressure-treated wood is a recommended choice. Softwoods, such as Douglas fir and pine, are commonly treated by injecting the wood with a preservative under pressure. Other softwoods are suitable for above-ground uses, provided they are made more durable by painting, staining, or sealing. Staining or painting will enhance the appearance and mitigate warping, but these additional steps also mean greater maintenance and cost. Many homeowners opt for pressure-treated wood due to its initial low cost and maintenance-free characteristics. When any exterior wood product is used above ground, changes in moisture content can lead to warping.

In addition to rot resistance, lumber is graded on strength and appearance, and divided into either select lumber or common lumber. Select lumber is used where appearance and finish are important. It includes grades A through D, where Grade A material contains no knots or defects, while Grade D has large blemishes. Common lumber is used for framing and structural members and is divided into grades (Grade 1–5) that denote different structural quality.

The type of fastener used for landscape projects is as important as the type of wood. Corrosion-resistant fasteners, such as stainless steel or galvanized, withstand environmental elements and increase the longevity of any project. Screws have replaced nails in most facets of outdoor wood construction, most notably for decking, and are preferred over nails for strength and permanence. Functional fasteners that are also decorative can be used as a type of landscape embellishment to mimic an architectural style such as arts and crafts or mission. In addition, specific, repeated patterns of fastener placement (number, distance from board edge, amount of stagger for multiple fasteners, etc.) can unify an outdoor project and ensure structural integrity.

Composite Wood Products. Manufacturers have developed composite wood products that are moisture and insect resistant and do not require sealing, painting,

FIGURE 7-35

Composite wood decking is a good choice around this hot tub. *(Ann Marie VanDerZanden)*

or staining. **Composite wood** is made from a combination of recycled plastic, wood products, and glue (resin); and the ratios vary by manufacturer. Occasionally, a small amount of virgin material, compared to the volume of recycled plastic and wood, is added to increase strength and wear resistance. The plastic and wood combination resists ultraviolet light (sunlight) damage and does not warp, bow, or fade over time. It is also splinter free and skid free, which makes it a safer alternative to wood and particularly useful around swimming pools and hot tubs (Figure 7-35).

The cost of composite wood products compared to wood is slightly more. But composite products do not require frequent maintenance such as painting or staining, and most products have a 25-year replacement warrantee for cracking, warping, splintering, etc. The lower maintenance cost and longer life further its value as an alternative to wood.

Composite wood is available in the same sizes as other dimensional lumber (e.g., $1'' \times 4''$; $2'' \times 4''$). It is also milled into prefabricated decorative elements such as **balusters**, chamfered handrails, rail posts, and post caps (Figure 7-36). Composite materials are available in a range of colors, including light and dark tan, gray, various shades of brown, and muted shades of green and brick red (Figure 7-37). This allows the designer to customize the hardscape so it complements the surrounding landscape and house.

baluster: An upright, usually rounded or vase-shaped support of a balustrade or handrail.

HARDSCAPES IN THE VERTICAL PLANE

Fences and walls are the most common examples of hardscapes in the **vertical plane** of a landscape. Both elements are used to create privacy, a sense of security, and direct circulation patterns and even to block undesirable views. Walls also mitigate topographical challenges. In addition to their functional roles, fences and walls can serve as unique features in the overall landscape. As in the ground plane examples described previously, the vertical element's function in the landscape will determine which material is most appropriate.

FIGURE 7-36

Deck rail systems made from composite wood are decorative and durable. *(Steven Rodie)*

FIGURE 7-37

Composite wood products are available in a range of colors, and many have a texture that mimics wood grain. *(Ann Marie VanDerZanden)*

Fences

Some fences are used to delineate a boundary or property line, while others direct traffic or create privacy. Those that define a boundary or direct traffic can be constructed from a variety of materials, including metal, vinyl, chain link, post and rail, split rail, or other materials that do not create a solid façade (Figure 7-38). Because privacy is not a main function of these fences, they tend to range in height from 3 to 5 feet. The size and material used should be in scale with the house and surrounding landscape, and should complement the architecture of the house or neighborhood if possible.

A privacy fence, also sometimes called a stockade fence, should be at least 6 feet tall and solid in order to maximize privacy (Figure 7-39). A fence this tall and

FIGURE 7-38

This fence defines the property boundary without being obtrusive. *(Ann Marie VanDerZanden)*

FIGURE 7-39

The woven effect of this privacy fence gives it texture while still functioning as a physical and visual barrier. *(Ann Marie VanDerZanden)*

dense, however, can be imposing. To soften its impact in the landscape, include groupings of shrubs and perennials in front of it or partially obscure the fence with a climbing vine. Many communities require a building permit for privacy fences and have zoning laws that determine fence heights. In addition, most swimming pools and large water features are required to have secure fencing to limit access for small children. As a precaution for most fence projects, you should check local building codes before designing and installing the fence.

Walls

Landscape walls are categorized as either freestanding or retaining. **Freestanding walls** are similar to fences; they define a boundary, direct traffic, or obscure a view. They can also be decorative landscape features that create an attractive background or serve as a focal point. Even though the primary function of a **retaining wall** is structural, ideally it will also look good. From a structural standpoint, a retaining wall must be designed and constructed to withstand the weight of the retained soil, as well as the pressure of the water contained in that soil. Consulting a civil or

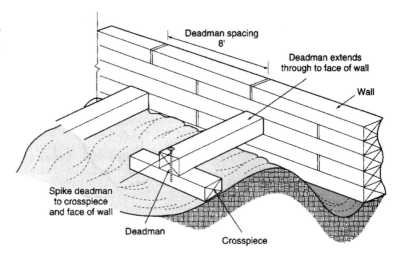

FIGURE 7-40

Typical deadman construction for a timber wall. *(Thomson Delmar Learning)*

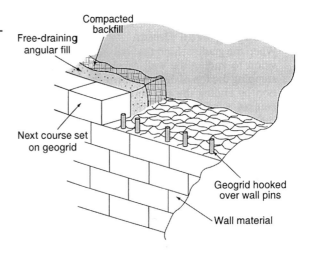

FIGURE 7-41

Example of geogrid installation. *(Thomson Delmar Learning)*

structural engineer, landscape architect, or other landscape professional is essential to safe wall design and successful installation, especially for walls over 3 feet high.

DEADMEN IN RETAINING WALLS. A **deadman** is used to further the structural integrity of retaining walls (Figure 7-40). This technique is commonly used with timber retaining walls and involves anchoring a T-shaped structural member deep into the soil behind the wall. This helps the wall withstand the pressure that builds up behind it. From a design standpoint, the size and location of the deadmen must be considered, and there must be adequate room behind the wall for them to be installed. If there is not enough room, then the wall may need to be redesigned, and the overall landscape design may be affected. Consult a reference (some are listed at the end of this chapter) on constructing retaining walls for deadmen details. Walls built from precast concrete units can also be anchored into the backfill for additional strength. Instead of using deadmen, **geogrids** are layered between block courses in the wall and layers of backfill to lock the backfill and wall together (Figure 7-41). This method allows extremely tall engineered walls to be safely built.

geogrid: Open-weave fabric used to stabilize retaining walls.

FIGURE 7-42

This timber wall ties back into the slope it is retaining, thus anchoring it into the landscape. *(Ann Marie VanDerZanden)*

WALL DESIGN CONSIDERATIONS. A well-designed and properly installed wall can be a distinctive landscape feature. Yet sometimes only minimal thought goes into how to make the wall more than just a functional element in the design. The result can be walls that look out of place in the overall landscape or walls that create maintenance problems for the homeowners. Following are examples of common dilemmas with wall design and strategies to avoid them.

Is a Wall Really Necessary? Before committing to a wall, consider its design purpose and possible alternatives: Is a truly flat area really needed above the wall? Or would a shorter wall that left a slight slope still be acceptable? Is a wall really needed to level off a difficult-to-maintain turf slope? Perhaps the slope could be planted with groundcover or other well-adapted plants. Although precast concrete unit walls have significantly broadened the flexibility and affordability of wall design and installation, a wall can still be one of the most expensive hardscape elements in a landscape. This cost should be weighed against the benefit the wall would bring to the design.

The Floating Wall. To avoid the illusion of a wall not attached to the landscape, design retaining walls so wall ends tie back into the slope being retained (Figure 7-42). Walls that do not have a definite end appear to unravel in the landscape and typically create a maintenance problem with drainage, soil erosion, or difficult mowing at the wall ends.

The Oversized Wall. When a relatively tall wall (over 4 feet) is required, consider using a series of lower walls stepped back into the grade. Terracing multiple shorter walls will create more landscape interest and may allow for additional planting areas next to the walls. These planting areas help soften the visual impact of the walls and provide unique planting opportunities (Figure 7-43).

FIGURE 7-43

Using two shorter walls makes this slope manageable and provides additional planting areas. *(Steven Rodie)*

FIGURE 7-44

No turf areas are adjacent to this stone wall, making maintenance easier. *(Thomas W. Cook)*

Minimizing Maintenance. If possible, wall design should provide for mulched landscape beds, rather than turf, above and below retaining walls. In addition to reducing the need for string trimming along the wall, these plantings can help create a safety barrier (visual or physical) for people using the landscape areas along the top of the wall (Figure 7-44).

A Wall with Multiple Uses. Walls that are 1.5 to 2 feet tall and approximately 10 inches deep can also be used for casual seating. To make the wall more decorative, you can use a different material for the slab on top (Figure 7-45). This combination of form and function can greatly enhance a landscape, particularly if space is limited.

MATERIALS SYNOPSIS. Materials used to build fences include wood, composite wood, wrought iron, vinyl, and chain link. Common wall materials include

FIGURE 7-45

This stone wall with a slab on top creates a seating area at the edge of this deck. *(Ann Marie VanDerZanden)*

brick, natural stone, precast concrete units, and landscape timbers. Decorative effect tends to be a deciding factor in selecting fence material, while the appropriate wall material is contingent on structural criteria and height in addition to decorative appearance. Table 7-5 summarizes uses, relative prices, and maintenance requirements for a number of hardscape materials.

Metal. Metal fencing can be sleek and simple or intricate and ornate (Figure 7-46). Choosing a style that complements the house's architecture or landscape style is an important design consideration. Wrought iron has been used as a fencing material for centuries. Newer products made from aluminum or pressed steel are a relatively cheaper alternative to wrought iron, but they are still expensive compared to some of the other fencing materials. Metal fencing tends to be expensive to install because of the expertise required to do it well, and it may require periodic maintenance. These expenses should be discussed with the client before selecting this material.

Vinyl. Vinyl fencing continues to gain popularity among homeowners because it is relatively inexpensive, it is easy to install, and it has minimal maintenance requirements (Figure 7-47). It can be used to achieve a similar effect as a wood fence without the additional time and expense of regular painting.

Chain Link. Although arguably not the most attractive fencing material, the functional merits of chain link cannot be questioned. Chain link is a woven metal material that is stretched between metal posts and secured with fasteners to the posts. It is available in a variety of heights and is an effective physical barrier. Traditionally available only in galvanized grey, it is now available as vinyl-coated or painted in black, green, and brown. These colors help the fence blend into the surrounding landscape.

TABLE 7-5

Summary of hardscape materials use, relative price, and maintenance requirements.

HARDSCAPE MATERIAL	PATIO	DECK	WALKWAY	DRIVEWAY	STEPS	WALLS	FENCE	PERGOLA/ARBOR	RELATIVE PRICE — Materials	RELATIVE PRICE — Installation	MAINTENANCE
Gravel			◆	◆					Low	Low	High
Natural stone	◆		◆	◆	◆	◆			High	High	Medium
Brick	◆		◆	◆	◆	◆			High	High	Low
Concrete pavers	◆		◆	◆	◆				High	High	Low
Poured concrete	◆		◆	◆	◆	◆			Medium	Medium	Medium
Wood		◆			◆	◆	◆	◆	Medium	Medium	Medium
Composite wood		◆			◆	◆	◆	◆	High	Medium	Medium
Wrought iron							◆	◆	High	Medium	Low
Chain link							◆		Low	Low	Low

339

FIGURE 7-46

Black metal fencing is durable and creates a physical barrier in this landscape. *(Steven Rodie)*

FIGURE 7-47

White vinyl fencing has minimal maintenance. *(Kathleen Wilkinson)*

Brick. Perhaps the biggest advantage of brick is its versatility. Bricks come in a range of colors, and when color is combined with a variety of mortar colors and thicknesses, the possibilities are nearly endless. Mortar, used to bind bricks, can influence the wall design. For example, using a mortar that is significantly lighter than the brick and layering it 1 inch thick will accentuate the building pattern (Figure 7-48). Another way to emphasize the pattern, and add texture to the wall, is to allow thick mortar to ooze out between the bricks (Figure 7-49). Yet another design advantage of brick is that a freestanding wall can be viewed on multiple sides because all brick surfaces are finished.

Natural Stone. Whether dry-stacked or built using mortar, stone such as flagstone or fieldstone makes a beautiful and natural-looking wall. The artful construction and quality craftsmanship of dry-stack stone walls built more than 100 years ago are still evident in many New England and Midwestern locales. Provided they are

FIGURE 7-48

Light mortar contrasts with darker bricks and accentuates the wall pattern. *(Ann Marie VanDerZanden)*

FIGURE 7-49

Thick mortar protruding between bricks creates a rustic look. *(Ann Marie VanDerZanden)*

built correctly, both dry-stack and mortared walls can be used as retaining walls. Just as with brick walls, freestanding stone walls can be striking and may be constructed to be viewed from multiple sides.

Dry-stack walls are often built from either cut stone or fieldstone. Since cut stone has relatively flat surfaces and comes in modular lengths and widths and a uniform thickness, it is relatively easy to build with this material (Figure 7-50). In contrast, the term **fieldstone** refers to stone that has not been milled and has retained its natural, uneven shape and edges. Fieldstone is generally sold by the pallet or by the ton (Figure 7-51). Walls built using fieldstone tend to take more time than cut stone walls because of the variable sizes and shapes.

footing: A structural element used to provide stability to a wall or other hardscape; often made of concrete.

veneer surface: A decorative surface applied to a wall; typically stone or brick.

Hidden Costs. In most situations where mortared brick, and sometimes stone, walls are desired, there is an additional cost that must be considered. This type of wall usually requires an internal structure of poured concrete or concrete block placed on a poured **footing** below the frost line. This internal core provides rigidity and structural support, particularly for retaining walls. Brick or stone is then added as a **veneer surface** to the concrete core. This configuration can add significant costs to a landscape budget. A designer should weigh the aesthetic benefits of rigid wall construction with the typical cost trade-offs compared to a dry-laid stone or segmental block wall, which remains somewhat flexible because it is built on a compacted gravel base.

FIGURE 7-50

This natural stone retaining wall was built using stone of different sizes but each piece has flat sides. The effect is more informal than if all of the pieces were a similar size. *(Thomas W. Cook)*

FIGURE 7-51

Pallets of fieldstone in wire baskets at a hard good supply yard. *(Ann Marie VanDerZanden)*

Precast Concrete Units. Precast concrete units, also referred to as **segmental wall units,** can be used for both freestanding and retaining walls. Rather than having only one decorative façade, recent modifications in the manufacturing process have produced units that are finished on two sides (Figure 7-52). Manufacturing modifications have also made this product available in a range of decorative colors. These units are relatively lightweight (30–70 pounds each; some of the specialized block designed for commercial applications weighs more than 100 pounds); they are uniform in size and color and easy for professionals and homeowners to install. Precast concrete segmental walls are constructed on a compacted gravel base (Figure 7-53) and do not require a poured foundation or concrete footings. The resulting wall creates a functional and formal landscape feature.

Hardscapes: Selection and Use 343

FIGURE 7-52

Segmental wall units create a formal-looking wall and sometimes are finished on two or three sides. *(Ann Marie VanDerZanden)*

FIGURE 7-53

A compacted gravel base is required for building a segmental wall. *(Thomson Delmar Learning)*

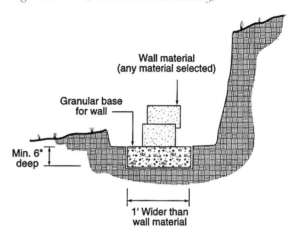

Landscape Timbers. Landscape timbers are an attractive alternative to wooden railroad ties; they make a rustic-looking retaining wall that blends into the landscape (Figure 7-54). These treated timbers are easier to work with than railroad ties because of their uniform size and length, lighter weight, and absence of sticky creosote (a wood preservative) and other debris like nails, spikes, and rocks often found in railroad ties. Landscape timbers used for retaining walls should be at least 6″ × 6″ or 7″ × 7″ to provide the necessary structural strength. Timbers are easy to install on a compacted gravel base, and they are typically pinned together with steel spikes or rebar. Using appropriately sized fasteners, installing properly designed deadmen, and allowing for adequate drainage behind the wall will ensure wall longevity.

HARDSCAPES IN THE OVERHEAD PLANE

The **overhead plane** or ceiling of an outdoor room can be defined by hardscape elements such as pergolas, arbors, and gazebos. When added to a landscape, these structures create a sense of enclosure and a more intimate space within the overall scheme of the landscape. When these elements are combined with patios, walls, fences, and plant materials, outdoor rooms are formed, both visually and physically, creating a useful and good-looking space.

FIGURE 7-54

Landscape timber walls can blend in with the landscape. *(Joseph E. VanDerZanden)*

Overhead structures, especially those that are free-standing, can be a significant safety concern if not properly designed. Appropriate sizing of structural members, fasteners, footings, and posts, together with sufficient sway stability and support for wind- and snow-loading, must be considered in overhead structure design. Landscape designers should seek professional assistance with structural design issues to ensure the safety and longevity of landscape structures.

Pergolas and Arbors

Pergolas and **arbors** are open, airy, architectural structures generally constructed from wood or metal. They can have posts with cross beams and sometimes a second layer of cross beams to create a crisscross pattern. Sometimes, metal is added to the overhead section of the structure for ornamental appeal or structural stability. Arbors are generally smaller than pergolas (Figure 7-55). Pergolas, on the other hand, often cover patios, decks, or long walkways and help define a garden space (Figure 7-56). Large pergolas can also be used to create a transition between two buildings such as a house and detached garage. Both structures are often covered with climbing vines to create an attractive ceiling that provides both shade and shelter from the elements.

Keeping arbors and pergolas in scale with the rest of the landscape and the buildings is essential. When these structures are too large in comparison to the other elements in the landscape, they will dominate the visual and physical space. Those that are too small will look out of proportion. When pergolas are constructed near a house, they should have substantial mass and an architectural style that complements the house (Figure 7-57). One way to achieve this is by using adequately sized columns such as 6" × 6" posts. Additionally, wrapping commonly used 4" × 4" posts in wood casings, stone, or brick can give the columns additional visual mass and keep the overall structure in proportion with the house.

ARBOR AND PERGOLA DESIGN NUANCES. When designing the overhead portion of a pergola or arbor, consider the patterns created by the crossbeams. Beams that overlap and have angled or decoratively cut ends will create interesting

FIGURE 7-55

Arbors can be built to reflect a variety of landscape styles and are a stunning way to define a landscape entrance. *(Ann Marie VanDerZanden)*

FIGURE 7-56

Pergolas are a great way to extend the living space to the outdoors, but they need to be appropriately sized to the landscape. *(Ann Marie VanDerZanden)*

patterns and shadows. For example, overhead members with a north–south orientation will cast a shade pattern that moves during the day, while an east–west orientation casts shadows that will seasonally change as the sun moves from a high summer sun angle to a low winter sun angle. Choosing the right vine for the structure also enhances the design. A wisteria vine can twine on thick crossbeams and requires substantial support for its heavy weight, while a clematis vine has less weight and requires a tighter, smaller weave of crossbeams to better twine through and attach its tendrils for support.

FIGURE 7-57

This structure has large posts to keep it in scale with the house. *(Ann Marie VanDerZanden)*

FIGURE 7-58

Gazebos can be simple or intricate and often function as a focal point in the landscape. *(Ann Marie VanDerZanden)*

Gazebos

The term gazebo originates from the blending of two words: Gaze and the Latin term *ebo,* meaning to see or look. As such, they are often used as vantage points (not focal points) from which the landscape can be viewed. This is not to say, however, that a beautiful gazebo cannot be a focal point in its own right. Gazebos are different from arbors and pergolas since they have a definite roof and at least partial walls (Figure 7-58). They can be constructed from wood, metal, or a combination of these, and other materials such as brick or stone.

Whether formal or casual, for a gazebo to work in the landscape it must be inviting and accessible. Placing a simple table and comfortable chairs in the gazebo provides a place for the homeowner to relax and enjoy the garden. Creating walls with a fine mesh screen to keep insects out can also make the gazebo a comfortable place on summer evenings. Paths, decks, or open lawn areas that make it easy to get to the gazebo further enhance its value in the design and increase the likelihood it will be used.

LANDSCAPE LIGHTING

With the flip of a switch, a landscape can be transformed from a dark, shadowy space into a beautifully illuminated nightscape. Night lighting enables homeowners to enjoy their outdoor living space well into the evening and extends the time they are able to spend in the landscape earlier into the spring and later into the fall.

Advances in landscape lighting technology (low voltage, solar) and materials (copper, ceramic, metal, etc.) give designers many options to choose from. Many of these lighting elements are safe and relatively easy to install, and they contribute to the beauty of the landscape. They can reinforce a style (e.g., arts and crafts, modern, nautical, retro) already present in the design or showcase architectural features such as a column, roof line, or railing system. Quality lighting design can significantly enhance the look and feel of a landscape, but it should be considered during the overall design process and not as an afterthought. It can be tempting to use lighting in a wide variety of locations and applications in the landscape, but be aware that the artistic effect is lessened when it is used in excess.

What Lighting Adds to a Landscape

Landscape lighting can be classified as lighting for functional purposes and lighting for purely aesthetic purposes. Functional lighting illuminates driveways, walkways, and access points. It also provides security. Aesthetic lighting showcases architecture or plant materials and adds drama to a landscape. In some cases, functional lighting can also be aesthetic, such as an up-lit cluster of palm trees that also illuminates the front walkway of a Mediterranean-style home. Using a variety of lighting fixtures with different angles or light sources can meet these functional and aesthetic needs.

FUNCTIONAL LIGHTING. Functional lighting centers on illuminating the landscape enough for people to move through the space easily and discouraging unwanted people from entering the landscape. Lighting along walkways and driveways provides a clear path for people to walk and guides them from one area of the landscape to another (Figure 7-59). Primary pathways should be illuminated if they will be used at night. For added safety, tuck lights into stair-step risers (Figure 7-60a) or use rope lights under the lips of stairs (Figure 7-60b) to make a clear distinction of where the tread stops. For security purposes, add motion-sensing lights with high-wattage bulbs to discourage intruders, enhance safety in the landscape, and save on energy costs.

AESTHETIC LIGHTING TECHNIQUES. Four of the most common landscape lighting techniques are: uplighting, spotlighting, backlighting, and downlighting. Each of these techniques results in a unique visual effect in the night landscape and may serve functional roles as well. For example, an uplight by a front entrance provides safety for a visitor walking to the door, security for the homeowner looking out the door, and the aesthetic enhancement of illuminating a tree silhouette in that area. The techniques can be used singly or in combination; novice designers can make use of the expertise of lighting contractors.

FIGURE 7-59

At night these lights will illuminate this primary walkway. *(Ann Marie VanDerZanden)*

FIGURE 7-60

Lights installed in the step (a) or rope lights under the front edge of a step (b) prevent tripping at night. *(Ann Marie VanDerZanden)*

(a)

(b)

FIGURE 7-61

Uplighting highlights the peeling bark of the paperbark maple. *(Steven Rodie)*

Uplighting. This effect is achieved when a light source is positioned near the base of an object, and the light beam is directed upward. Uplighting draws attention to surface details of trees, shrubs, or other landscape objects (Figure 7-61). Trees or shrubs with undulating or crinkly leaves result in strong light-and-shadow contrasts on the leaf surface. Plants like paper bark maple (*Acer griseum*) with peeling bark or cornelian cherry (*Cornus mas*) with fissured bark are striking when an uplight is focused on the trunk. Hardscape features such as the columns of a pergola can be up-lit to emphasize attractive building materials such as stone or copper and to highlight unique architectural details.

Spotlighting. A spotlight has a strong beam of light that is positioned some distance away from the object being lit. The beam of light is directed at the object, and it becomes a focal point because its bright surface contrasts with the surrounding darkness. Choosing objects with interesting texture to spotlight and altering the angle of the beam so shadows are created across the face of the object result in a dramatic nighttime focal point.

Backlighting. Backlighting is created when a light source is placed close behind an object. Using a light with a lower wattage than an uplight or spotlight creates a

FIGURE 7-62

Downlighting makes the landscape usable at night. *(Ann Marie VanDerZanden)*

more subtle effect and gives the illusion that the object is glowing against the darkness of the night. Backlighting a row of evergreen shrubs along a driveway can create a stunning entrance to a property. A similar effect can be created when this is done to a row of small shrubs flanking the walk to the front door.

Downlighting. Downlighting is used to mimic moonlight and to cast ground shadows. This type of lighting is more restrained than the other types of lighting described earlier and is an effective way to provide just enough light to make the landscape pleasant. Low-voltage lights (usually 12-volt) are useful in creating this subtle effect because they create a soft glow of light that bathes an area with gentle illumination (Figure 7-62). Low-voltage lighting is relatively easy to install, and this can be done by the homeowner.

Lighting Water Features. A water feature can become totally transformed at night with a few well-placed landscape lights. In addition to backlighting or uplighting the plants or stones around a water feature, waterproof, low-wattage lights can be placed in the water or in adjacent landscape boulders (Figure 7-63). Well-placed lights that illuminate falls and/or the reservoir result in a stunning effect.

GARDEN FEATURES AND EMBELLISHMENTS

Once the framework of an outdoor room is in place, you can select and position garden features and embellishments such as water features, fire pits, statuary, and containers. Some of these garden features will be strictly utilitarian; others may be all whimsy, and some may be both. Their role in the landscape often influences their location. For example, situating a potting shed so it is easily accessible and softening its appearance with plants can result in a functional feature that does not

FIGURE 7-63

A light placed in the boulder illuminates the water feature at night. *(Steven Rodie)*

FIGURE 7-64

The plantings around this shed soften its visual impact in the landscape. *(Thomas W. Cook)*

dominate the landscape (Figure 7-64). Another example is the location of a water feature. If the client wants to hear the soft murmur of water, then a trickling fountain should be located near a deck or patio. Locating that feature in the far corner of the landscape lessens its visual and aural impact.

When the landscape area is large or if the space has been divided into discrete rooms, more features can be used than in a smaller area. Too many features in a small space will compete with each other and result in a visually disordered landscape. Just as with other design considerations, features located closest to the house should be more formal and architectural in their appearance. Features positioned farther away from the house, where the character of the overall landscape may become more informal, can be more rustic and casual.

FIGURE 7-65

This flexible liner makes it easy to install multiple falls in this water feature. *(Ann Marie VanDerZanden)*

Water Features

Water feature is a broad term that includes ponds, waterfalls, fountains, streams, cascades, swimming pools, and even bird baths. Whether experienced in close proximity or viewed from a distance, a well-placed water feature can be the major focal point in a residential landscape. Water adds light and movement to a landscape and stimulates the senses like no other landscape feature. Water features are available in an endless variety of shapes, sizes, and sounds, and they can easily be customized to meet the client's needs. These characteristics allow the designer to create a unique landscape for a client.

Water, and the ability to manipulate it, has historically been associated with power and wealth. The ancient Romans were some of the greatest builders in history; they constructed elaborate systems (aqueducts and fountains) that provided water for the population in the form of baths, public laundries, and spectacular fountains. During the French Renaissance, King Louis XIV capitalized on new water pumping technology and commissioned numerous large and complicated water features at Versailles. Records show that the sheer number of fountains at his estate made it impossible to run them all simultaneously because there was not enough water pressure.

Today, many clients are requesting water features for their own backyard. Advances in technology and materials make water features an affordable addition to most any landscape. Instead of using concrete to construct the water reservoir, newer water features use either rigid or flexible liners. Rigid plastic liners are available in many sizes and shapes and are best suited for smaller features. Flexible liners are commonly available in polypropylene, polyethylene, or synthetic EPDM (ethylene propylene diene monomer) rubber. The EPDM material is very durable (often with a 20-year guarantee) and is not toxic to fish. The flexible liners can be used to create large water features and make it easier to install multilevel water features (Figure 7-65). Small recirculation pumps and flexible plastic tubing are quick and easy to install, particularly compared to the way the Romans did it (Figure 7-66).

FIGURE 7-66

Cross-section of a water feature showing components. *(Thomson Delmar Learning)*

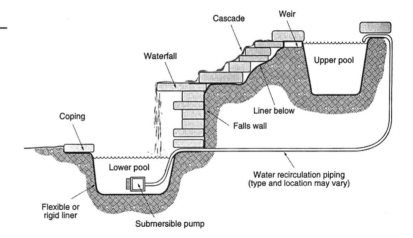

FIGURE 7-67

Benches surrounding this reflecting pool allow for quiet contemplation. *(Steven Rodie)*

TYPES OF WATER FEATURES. Water features fall into one of two categories: those with static water or those with moving water. A pool where the water is contained and does not move is the most common static water feature. This type of feature often takes the form of a swimming pool or a reflecting pool. Swimming pools have evolved from the traditional bright-blue, in-ground rectangle to cutting-edge designs that are as much art pieces as they are swimming pools. Reflecting pools come in a variety of shapes and sizes to fit the landscape and add a calming element to the landscape (Figure 7-67). Their reflective capability is improved by painting the interior surface a dark color.

Water features with moving water consist of cascades, waterfalls, fountains, or some combination of the three. Cascades are created when a stream of water flows from a higher elevation to a lower elevation. The stream of water stays in a shallow channel as it moves over cobbles or large stones (Figure 7-68). Once the water reaches the bottom of the slope, it is deposited into a lower reservoir and is recirculated back to the top of the feature.

FIGURE 7-68

This cascading water feature uses a variety of natural stone and is the major focal point for this landscape. *(Thomas W. Cook)*

FIGURE 7-69

This multifalled water feature includes fanciful garden art that is also a fountain and a bottom reservoir. *(Ross and Beth Brockshus)*

Waterfalls are different from cascades because the water does not move in a stream. Instead, a sheet of water drops from a higher elevation to a lower elevation. A water feature may have a single waterfall or multiple falls separated by reservoirs of water (Figure 7-69). The waterfall may be a short drop that is relatively quiet, or a tall, loud drop. Stones jutting through the waterfall also affect the sound the water makes as it splashes to the lower elevation. These features also have a recirculating pump submersed in the lower reservoir to pump water back up to the top.

Fountains will vary in expense, depending on how elaborate they are and what material they are made from. They can range from a simple bubbler fountain (Figure 7-70) that creates a trickling sound to a multinozzled feature that sprays water out of gargoyle mouths. The location of the fountain will influence the style used. Some fountains are visible on all sides, while others may only be viewed from

FIGURE 7-70

A bubbler fountain next to this entrance creates a soothing sound for arriving guests. *(Ross and Beth Brockshus)*

one angle. Fountains generally have a reservoir, either above or below ground, at the base of the fountain where a submersible pump forces water through the feature.

Pondless water features are another option for homeowners to consider. A **pondless water feature** differs from the traditional water feature because it does not have a collection reservoir above ground. Instead, after the water goes over the cascades or falls, the water falls onto a bed of rocks (usually), filters through the rocks, and collects in an underground reservoir (Figure 7-71). A submersible pump then circulates the water back to the upper pool. Without the above-ground reservoir, less debris accumulates in the water, so there is less maintenance cleaning filters.

Water Feature Maintenance. In general, water features tend to have a significant maintenance requirement. Depending on winter temperatures, some ponds may need to be drained in the fall and refilled in the spring. Even if this annual maintenance is not required, there are maintenance issues associated with aquatic plants and fish in the pond. The mechanical equipment such as the filters, skimmer, and pump also need regular maintenance. It is important to discuss these maintenance issues with your client before he or she decides to include a water feature in the new landscape.

WATER FEATURE FUNCTIONS IN A LANDSCAPE. From a design standpoint, it is important to decide what role the water feature will perform in the landscape. Because it stands out from the rest of the landscape, a water feature naturally functions as a focal point. Aside from that, the homeowner may want to include a water feature to mask urban noises such as traffic and neighbors. In the case of mitigating noise, a feature with cascades, falls, or a fountain should be used. The noise that the homeowner wants to mask will determine how far the water should fall and the volume of water used. In cases where the client wants to hear water gurgling as a gentle background noise, a gently cascading feature or a fountain may be more appropriate. Another reason to include a water feature is to showcase fish and aquatic plants. Bubblers and fountains can enhance oxygen levels in a garden pond, which creates a healthy water environment for both fish and plants.

FIGURE 7-71

Pondless water features do not have an above ground collection reservoir. *(Josh Sutton)*

Designing with the client's needs in mind will ensure that the water feature meets the homeowner's objectives and looks good too.

Fire Pits and Outdoor Fireplaces

Fire pits and outdoor fireplaces have become popular additions to home landscapes in the past few years. Maybe it is the sound of crackling flames, the thought of warm s'mores, or the soothing scent of burning wood that conjures up memories of outdoor adventures. Whatever the case, fire pits and outdoor fireplaces are wonderful gathering points in the landscape that entice homeowners and visitors to linger longer in the evening landscape.

This design feature takes a little more planning than some of the others discussed in this chapter. First, you will need to check with local ordinances to see if open fires are allowed. The feature needs to be located well away from structures and tree canopies to prevent the fire from spreading (Figure 7-72). Portable fire rings are available and in some situations may provide the flexibility necessary for safe and legal use.

Garden Embellishments

Among the last additions to a home landscape are garden statuary and other embellishments. This ornamentation should complement the style or theme of the outdoor room where it will be placed. A rustic pot or antique iron gate will be attractive in an informal, English cottage-style composition, while statues or

FIGURE 7-72

A variety of materials can be used to build fire pits, which can either be built into the ground or sit on top. Regardless of type, it is important that they are surrounded by fireproof materials. *(Matt Helgeson)*

containers with clean, sleek lines will complement a contemporary space. As a designer, you might provide guidance on material, function, shape, and location in the landscape, but the client should be allowed to select pieces that reflect his or her personality and style. This results in an outdoor living space that reflects a homeowner's unique personal taste.

Garden embellishments can be categorized as statuary, containers, furniture, and special effects. Embellishments come in a variety of materials, sizes, shapes, and styles. They look best when you group them in odd numbers and when you use items of similar materials, shape, or color. Since these elements often become focal points, it is also important to locate them where they will look most appealing.

STATUARY. The ancients were renowned for their use of statuary in the landscape, yet not all garden statuary needs to be in the form of a 15-foot tall Mayan godhead. Statues can be carved or cast to represent almost anything, including people, animals, geometric shapes, and natural elements. Whether made from natural (stone, wood) or synthetic (concrete, metal, plastic, or fiberglass) materials, each brings a subtle nuance to the composition (Figure 7-73). Some materials will even change in color and texture as they age and weather over time.

obelisk: A tall, four-sided shaft of stone, usually tapered and monolithic; rises to a pointed pyramidal top.

Geometric shapes like an **obelisk** and statues of the human form (Figure 7-74) tend to have a formal effect; oversized animals and garden gnomes (Figure 7-75) bring a sense of whimsy to the garden. The client should select statues that reflect his or her personality and lifestyle.

CONTAINERS. The combination of size, shape, and material makes for abundant container choices for a designer (Figure 7-76). Similar to statuary, containers are made from metals, stone, concrete, fiberglass, plastic, and wood. Newer manufacturing techniques make some of the synthetic materials like concrete and fiberglass more representative of natural materials like stone and wood. For the client who

FIGURE 7-73

A crane, which symbolizes longevity, is commonly found in Japanese style gardens. *(Ann Marie VanDerZanden)*

FIGURE 7-74

This statue of a female sits in quiet repose under a bridge. *(Ann Marie VanDerZanden)*

wants to move containers to different locations in the landscape, consider using some of the synthetic materials that are much lighter and easier to pick up and relocate.

Containers can be planted or left unplanted, used singularly or in groups. For a subtle effect, planting compositions in the containers can be similar to the surrounding landscape; or for a more striking effect, the plants can contrast with the landscape. Large containers with a combination of woody plants and annuals provide an opportunity to change the composition each year, depending on the client's wishes.

Each type of container lends a unique feeling to the landscape. Urns are a type of container with a distinctive, flared top and often a footed base. Urns are often used in

FIGURE 7-75

A jaunty garden gnome peeks out curiously from underneath a hedge. *(Ann Marie VanDerZanden)*

FIGURE 7-76

Containers are available in a range of materials, sizes, and styles. *(Ann Marie VanDerZanden)*

pairs to mark an entrance or pathway and set a formal tone (Figure 7-77). Terracotta and ceramic containers come in attractive colors but must be handled carefully and brought inside during cold winter months to avoid cracking.

An updated spin on a traditional technique that has recently become popular in the United States are trough gardens or **hyper-tufa** containers (Figure 7-78). A mainstay in English gardens for centuries, troughs come in a range of sizes and shapes and are often planted with dwarf conifers and alpine plants.

Although container plantings are relatively easy to incorporate into the landscape, they do require regular and frequent maintenance. Container plants need consistent moisture and fertilization in order to thrive, and some plants need regular deadheading in order to keep blooming. Because of these maintenance issues,

FIGURE 7-77

Urns are often used in pairs and add a formal element to this entrance. *(Ann Marie VanDerZanden)*

FIGURE 7-78

Trough gardens planted with dwarf conifers are a great addition to this backyard. *(Paula Flynn)*

containers should be placed with accessibility and aesthetic impact in mind. Container plantings also require proper drainage. Pots and tubs placed on surfaces that are easily stained or damaged may need to be slightly elevated or have a saucer placed underneath to catch excess water.

FURNITURE. Adding chairs, benches, and tables can make an outdoor room very livable. If the client is planning to use the furniture for relaxing or entertaining,

FIGURE 7-79

Metal mesh furniture creates a comfortable sitting area on this patio. *(Ann Marie VanDerZanden)*

as opposed to just for ornamentation, it is essential that the furniture be comfortable. Although some garden furniture looks great, it isn't necessarily useable. Cast- and wrought-iron furniture pieces are beautiful, and they are heavy and will not blow over in the landscape, but they are incredibly uncomfortable to sit on unless cushioned. Molded metal and mesh chairs tend to be more comfortable and look contemporary (Figure 7-79). Wooden furniture can be rustic, elegant, or anywhere in between.

Built-in furniture, such as deck benches, flat sculpted boulders, short walls, and broad deck, or patio steps, should also be considered in the mix of furniture provided. This type of furniture provides additional seating choices and is particularly useful when an outdoor space is filled beyond its normal capacity (e.g., a really big garden party) or when portable furniture is in storage. This built-in furniture can also help define the edges and boundaries of outdoor spaces. For example, a built-in bench can provide seating as well as a safety barrier at the edge of a low deck drop-off, and it can eliminate the need for a taller railing, which would block landscape views from the deck (Figure 7-80).

FOUND OBJECTS. To some people *found objects* is just another term for junk! To others, found objects are those delightful seashore, flea market, garage sale, or heirloom items that add a unique twist to a landscape composition. Some large objects will become focal points, while others may be small and more restrained. Regardless, these objects tend to attract viewer interest in the landscape and typically hold special meaning for the homeowner. It makes sense to place them in close proximity to garden areas that are accessible and visible. Placing found objects randomly throughout a landscape tends to dilute their importance, and they may be too far away to be appreciated.

fresco: A painting done on fresh, moist plaster with pigments dissolved in water.

SPECIAL EFFECTS. Special effects like murals or **frescos** and mosaics made from recycled china, glass, ceramic, or stones can personalize the floors and walls of

FIGURE 7-80

The built-in benches on this deck provide seating areas and prevent people from accidentally stepping off the deck. *(Steven Rodie)*

any outdoor room. A mural can be used to make a small garden look larger or to transport the garden visitor to a far away place. Mosaics can cover floors, walls, and furniture such as chairs, benches, and tables, or be much smaller like stepping stones. They can be subtle and made from shades of the same color, or they can be a vibrant combination of complementary colors.

SUMMARY

Hardscapes enable a person to move through the landscape, create privacy, or simply provide a place from which to enjoy the landscape. They are an integral part of many residential landscapes, and often their sheer size makes them difficult to ignore. The stone path, copper archway, or garden gazebo will grace the landscape all year long. Because of this constant presence in the landscape, hardscapes are always in view and often a focal point. Designing attractive hardscapes that meet the homeowner's personal needs can determine whether a landscape designer is successful.

KNOWLEDGE APPLICATION

1. Define the following terms:
   - deadman
   - retaining wall
   - fieldstone
   - flagstone
   - primary walkway
   - stamped concrete
   - composite wood

2. Describe three examples of hardscapes in the following planes:
   - ground plane
   - vertical plane
   - overhead plane

3. For each of the examples in question 2, list and describe appropriate materials used to construct them.

4. Differentiate between uplighting, backlighting, downlighting, and spotlighting.
5. What is the standard equation to calculate step dimensions?
6. Differentiate between these water features: reflecting pool, cascade, waterfall, fountain, bubbler fountain.
7. Describe four examples of garden embellishments.

REFERENCES

American Society for Testing and Materials (ASTM). ASTM International. West Conshohocken, PA. <http://www.astm.org>

The Brick Industry Association. Reston, VA. <http://www.gobrick.com>

Britannica Concise Encyclopedia. (2006). Encyclopedia Britannica Premium Service. <http://www.britannica.com>.

Free Dictionary by Farlex. (2006). <http://www.thefreedictionary.com>

Sauter, D. (2005). *Landscape construction.* Clifton Park, NY: Thomson Delmar Learning.

Seferian, H. (2004). *Hardscaping: High style, low maintenance outdoor spaces.* New York: McGraw-Hill.

Stevens, D. (1995). *The garden design sourcebook.* London: Conran Octopus Limited.

Wirtanen, D. (2006). Personal communication.

CHAPTER 8

Pricing the Proposed Project

The landscape design process and the associated sales process outlined in Chapter 5 have multiple steps and can proceed in many different ways (see Figure 5-1). This chapter will focus on how to proceed after the client has retained your design (and possibly installation) services. The chapter defines business terms, explains pricing strategies for design and installation services, lists the components included in a landscape installation estimate, and provides examples of how to account for uncertainties associated with landscape installation.

OBJECTIVES

Upon completion of this chapter you should be able to

1. describe pricing strategies for landscape design services.
2. differentiate between cost and price.
3. differentiate between an estimate and a bid.
4. describe the process of pricing a landscape installation project.
5. describe the most common uncertainties that affect pricing a landscape installation project.
6. complete the basic calculations necessary to develop an estimate.
7. prepare an estimate for a landscape installation project.

KEY TERMS

design contract	direct cost	estimate	labor cost
retainer fee	indirect cost	bid	production rate
cost	landscape contractor	lump sum bid	direct job overhead
profit	price	proposal	general overhead
break even	contingency	materials cost	change order

CHARGING FOR YOUR EXPERTISE

There are a number of ways to generate income from horticultural knowledge and design expertise. Creative marketing strategies can lead to multiple income streams. Client services should not be limited to just design; consider developing a nursery pull-list and completing research on plant or hardscape materials. The pull-list includes the plant names (botanical and common), specific cultivar, quantities and size (e.g., 1 gallon, 3 gallon, 24-inch B&B) for each plant in the design. Some designers will also include reasonable alternatives, in case the nursery or garden center does not carry a particular plant. The client can take the pull-list to his or her local nursery or garden center and know exactly which plants, how many, and what size to purchase.

A similar service is calculating hardscape materials for the project. Included in this information might be how many cubic yards of crushed limestone base or sand is needed for a paver patio, the number of pavers or tons of flagstone (limestone, bluestone, etc.), and even the amount of edging material for between planting beds and turf areas, based on a calculation of lineal feet. For some clients, particularly those interested in doing the installation themselves, providing these shopping lists facilitates efficient use of their time and money.

Maintenance consultation on a new or existing landscape is another income generator. Many times, clients need to be educated on the initial establishment and care of their new landscape. Identifying plant materials and creating a maintenance plan with information on annual care—including a fertilization schedule, pruning schedule, and maybe even an integrated pest management (IPM) approach for the site—can be valuable to the homeowner (Figure 8-1). Additionally, some clients who enjoy working in their landscape may just need a little direction and expert advice. Suggestions on how to approach a landscape dilemma, but not a formal design, will aid these customers a great deal. This type of consultation can be an important source of income and could result in future work when the designer is perhaps called on later to create a formal design and possibly complete the installation.

FIGURE 8-1 Providing a landscape maintenance plan to clients can be an additional income stream for a landscape designer. *(Ann Marie VanDerZanden)*

Basic maintenance plan

	April	May	June	July	August	September
	Fertilize trees and shrubs if needed					Turf fertilizer and weed control
	Annual pruning					
	Turf fertilizer					
	Cut herbaceous perennials to ground					
	Fertilize perennials (slow release)		Fertilize perennials (slow release)		Fertilize perennials (slow release)	
		Fertilize annuals (slow release)		Fertilize annuals (slow release)	Fertilize annuals (slow release)	
		Annuals: Apply soluble fertilizer (bi-weekly)			→	
		Annuals: Deadhead as needed (weekly)			→	

PROCEEDING AFTER A SUCCESSFUL FIRST MEETING

A successful first meeting with a potential client is one in which the designer is hired. Client design expectations can range from those who want a rough sketch of possible design solutions to others who need to experience the whole design process from client interview to a rendered final plan. Most others will fall somewhere along this continuum of customer expectations. Once hired, a designer can spend a significant amount of time on the project. Because of this, it is important to separate the design fee from charges associated with the installation.

PRICING THE DESIGN ONLY

Often, residential design projects comprise the following steps: three client meetings, site measuring and site analysis, preparation of a base map, research on plant and hardscape options, concept development (includes functional diagrams and alternative design solutions), development of a preliminary design(s), and production of a final design (Booth and Hiss, 2005). For clients who want the whole package, designers can combine items into a single design fee. The size and scale of the project will influence how long it will take to complete the design process. Typical residential designs take anywhere from 8 to 20 hours to complete, and the price can range from $300–$3000 (Booth and Hiss, 2005; J. Mason, personal communication).

In some cases, a client may only want a couple of meetings, a site visit (including a walk through the existing landscape while discussing possible design ideas), and some rough sketches or plans that illustrate those design ideas. In this situation, it may be more practical to charge by the hour for design services. Just as the total price for a design can vary, so can the hourly rate charged by a designer. Experience is one factor that significantly influences a designer's rate. For example, a designer new to the field may charge $25–$30 per hour, while a designer with an established practice may charge $150, or more, per hour. The local market is another factor that influences the hourly rate. For example, the landscape design market in suburban southern California will likely allow for a higher designer fee than would be realistic in a small rural community in the Midwest.

Ultimately, pricing a project will depend on the **business model**. If the business does only landscape design, the project may be priced differently than if the business does installation too. If installation services are provided and the client hires the designer to install the project, a portion of the design fee may be credited toward the installation price. On the other hand, if the client wants to install the project himself or herself or wants to hire a different landscape contractor, the fee may be slightly higher. As with all business ventures, it is important to charge a price that allows for an adequate profit margin. At the end of this chapter is a description of how Country Landscapes, Inc., a landscape design/build company that has been in business for 25 years, approaches the designing, bidding, and installation processes.

business model: The mechanism by which a business intends to generate revenue and profits; a summary of how a company plans to serve its customers; involves both strategy and implementation.

Developing a Contract for Design Services

Once the client decides to continue with the project, the designer and the client will enter into a **design contract**. As with any legal document, there are certain details that the contract should contain, including: names and addresses of involved parties, scope of work, types of drawings/products the client will receive, number of client meetings, time schedule, and payment (Booth and Hiss, 2005). Figure 8-2 is a sample contract template that can be adapted for other projects. Consult a legal professional when developing an initial contract to ensure that the necessary information has been included.

NAMES AND ADDRESSES.   In order to facilitate communication between the designer and client, the contract should include the names and contact information, including address, phone number(s), and e-mail, for parties involved in the contract. In some cases, the contact person is different than the person signing the contract. Sometimes, the person paying the bill is a different person. Having this information clarified and readily available decreases possible misunderstandings between parties during the project. It also makes contact easier once the project is completed and follow-up communication is needed.

SCOPE OF WORK.   Outlined in this section of the contract are the specific tasks the designer will complete as part of the project. A typical list might include: completing a site visit for analysis and measurements, developing the base map, researching plant and hardscape options; creating functional diagrams and alternative design solutions; developing a preliminary design for review by the client; and generating a final design for the project. This section should also list what is not covered in the contract. Examples of items not included in the contract might be installation of the project or detailed construction drawings for features like patios, decks, fences, or water features. It is important that these exclusions be outlined so the client understands exactly what he or she is getting for the price he or she pays.

PRODUCTS OR DRAWINGS.   This section of the contract should describe the specific products or drawings the client will receive, including the type (black and white, color rendered), size, and how many. For most residential designs, this will include a preliminary design(s) and the final design. Additional drawings might include sections or elevational renderings of different parts of the landscape (Figure 8-3). Some clients may also want detailed information (e.g., unique characteristics, maintenance requirements, and a maintenance schedule) on the plants selected for the project. If this is the case, the designer and client should discuss the matter ahead of time and add items to the scope of work section. Since compiling this information will require additional time, a charge should be associated with it, either on an hourly basis or as a lump-sum addition to the design fee.

CLIENT MEETINGS.   This section outlines the number of meetings the client will have with the designer during the design process. For most projects, there will be three meetings: the initial information gathering meeting and a discussion of how to proceed, a second meeting where the preliminary plan is presented and the client can provide feedback, and a final meeting where the landscape professional presents the final design to the client (Figure 8-4). For large or particularly complex projects, additional meetings may be needed. The time spent preparing and conducting these additional meetings should be accounted for in the design fee.

FIGURE 8-2 Sample contract for design services. *(Tigon Woline)*

Contract for Design Services

Harmison's Gardening Services, L. L. C.
500 Maple Lane
City, State

Client: _____

Address: _____

Phone: _____

1. The purpose of this Agreement is to set forth the terms and conditions under which Harmison's Gardening Services, L.L.C. (hereinafter called "Contractor") will provide design service for Client at the above address.

2. Contractor agrees to perform the following design services.
   Complete site analysis and inventory
   Complete necessary site measurements to prepare a base map
   Prepare two preliminary designs for review by Client
   After review of preliminary designs by Client, Contractor will refine design concepts to develop a final plan.

3. Contractor will furnish labor and equipment necessary to perform the above design services.

4. Contractor will provide Client the following products:
   Two preliminary landscape plans including some specifications on plants and hardscapes.
   One final plan with detailed planting key, hardscape specifications including materials, and paving pattern (if appropriate).

   Additional products such as elevational or perspective drawings, maintenance plan or detailed construction details for wood or metal features are not included in this contract. These products are available upon request for an additional fee.

5. Contractor will provide Client with preliminary drawings within 21 days after signing this contract. A meeting between Client and Contractor will be arranged to determine how to proceed to the final plan. Contractor will present the final plan to the Client 10 days after this meeting.

6. Client shall pay to Contractor at the rate of (insert amount here) for the service herein agreed to be performed. Contractor will bill Client and Client shall make payment within ten days of billing date. Client agrees to pay a service charge of $10.00/10 days overdue for all payments not made when due.

7. The terms of this Agreement shall commence on_____and shall continue in full force and effect thereafter until it is terminated by thirty days written notice by either party to the other.

8. This Agreement shall be governed by the laws of the State of (enter state) and constitutes the entire agreement between the parties regarding its subject matter.

Harmison's Gardening Service   Client

_____   _____
Signed name   Client signature

_____   _____
Printed name   Client printed name

_____   _____
Date   Date

FIGURE 8-3

The client requested an elevational rendering of this area of the landscape in addition to a fully rendered final design. *(Courtesy of Darin Froelich. Ann Marie VanDerZanden)*

FIGURE 8-4

This designer is explaining the details of the final design with the client. *(Ann Marie VanDerZanden)*

TIME SCHEDULE. The time schedule section will describe when the design process will start and when the preliminary and final designs will be completed. Providing some specifics on the planned schedule is important, but you should allow for some flexibility. Giving a client precise completion dates for the different phases is risky, since circumstances may arise that prevent deadlines from being met.

PAYMENT. The fee for design services and the payment schedule can vary, depending on the type of business. For most projects, design fees can be divided into three categories: a retainer fee, a partial completion fee, and a final completion fee (Booth and Hiss, 2005; J. Mason, personal communication).

The **retainer fee** is often 10 to 20 percent of the total design fee, and it is required before work on the project starts. When the designer presents preliminary plans, the client usually pays the partial completion fee, which is 40 to 60 percent of the total design fee. The final completion fee consists of the remaining amount and is collected when the final design is presented to the client. In some cases, the entire design fee is collected up front; in other cases, only a small percentage is collected initially, with the bulk collected at completion. The payment schedule should be created to match the business model and cash flow needs.

A FINAL MEETING WITH THE CLIENT

In some cases, presenting the final design to the client will be the end of the designer's role in the landscape project. In these situations, the client may then hire a landscape contractor to do the installation, or he or she may do the installation himself or herself (see Figure 5-1). If the client is planning to hire a landscape contractor, he or she may ask you for a recommendation. Directing him or her to qualified and reputable contractors who can translate a two-dimensional landscape drawing into a three-dimensional, living landscape is an essential, final step in making the project come to life. A critical component of a designer's business is to develop a professional network for client referrals. A number of specialists should be included in this network, such as stone masons, water feature experts, lighting consultants, and electricians.

Important Business Definitions

In order to be conversant in the landscape industry, it is important to understand the terminology used for various aspects of business in general. Some common terms are defined below.

Cost: What it will cost the landscape contractor to complete the job, including materials, labor, equipment use, and overhead.

Profit: Financial gain realized on the project after all job costs have been accounted for.

Break Even: The point where all costs associated with the project have been recovered, but no profit has been gained.

Direct Cost: Costs that are directly tied to a particular landscape project. Most often, these include labor, materials, and equipment costs required to complete the project.

Indirect Cost: Costs associated with running a business. (e.g., office rent, utilities, insurance, etc.) that cannot be assigned to one specific job. These costs are recouped over multiple jobs.

LANDSCAPE CONTRACTING TERMINOLOGY

A business person who provides landscape installation services is termed a **landscape contractor**. In some cases, the business might also provide design services, in which case the business is referred to as a design/build firm. Landscape contracting is similar to other building trades, and it is important to understand fundamental business terms (cost and price) as well as terms specific to the profession (bid, estimate, proposal, and contract).

Differentiating Between Cost and Price

Landscape business professionals differentiate between cost and price in order to have a profitable business enterprise. Without making this critical determination, even the most talented landscape professionals may see their business fail. In purely economic terms, cost is the financial recovery of expenditures (Ingles, 2004). For a landscape contractor, the cost of a project includes labor, materials, equipment, overhead, and sometimes the subcontractor costs needed to complete the project. All of these expenses are assumed by the landscape contractor during the construction of the project. Just to break even on a project, a contractor would charge only these costs. Breaking even is seldom desirable because there is no profit!

Essentially, **price** is cost plus profit. It is what a contractor will charge a client to complete a landscape project. In addition to the project costs listed above (labor, materials, equipment, subcontractor, and overhead), the price will also include contingency and profit. Because landscape installation can be affected by so many variables—including weather delays, installation crew inefficiencies, miscalculation of materials to name just a few—a **contingency** allowance is included in the price. This allowance takes into account the unpredictable nature of landscape installation and helps preserve the contractor's profit. Calculating contingency allowances will be addressed later in this chapter.

To calculate a profit margin, some contractors will determine the direct job costs for a project (labor, materials, equipment, and direct overhead) and multiply that by a factor (percentage) based on a profit margin set forth in their business plan. That value is added to the cost of the project and will determine the price (Figure 8-5).

As any seasoned landscape professional will attest, no two landscape projects are alike. Each presents its own unique set of circumstances that need to be considered when calculating costs and determining the price. Experience with estimating the costs associated with labor, materials, and equipment use is critical to accurately calculate the overall project costs. In addition, understanding the market and what it will bear in regard to profit and contingency is essential when determining the price. Calculating a large profit margin and including too large a contingency may price a landscape contractor out of a project.

Differentiating Between Estimate, Bid, Proposal, and Contract

The terms estimate, bid, proposal, and contract are used throughout this chapter. These terms have related, but different, meanings. What follows are comparisons of how these terms are typically used in the landscape industry.

FIGURE 8-5

For this basic, straightforward job, profit was calculated by multiplying all direct job costs by 15 percent. *(Ann Marie VanDerZanden)*

Schuh Account 1421 30th St.	
Materials	
Plant	$ 5,340
Hardscape	530
Irrigation	3,530
Total[a]	9,400
Labor	
Landscape	3,530
Irrigation	1,820
Total[b]	5,350
Subcontractor	530
Total Direct Job Costs	15,280
Profit (15% × Direct Job Costs)	2,280
Total Bid Price	**$17,560**

[a] Total materials cost includes overhead.
[b] Total labor cost includes overhead and labor burden (payroll taxes, insurance, etc.).

ESTIMATES AND BIDS. The terms estimate and bid are very different. An **estimate** is an approximation of what price a client will pay for the landscape project (i.e., design and installation). It is important to clarify with the client that the actual price may be higher or lower, depending on a number of factors that could affect the project. For example, after the design has been completed, the designer could generate an estimate for installing the plants and hardscape. This estimate is a best guess of what it will cost the client to have the project installed. However, if during construction there was a problem excavating the patio and it took the crew an extra 5 hours to complete that part of the project and it required four more cubic yards of crushed base to fill in the larger hole, the client would be charged for this additional expense.

In contrast, a **bid** is an offer to provide services, such as the design and installation of the project, for a set price. In the scenario described previously, in which problems associated with excavating occurred, if the client accepted a bid for the project, he or she would only pay the bid price. This is a case in which the landscape contractor would absorb the additional cost for the project, resulting in reduced profit.

For large landscape projects, clients will often solicit multiple estimates before they decide on a landscape professional. Estimates can be developed relatively quickly and accurately to give a client a sense of what the project will cost. In some cases, if the client accepts the estimate, the estimate is then made into a formal bid, and a contract is signed to proceed with work. Although all of the estimate information is included in the bid (Figure 8-6a), some contractors do not include the specific details in the bid version the client sees (Figure 8-6b). In situations in which the work

FIGURE 8-6 Example of a bid presented to a client (a) compared to an estimate (b), which includes details for different components of the project and is used for internal purposes by the company. (Jim Mason)

BID

ITEM	QTY	SIZE	PRICE	MAT TOTAL	LABOR	
BARBERRY, 'CRIMSON PYGMY'	396	#3 CNT	$22.50	$8910.00		
SUMAC, 'GRO-LOW'	712	#2 CNT	$15.35	$10929.20		
SHREDDED OAK MULCH BULK (16+YD)	185	CU YD	$19.00	$3135.00		
			SECTION TOTALS:	$22,974.20	$9,759.64	$32,733.84

Remarks:

We PROPOSE to furnish material and labor in accordance with the above specifications for the sum of $35,025.21
A minimum deposit of $8,756.00 is due with signed proposal before work can be scheduled.
In the event that the specified varieties or sizes of nursery stock are not available, we reserve the right to make substitutions of variety and size at no additional costs when such substitutions do not compromise the integrity of the design. If no substitutions are available, the items will be backordered to be installed and invoiced as a separate job.

MATERIALS	$22.974.20
LABOR	$9,759.64
(ADJUSTMENT)	$0.00
SALES TAX	$2,291.37
TOTAL	$35,025.21

AUTHORIZED SIGNATURE _____ DATE _____

ACCEPTANCE OF PROPOSAL - The above prices, specifications, and conditions (including those in Attachment A) are satisfactory and are hereby accepted. Country Landscapes is authorized to do the work as specified.
_____ is hereby paid as a deposit with Check # _____
The balance will be paid within 10 days of job invoice date.
A finance charge of 1 1/2% per month, which is an annual percentage rate of 18%, will be charged on accounts over 30 days past due.

Date of acceptance _____
Signature _____
Signature _____

(a)

BID

ITEM	QTY	SIZE	PRICE	TOTAL	LABOR	TOTAL	ITEM TOTAL
BARBERRY, 'CRIMSON PYGMY'	396	#3 CNT	$22.50	$8910.00	$5.83	$2308.68	$11,218.68
SUMAC, 'GRO-LOW'	712	#2 CNT	$15.35	$10929.20	$5.83	$4150.96	$15,080.16
SHREDDED OAK MULCH BULK (16+YD)	185	CU YD	$19.00	$3135.00	$20.00	$3300.00	$6,435.00
						SECTION TOTAL:	$32,733.84

Remarks:

We PROPOSE to furnish material and labor in accordance with the above specifications for the sum of $35,025.21
A minimum deposit of $8,756.00 is due with signed proposal before work can be scheduled.
In the event that the specified varieties or sizes of nursery stock are not available, we reserve the right to make substitutions of variety and size at no additional costs when such substitutions do not compromise the integrity of the design. If no substitutions are available, the items will be backordered to be installed and invoiced as a separate job.

MATERIALS	$22.974.20
LABOR	$9,759.64
(ADJUSTMENT)	$0.00
SALES TAX	$2,291.37
TOTAL	$35,025.21

AUTHORIZED SIGNATURE _____ DATE _____

ACCEPTANCE OF PROPOSAL - The above prices, specifications, and conditions (including those in Attachment A) are satisfactory and are hereby accepted. Country Landscapes is authorized to do the work as specified.
_____ is hereby paid as a deposit with Check # _____
The balance will be paid within 10 days of job invoice date.
A finance charge of 1 1/2% per month, which is an annual percentage rate of 18%, will be charged on accounts over 30 days past due.

Date of acceptance _____
Signature _____
Signature _____

(b)

is more complicated or subcontractors are required, it may be necessary to modify the estimate to develop a more detailed bid that will more clearly explain the various prices associated with the project.

WHAT THE CLIENT SEES IN A BID. To create a landscape bid, the designer completes a detailed estimate for the project. Oftentimes, this estimate is used as an internal document within the company, and a bid is generated based on total figures for the categories of materials and labor. Embedded in those figures are additional amounts to cover overhead, profit, and contingency. (Each of these will be discussed in more detail later in the chapter.) Clients do not need to know how much of the company's overhead is being charged to their project, or how much profit the contractor is planning to make on the project. Because of this, some contractors present their clients with a **lump sum bid**, in which the client is given one all-inclusive figure for the project. Other contractors provide the client with an itemized or partially itemized bid. In this case, the project is divided by category (e.g., patio installation, seeding of a turf area, shrub installation), and for each category the materials and labor are included. Some contractors itemize the materials costs but provide a lump labor figure associated with the installation of those materials (Figure 8-6b). Using a lump sum for the labor gives a contractor flexibility, in case one part of the project takes more or less time to install than expected. Bid formats vary and depend a lot on personal preference. Most contractors would agree that you want to provide the client enough detail so he or she knows what the different components of the project are going to cost, but that you provide yourself enough flexibility to cover expenses of the installation.

Developing an Estimate. The accuracy of an estimate can vary greatly, depending on how much detail is included in the document. A combination of experience with both landscape installation and landscape estimating is essential in generating an accurate estimate.

A landscape estimate includes the costs of the various components required to complete the project. Most estimates will include the price for design services, materials, labor, overhead, contingency, and profit. Complex projects may have many phases for which costs must be estimated, while other projects may be more simplistic and include relatively few components. One factor that is common across the board for landscape installation projects is the number of variables that can affect the ability of the contractor to complete the project for the estimated price. Examples of these variables include availability of plants and hard goods, the actual costs of these materials (which can vary throughout the season), transportation costs, labor costs, unforeseen problems that arise during the installation, and weather; and the list goes on. A good estimate will consider all of these elements so the landscape contractor can make a profit and have a viable enterprise.

Proposal. Sometimes, the terms bid and **proposal** are used interchangeably. Generally, they mean the same thing and are an offer to complete landscape services for a set price.

Contract. In some cases, a proposal is used as a contract when it is signed by both the client and contractor. Until it is signed by both parties, it just remains a document that outlines the project and has a price associated with it. A contract is a legally binding document, and when it is signed, it signifies to the contractor that the client has given his or her approval to proceed with the project, and it will be completed for a fixed price.

THE DESIGN/BUILD BUSINESS

Commonly, landscape companies offer both design and installation services. Combining these two services allows a client to work with a single company from start to finish on the project. This is a benefit to the customer because he or she will have one person, or team, whom he or she will communicate with for the duration of the design process and the entire installation phase. In this type of business, there should be an established line of communication between the design and installation components when it comes to project scheduling. This should make for a seamless transition between the design and installation phases, and it should ensure that the installation crews have a complete understanding of the design concept and installation requirements.

This combined offering of services is also beneficial to a design professional since knowing the skills and unique expertise of the installation crew(s) can help shape some design decisions. For example, if there is a crew with extensive experience installing water features, you may consider adding that feature to a design since you know it will be expertly installed. If the crew has installed a number of concrete paver projects with complicated patterns, you may consider selling the concept of a one-of-a-kind patio to a client and be able to charge a premium price. The result is a well-designed and well-installed landscape that is a valuable long-term investment for the client.

Designers Can Benefit from Collaborating with Installation Professionals

Good designers have learned to balance their design biases and desires for having a finished design their way, with a willingness to consider insights from experienced landscape contractors and installation crew members. Obviously, compromises in design intent and quality should not be automatically approved by a designer. But if lines of communication are well established between a designer and experienced construction professionals, discussions can lead to benefits. Less time may be spent on drawing specific installation details because they can be worked out on site, and an experienced installer may have suggestions that result in a better design product at a reduced cost.

 ## PRICING FOR DESIGN AND INSTALLATION SERVICES

The design/build business model influences the approach to pricing a landscape project. Offering a complete menu of design and landscape services allows a design/build business to modify how it would price the services if they were offered separately. In some cases, it provides flexibility in what to charge for and how much to charge. It provides multiple opportunities where profit margin or contingency allowance can be accounted for, thus avoiding the appearance of a large single-line item for the two components that might otherwise be dissatisfying to a client. The following sections focus on pricing a project that includes design and the installation costs.

Design Services

The client may be charged a flat design fee or an hourly rate, depending on what the designer provides. Dividing the flat fee by the number of hours to complete the

FIGURE 8-7

If using the flat fee design pricing method, consider calculating your hourly rate occasionally to make sure you are charging a reasonable fee for time spent on the design.

Design Fee Comparison

$300 Design Fee

10 hours to complete – design rate is $30/hour

20 hours to complete – design rate is $15/hour

design gives an hourly design rate. That price can be compared to typical hourly design rates ($25 for minimal experience; $150 for an experienced designer). If calculating the flat fee into an hourly rate results in a low hourly rate but a designer still wants to just charge a flat fee, increase the flat fee to reflect a suitable hourly rate (Figure 8-7). The design fee estimate should accurately reflect the amount of time spent on the design so the client is charged accordingly.

Materials Costs

Materials costs are the actual cost of the materials used to create the landscape. This cost includes unit items like plants, pavers, or natural stone; and bulk items such as topsoil, mulch, and crushed limestone and sand for a patio base. In some cases, quantity can influence cost, and a lower price may be available when larger quantities are purchased. Oftentimes, there is a delivery charge from the wholesaler when materials are delivered to the building site. The charge may be associated with the use of a forklift on site, a **pallet fee**, or simply the cost in using a vehicle and employee to deliver the product. Delivery charges generally range from $25 to $50, and this should be accounted for as part of the materials charge. Although $25 to $50 does not seem like much, multiple delivery charges added together for a single project can quickly eat into expected profit.

pallet fee: Fee (typically $20–$30 per pallet) charged for delivery of goods on a pallet (e.g., pavers, stone, brick); generally refundable when pallet is returned to distributor.

Labor Costs

Wages and benefits of the landscape crew make up the majority of **labor costs** and are greatly influenced by the efficiencies or inefficiencies of the installation crew. Inexperienced or poorly managed crews and those without the most suitable equipment may take longer to complete a task than properly trained, equipped, and experienced crews. Crews that are too small or too large for a project can be inefficient. For example, if the crew is too small, the flow of work may be interrupted because individuals have to wait for one part of the project to be completed before moving on to the next step. In contrast, when crews are too large, employees may interfere with each other as they move around the work area, or there may not be enough work to keep all crew members busy. The result of these crew inefficiencies is a low production rate.

Production rate is the time required to complete a specific task. For example, how long it takes to plant a 1-gallon shrub; how long it takes to spread 3 inches of mulch over a 2000 square feet planting bed; or how long it takes to install 1000 square feet of sod. A number of resources are available that provide general production rates for the wide variety of tasks associated with landscape installation. (See references at the end of this chapter.) These references are particularly helpful for beginning estimators in determining how long a job will take, what equipment expense will be associated with it, and what the labor cost will be. Experienced estimators can also benefit from referring to these sources as a means to check how efficient their landscape installation crews are functioning.

FIGURE 8-8

Comparison of cost for each step of seeding a turf area compared to the lump amount used as the industry standard. *(Ann Marie VanDerZanden)*

Description Turf Area = 1000 ft^2	Quantity or Unit	Unit cost[a]	Total cost	Total cost including overhead and profit	2002 Industry standard for seeding 1000 ft^2 turf area
Scarifying subsoil (skid steer)	1	15.75	15.75		
Grubbing/removing boulders	1	24.80	24.80		
Spreading planting medium, 4" deep (skid steer)	1	241.60	241.60		
Spreading limestone (1#/SY) (push spreader)	111.1	0.11	12.22		
Spreading fertilizer (.2#/SY) (push spreader)	111.1	0.09	9.99		
Tilling planting medium, 4" deep (26" rototiller)	111.1	0.36	39.99		
Raking planting medium (by hand)	1	18.75	18.75		
Rolling planting medium (push roller)	1000	0.09	90.00		
Seeding turf; Bluegrass 4#/MSF (push spreader)	1	37.90	37.90		
Straw mulching 1" deep	111.1	0.69	76.66		
Total			$ 567.66	$ 663.30	$ 585.00

[a]Unit cost data are the bare costs including material, labor and equipment from *Means Site Work & Landscape Cost Data*, 2002. Data does not include overhead and profit.

scarify: To break up the soil surface.

grubbing: To remove existing vegetation prior to new construction.

PRODUCTION RATE EXAMPLE. Landscape installation is a sequential process best completed in a specific order. Interruptions in work flow often result in low production rates, and these crew inefficiencies can have a major impact on profit. Many residential landscapes include seeding a lawn. Since this is such a common installation procedure, an industry standard quantifies how long it takes an average crew to seed a 1000-square-foot area. A generally accepted price is $585/1000 square feet (Angley et al., 2002; Means Site Work 2006).

This $585/1000 square feet is based on completing the following steps: **scarifying** subsoil (skid steer), **grubbing** and removing boulders, spreading planting medium (skid steer), spreading limestone and fertilizer (push spreader), tilling planting medium (rototiller), raking planting medium (by hand), rolling planting medium (push roller), seeding turf (push spreader), and straw mulching (Figure 8-8). The optimum-sized crew working at optimum efficiency can seed a 1000 square foot area for the $585 price. Using this inclusive figure makes estimating easier because an estimator can use a figure based on the area to be seeded, rather than projecting individual costs for each step of the installation process. This results in one line on the estimate, rather than nine individual entries that add up to the same amount.

Labor costs also include items such as worker's compensation, state and federal withholding taxes, and unemployment taxes. Other factors to consider are unproductive time (e.g., lunch breaks), mandatory state and federal paid holidays, and paid vacation or sick days. All of these additional costs must be accounted for and included as part of the actual cost of an employee's time.

Overhead

Overhead for a landscape job can be divided into two categories: direct job overhead and general overhead. **Direct job overhead** costs are those that can be directly attributed to a specific landscape project, such as a building permit fee for that site, dumpster rental fee, or landfill tipping fee for disposal of removed landscape waste from a site. These costs are not actually part of the installation of the landscape, but they are still required for the project.

General overhead costs are those expenses associated with operating a business that cannot be directly attributed to a specific landscape project. (Sometimes they are also referred to as indirect costs). Examples of these costs include office rental fees, electric, gas and water for the office, insurance premiums, and business license fees. The total amount of these general overhead costs needs to be recouped annually.

How general overhead expenses are allocated across all the landscape jobs completed during the year varies. In some cases, a fixed percentage of the total overhead costs is allocated to each job. In other cases, a percentage of the total overhead is assigned to a job, based on the size of the project and the time required to complete it. This second method more accurately reflects the amount of the business' overhead that is related to the individual projects. Larger projects that require more time and administration to complete will be charged a greater amount than a small job that was completed in a short amount of time and needed minimal oversight. Regardless of the approach used, it is vital these overhead costs are included as part of the estimate. Excluding these costs from an estimate can negatively affect the actual profit realized on a project.

Contingency

As described earlier, a contingency allowance can be viewed as extra insurance for the landscape contractor. It is a safety net that helps to account for unknowns associated with a specific project and the unpredictable nature of landscape construction in general. Examples of contingencies for a specific project might include landscape plans or specifications that are either unclear or overly specific, unavailability of plant materials and hard goods, transportation difficulties getting materials to the site, working on a confined site where access to the work area is limited, or delays related to dealing with a difficult designer or client.

There is no set formula to determine a contingency allowance, and the amount will vary depending on the size, seasonal timing, and complexity of the landscape job. Multifaceted and complex projects, or those scheduled when installation will likely be hampered by inclement weather, tend to have a high contingency allowance. In contrast, a simple project scheduled for installation during ideal weather conditions will have a relatively low allowance.

Some estimators multiply the entire project cost by a specific percentage, regardless of the project. Using years of experience, they are able to derive a percentage needed to protect the profit. This makes the contingency a lump sum, and the calculation is easy. A drawback is the need for experience with the estimating and installation process to calculate a suitable percentage. In some cases, this percentage may be too high when compared to competitors' bids for the same project, and the contractor may be priced out of the project.

Other estimators will set the contingency allowance relative to the complexity of the project. This approach more accurately reflects the likelihood of needing to recoup a contingency allowance. The larger and more complicated a project, the more likely there will be unforeseen complications that could compromise the overall profit. Small straightforward projects, such as installing a 20 foot × 20 foot paver patio, in an essentially flat, accessible yard of a newly constructed home, would require a small allowance. The contractor has done the same project a hundred times, with the same crew, and knows what to expect.

Yet other estimators will include contingencies on just a few specific items that are part of the project. For example, if the design calls for a new paving product that the installation crew has not used before, the estimator may assign a contingency to that part of the project. Or if the crew is installing a retaining wall that will require some difficult and tedious grading to be done beforehand, a contingency might be included. In some cases, estimators assign a larger contingency at the beginning of the construction season as newly hired crew members are familiarizing themselves with the products and process (D. Canova, personal communication). As the crew develops skills and increases their installation speed during the season, the earlier contingency allowance may be reduced and in some cases eliminated completely.

Using their years of training, experienced estimators will determine whether a contingency is necessary.

Contingencies

Rule number 27–2 United States Golf Association: A player is allowed to hit a provisional ball should the player believe the tee shot to be out of bounds or unplayable. If after approaching the initial tee shot the player discovers it is playable, that ball will be played. If, on the other hand, the initial ball is unplayable, the provisional ball comes into play. So how does this relate to estimating a landscape installation? The contingency allowance is your second ball; it is your what-if, your Plan B. This allowance accounts for the unknowns that are a huge part of landscape installation. Even the most experienced estimators cannot account for all of these variables, and most have at least one horror story about a project that had one glitch after another in which no contingency allowance would have been sufficient. The bottom line is that an estimator accounts for those unknowns so the profit is protected. The contingency allowance is a best guess at what those factors might be and what it might cost.

Profit

Profit is determined when the cost of the project is subtracted from the price of the project. When estimating the profit on a project, you must consider what the normal or typical amount of profit is in your area. Commonly, landscape contractors realize a profit between 10 and 25 percent of all the direct and indirect costs associated with the project (Angley, et al., 2002). In some regions of the country, the market will bear a higher profit margin than in other regions. Staying within the regionally accepted profit range may be important in order to be competitive with other landscape contractors.

Another consideration is gauging how much other contractors are likely to charge for the same job. This is not to say that superior work cannot demand a higher rate and still be competitive. Rather, knowing what the competition is charging and knowing the quality of work they are doing can help the estimator make better decisions when pricing a project.

A final consideration might be cash flow or work schedule. In some instances, the timing of a project may not be ideal, and fitting it into the overall installation schedule would be difficult. In this case, consider including a larger profit margin for the job, essentially making it worth the time. In other instances, the installation schedule may be slow, and it is necessary to get the project in order to keep the installation crews employed. In this case, a lower profit margin might be acceptable in order to keep everyone employed. However, estimating in this way is not desirable for the long-term success of the business.

Bidding low on jobs and sacrificing profit just to keep crews employed is not a sound business decision. Employee wages and benefits, or the number of employees, need to be evaluated. Working for lower profit margins just to stay busy may mean that projects with potentially higher profit margins are missed because crews are already committed. Calculating an adequate profit margin for each job, regardless of how large or how small, is essential to running a successful business.

DEVELOPING A BID

As defined previously, a bid is an explicit offer to provide services (design, installation, etc.) for a set price. The specifics of the bid are outlined in a contract that incorporates this information. After the involved parties, generally the client and landscape contractor, sign the contract, it becomes a legally binding document. If changes need to be made during the project, both the client and contractor must agree upon any changes, called amendments. As a result of these amendments, the price for the project may also change, and the client will pay this adjusted price.

Managing Change

As described earlier, landscape installation projects can be affected by many factors. As a result, the design itself may need to be altered; the materials used may need to be modified, or in some cases the installation process might need to be adapted to accommodate the change. The designer, the landscape contractor, or even the client, may initiate changes to the project. Changes implemented because of an unanticipated situation during installation are different from the changes described in this section and covered by a change order. The cost of unanticipated changes should be covered by the contractor's contingency allowance, while the cost of implementing a change outlined in a change order will be passed on to the client.

The party who initiates the change completes a change order. A **change order** describes in detail the amendment to the contract. It is important that the change order also include the change in price. Examples of changes a client might make include deciding he or she wants a different type of paving material (increase or decrease the cost depending on the material chosen), requesting a larger patio size than specified in the plan (which will usually increase the cost), or requesting additional plants in a planting bed. Depending on the requested change(s), the price will be adjusted up or down. Once the involved parties agree on the changes, the change order is signed, dated, and becomes part of the original contract.

Minor changes to the project may be easy for the installation crew to incorporate. However, larger changes that require additional or specialized equipment or a significant amount of labor can significantly increase the cost of the job. In addition to the type of changes requested, the timing of the request is equally important. If major components like grading, soil preparation, and hardscapes have already been completed, then changes related to these will be expensive because the crews will essentially be repeating the work. It is common for large commercial landscape projects to have multiple change orders. Even change orders to change orders are common, and this can drive a landscape contractor to distraction!

A final complication a change order presents a contractor is how the additional time required to complete the modified project will interfere with other projects on the installation schedule. The domino effect can delay current projects and postpone subsequent ones. In some cases, these delays will not be acceptable to clients, and they must be considered before signing the change order.

THE COUNTRY LANDSCAPES BUSINESS MODEL

Country Landscapes, Inc. is a thriving landscape design/build company. During their 25 years in business, they have developed successful strategies that complement their business model. Among these strategies is effective communication between the designer, installation crew, and client during all phases of the project. This seamless communication is essential to making the process run efficiently and having a satisfied client in the end.

Terms and Definitions Used by Country Landscapes

Each company has its own culture, and in some cases terminology, used to describe how and what it does. A few terms and definitions used by Country Landscapes are provided below.

Minimum deposit: This amount, generally 25 percent of the job's price, is required when the contract is signed and is not refundable. It helps ensure that the client is serious about following through on the project and is willing to part with his or her money.

Substitutions and backorders: In some cases, a particular plant cultivar may not be available (or perhaps backordered due to limited availability). Country Landscapes clients sign a waiver in the contract allowing the company to make appropriate substitutions when necessary, or to place a backorder on plants if a substitution is not available. The client realizes that a backordered product may significantly affect job completion. This substitution and backorder policy also applies to hardscape materials.

Work schedule: The operations supervisor signs a detailed work schedule that accounts for each installation project sold, crew availability, and equipment usage. The schedule dictates which projects are in process and the order in which others will be completed.

The Process

Because Country Landscapes designs every landscape it installs, it controls the project from start to finish. Once a potential client has contacted the company, an appointment is set for the customer and a designer to meet. During this meeting, the designer not only gathers information on the client's landscape needs and desires, but also explains the process that Country Landscapes uses when designing and installing a landscape project.

Once the client has accepted the design, an installation proposal (bid) is developed. This proposal includes a detailed list of the items to be installed as part of the project, contract terms stating the required minimum deposit, and agreement of the client to allow Country Landscapes to make substitutions or backorder materials if necessary without additional approval from the client (Figure 8-9). Once the contract is signed and the deposit has been collected, the project moves to the Operations Center, where it is added to the company's work schedule.

What the Client Sees

The company divides the proposal into the different project components, such as patio and stoop construction, path construction, and bed preparation (Figure 8-9). By organizing the proposal this way, the company allows the client to see the cost of each part of the project. Including the materials quantity, unit price, and total price shows the client the materials cost for the project and lets him or her know essentially what it would cost if he or she were to purchase the materials and do the installation himself or herself. The materials prices listed in the bid include the actual materials cost as well as an additional amount for overhead, profit, and contingency.

Labor is the only job cost not detailed in the Country Landscapes proposal. Keeping labor as a lump sum gives the company some control over how it will distribute these costs. For example, if the removal of a concrete pathway was estimated at 5 hours, but actually took the crew 7 hours to remove, Country Landscapes would not go back to the client and charge more (because the contract was signed, and the client did not agree to that additional cost). On the other hand, if the installation of a planting area estimated to take 4 hours took only 3, the client would not be refunded the labor price for that unused hour. Keeping labor as a lump sum allows for flexibility during landscape installation, where some things take longer than expected, and others take less.

Comparing Estimated to Actual Costs

At the completion of every job, the company analyzes a detailed comparison of actual to estimated man-hours (Figure 8-10). The principals at Country Landscapes feel that this is a critical step in a complete bidding and installation process. This audit highlights obvious discrepancies between actual and estimated hours. Since every landscape installation project is unique, it is important that each job be evaluated. They also check the accuracy of the bidding process by examining a large number of projects together.

A more comprehensive comparison is achieved when Country Landscapes does a thorough project cost analysis every quarter. The company spots developing trends and makes necessary adjustments to its prices. For example, when they began

FIGURE 8-9 Example of a proposal for a landscape design and installation. *(Jim Mason)*

01 STEPS

ITEM	QTY	SIZE	PRICE	MAT TOTAL	LABOR	
EXCAVATION (incl. haufing and landfill)	1		$10.00	$10.00		
BOULDERS	1	18 IN.	$22.00	$22.00		
BOULDERS	2	24 IN.	$52.00	$104.00		
			SECTION TOTALS:	$136.00	$181.57	$317.57

02 SOUTH BED & MAILBOX PLANTINGS

ITEM	QTY	SIZE	PRICE	MAT TOTAL	LABOR	
BED PREP	1		$20.00	$20.00		
INCLUDES REMOVAL OF FOUNDATION PLANTS. INCLUDES TRIMMING BARBERRIES & PRUNING LOWER BRANCHES OF TREE.						
PAVER EDGING (AUTUMN BLEND)	28	LN FT	$2.28	$63.14		
SHREDDED OAK MULCH • TOPDRESS	170	SQ FT	$0.12	$20.40		
SHREDDED OAK MULCH	80	SQ FT	$0.34	$27.20		
PREEN • PRE EMERGENT	250	SQ FT	$0.01	$2.60		
• *DELPHINIUM, MAGIC FOUNTAINS (DARK BLUE W) WHITE*		#1 CNT	$5.25	$0.00		
• PEONY SARAH BERNHARDT	1	#1 POT	$10.50	$10.50		
• SEDUM, AUTUMN JOY	2	#1 POT	$2.50	$5.00		
• KARL FOERSTER GRASS (FEATHER REED GRASS)	4	#1 CNT	$5.75	$23.00		
• RUSSIAN SAGE	3	#1 POT	$3.75	$11.25		
DAYLILY, STELLA D' ORO	10	#1 POT	$7.50	$75.00		
TRANSPLANTING • GAILLARDIA (no guarantee)	5	PLANT(S)		$0.00		
			SECTION TOTALS:	$258.09	$755.89	$1,013.98

Remarks: The italicized item(s) are on back order and were out of stock at the time we completed your job. We will schedule to install and bill these items as they become available. Please call if you wish to cancel or change any of the italicized items. Please go ahead and pay from this invoice for the work completed and Thank You for your patience.

We PROPOSE to furnish material and labor in accordance with the above specifications for the sum of $1,423.36
A minimum deposit of $356.00 is due with signed proposal before work can be scheduled.
In the event that the specified varieties or sizes of nursery stock are not available, we reserve the right to make substitutions of variety and size at no additional costs when such substitutions do not compromise the Integrity of the design. If no substitutions are available, the items will be backordered to be installed and invoiced as a separate job.

AUTHORIZED SIGNATURE _____ DATE _____

MATERIALS	$394.09
LABOR	$937.46
(ADJUSTMENT)	$0.00
SALES TAX	$91.81
TOTAL	$1,423.36

ACCEPTANCE OF PROPOSAL - The above prices, specifications, and conditions (including those in Attachment A) are satisfactory and are hereby accepted. Country Landscapes is authorized to do the work as specified.
 $373.00 is hereby paid as a deposit with Check # MC
The balance will be paid within 10 days of job invoice date.
A finance charge of 1 1/2% per month, which is an annual percentage rate of 18%, will be charged on accounts over 30 days past due.

Date of acceptance _____
Signature _____
Signature _____

using a new concrete paver product, they discovered it was taking about 10 percent longer to install compared to the paver they were using before. As a result, the profit on projects with the new pavers was substantially reduced. If they had not taken the time to do the quarterly analysis, the trend would have continued. After discussion with installation crews, they realized that the new pavers were denser and heavier, and took longer to transport on the job site. The density also increased the cutting time of individual pavers. Both of these factors translated into the additional labor cost.

Using Experience to Develop Estimates

Today, Country Landscapes has 25 years of experience developing production rates, assigning man-hours to a particular installation component, and compiling bids. The quarterly comparison of estimated and actual time spent on a project allows the

FIGURE 8-10 A comparison of estimated to actual man hours for the project references in Figure 8-9. *(Jim Mason)*

Code	Description		Unit	Qty	Actual		Hours
EVE	EXCAVATION (incl. haufing and landfill)			1	1		.55
BOU18	BOULDERS	18 IN.		1	1		.69
SR3	BED PREP			1	1		6.36
	INCLUDES REMOVAL OF FOUNDATION PLANTS. INCLUDES TRIMMING BARBERRIES & PRUNING LOWER BRANCHES OF TREE.						
PEAB	PAVER EDGING (AUTUMN BLEND)		LN FT	28	28		1.62
SOMTO	SHREDDED OAK MULCH - TOPDRESS		SQ FT	170	170		.77
SOMSF	SHREDDED OAK MULCH		SQ FT	80	80		.78
PREENS	PREEN - PRE EMERGENT		SQ FT	250	250		.36
DMFBW	DELPHINIUM, MAGIC FOUNTAINS (DARK BLUE W) WHITE		#1 CNT	0	0		0
PSBH1	PEONY SARAH BERNHARDT		#1 POT	1	1		.14
SCAJ1	SEDUM, AUTUMN JOY		#1 POT	2	2		.28
GRKFG1	KARL FOERSTER GRASS (FEATHER REED GRASS)		#1 CNT	4	4		.56
RS1	RUSSIAN SAGE		#1 POT	3	3		.42
DLSDO1	DAYLILY, STELLA D' ORD		#1 POT	10	10		1.39
BOU24	BOULDERS		24 IN	2	2		2.07
TRPP	TRANSPLANTING - GAILLARDIA (no guarantee)		PLANT	5	5		1.08

Job Notes SCHEDULE AFTER SEPT. 16

Job Evaluation

Est. Total Man Hours: 17.36
Actual Total Man Hours: 15.00

company to continually evaluate and refine its estimating process. New companies obviously will not have that experience to draw on, and when Country Landscapes was getting established in the industry its estimators referred to the Means Cost Estimating guides to generate estimates (referenced at the end of this chapter). These guides provide a solid basis from which estimates are developed. The comprehensive guides provide multiple options for estimating time and materials for the many components of a landscape project. For example, in the section on seeding lawns, there are 13 different types of seed mixes, and five different installation methods (push spreader, tractor spreader, stolens/sprigging, sodding, and hydroseeding).

COMPUTER SOFTWARE FOR ESTIMATING AND BIDDING

Estimating is perhaps the most tedious task performed by landscape contractors. Most contractors prefer spending time moving soil and planting trees, rather than calculating overhead costs and evaluating contingencies. Yet estimating, scheduling projects, and managing inventory are essential to a successful business. Finding ways to minimize time spent on estimating while still generating accurate and competitive bids will maximize a company's profit.

Many software programs streamline the estimating process and make it more accurate. Some programs are compatible with AutoCAD (as described in Chapter 1) or other graphic design software, while others are stand-alone estimating programs.

Most all of the programs use a spread sheet, where the estimator enters quantities, prices, and production rates (either their own rates or ones already calculated by the software) into an equation, and the calculations are completed. Although this type of estimating system does remove some potential for error, particularly related to the actual calculation, the end product is still only as good as the information entered into the software. If incorrect values are entered, or if the production rate is too high or too low, the final calculation will be inaccurate.

To increase the accuracy of the estimate, some systems provide a detailed checklist that confirms that all components of the project—as well as direct overhead, indirect overhead, contingencies (if necessary), and profit—are represented. In most cases, this checklist can be customized to reflect the unique features of individual companies. Below is a description of one computerized estimating system available to landscape contractors today.

EasyEst Estimating Software

Since EasyEst (CSC Software, Bonita, Calif.) is compatible with a number of software programs, it is a powerful tool, and it is useful in multiple areas of a landscape contracting business. EasyEst can import AutoCAD files as well as files generated in various computer-drafting programs. Using a digitizer and software, landscape areas can be measured quickly, as can individual materials (plants, lights, and other hardgoods) counts. These values are placed directly into EasyEst estimating software. In addition to calculating the site data described previously, this program can also account for site accessibility. For example, it will use different production rates for a frontyard with easy access compared to a backyard, where access will be limited and require additional time.

EasyEst provides an autoscheduling feature that allows the estimator to enter a project start date; and then, based on the project components (patio, retaining wall, planting beds, sod, etc.), an installation schedule is generated. This is used by managers or foremen when developing crew schedules.

EasyEst also provides tools that aid managers in monitoring their bottom line. The software provides a job-cost tracking feature, which can compare bid costs to final costs for a project. These differences in costs can be further broken down into each aspect of the job (patio, retaining wall, etc.) for easy comparison of predicted costs and actual costs. Estimators can then alter their bidding strategy, based upon this information. Another way to manage that bottom line is to have accurate accounting records. EasyEst is compatible with the QuickBooks (QuickBooks, Intuit, Inc., Mountain View, Calif.) accounting system. This allows for a seamless transfer of data from one program to another for generating price lists, invoices, and payments.

The system described above provides a unique set of features to make the estimating process more efficient and more accurate. There are many systems available, and each should be researched before you select a system to purchase. Unless correct and complete data are entered into these automated systems, the estimates will not be accurate.

FIGURE 8-11

Once the correct scale is entered into the system, the digitizer is held like a pencil and drawn along design areas to measure them. *(Rachel Cox)*

MANUAL MEASURING AND CALCULATING FOR ESTIMATING

Estimators traditionally calculate distance and area measurements on a landscape plan by using a scale (architect or engineer), measuring wheel, and planimeter. Scales are used to measure distances and lengths on drawings. A measuring wheel and planimeter are measuring devices with a rolling wheel calibrated to the landscape's scale. An estimator runs the wheel or planimeter along the perimeter of an area, and based on the measurements, it can calculate the area. Manually measuring a site can be a monotonous task, particularly on large landscape projects, and there is plenty of room for error. Using a measuring wheel or mechanical planimeter requires patience and consistency for accurate results; it is often best to measure distances or areas twice and take the average of the two calculations for more consistent results.

Estimators now have electronic tools, such as a digitizer and related software, which can make the process quicker and more accurate. A digitizer (Figure 8-11) is similar to a planimeter, except that it is an electronic instrument that automatically transmits measurements from blueprints into software estimating programs (provided you have the necessary software interface). If an electronic digitizer is used correctly, the mechanical element of a measuring wheel and planimeter is removed from the equation, and area measurements are highly accurate.

An estimator's job is further simplified when the landscape plans are generated in a program such as AutoCAD. With a few simple clicks of the mouse, an estimator can determine area measurements, lineal distances, and (in most cases) plant counts as well. Even though this provides accurate figures for bidding purposes, the benefits from having an actual person evaluate the project (site accessibility, accounting for inclement weather conditions, factoring in availability of hard-to-find plant materials or hardscapes, etc.) makes estimating a very personalized process.

Accuracy Is Essential

Regardless of measurement tools or techniques, it is critical that the estimator use the same scale in which the drawing was done. Today, it is easy to enlarge or reduce drawings through scanning or photocopying, and the scale of the drawing an

FIGURE 8-12

Even seemingly small differences between the scaled used to draw the design and the one used to estimate the design can result in major discrepancies when the full estimate is calculated. *(Courtesy of Adam Parrot. Ann Marie VanDerZanden)*

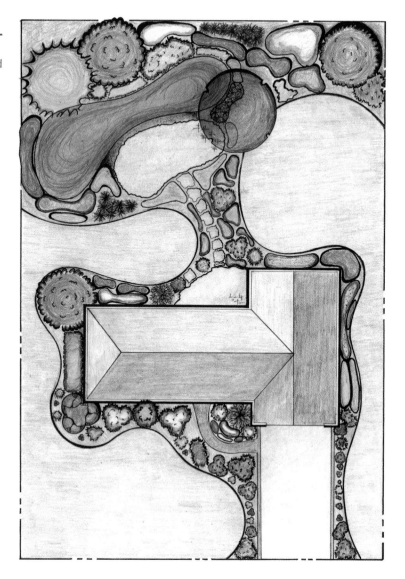

estimator is looking at may differ from the scale used to draw the design. If the difference is slight, it may go undetected, but estimation measurements may contain significant errors.

For example, in the landscape shown in Figure 8-12, the total area of the driveway is 665 square feet when measured using the scale of $\frac{1}{8}'' = 1'$, the scale the design was drawn in. If the copy of the plan the estimator used was reduced by 25 percent, then he or she would estimate the area to be roughly 500 square feet. A 165-square-feet difference of a hardscape surface can translate into a significant amount of money. Another example on this design is the total turf area. Using the correct $\frac{1}{8}''$ scale, the total turf area is 4884 square feet. If the estimator used the same reduced sized plan, the estimate would be 3663 square feet of turf. This means that the total area of turf to be seeded is underestimated by approximately 1200 square feet. Using the industry standard for seeding 1000 square feet of lawn described in Figure 8-8 means that the estimator would underestimate the cost to install the lawn area by close to $600. Because this is an error on the estimator's part, that amount will be subtracted from any profit on this project. (If the project was estimated with a contingency of $600, then the profit would be protected).

When in doubt, an estimator should compare the stated drawing scale to the measurement of a known distance on the drawing, either by using the graphic scale (if available) or by measuring a known hardscape dimension or a relatively standard drawing element (e.g., 4 feet for a front sidewalk width, or 16 feet for a standard two-car garage door).

SUMMARY

Creating a beautiful landscape design is typically only part of the designer's responsibilities in the design or design/build process. A designer must recoup money for time spent on the design, and that depends on the designer's level of experience and business model. For design/build businesses, once the design is drawn and accepted by the client, the next step is to develop an estimate on how much it will cost to install the project. From there, a bid is developed for the client so he or she knows how much he or she will be charged for installation. Accurate and complete estimates are critical to developing an inclusive bid that will generate adequate profit for the company and account for possible contingencies. Although a number of computerized estimating tools are available, none of them can compensate for experience in both the areas of landscape construction and in the art and science of landscape estimating.

KNOWLEDGE APPLICATION

1. Differentiate between the following terms:
   - cost and price
   - estimate and a bid
2. Define the following terms:
   - production rate
   - contract
3. Describe pricing strategies for:
   - design services only
   - design and installation services
   - generating income from your horticulture expertise
4. List and describe four common uncertainties that affect pricing a landscape.

LITERATURE CITED

Angley, S., Horsely, E., & Roberts, D. (2002). *Landscape estimating and contract administration*. Clifton Park, NY: Thomson Delmar Learning.

Booth, N.K., & Hiss, J.E. (2005). *Residential landscape architecture*. Upper Saddle River, NJ: Pearson Prentice Hall.

EasyEst Estimating Software. Construction Software Center Software. 3510 Valley Vista Road, Bonita, CA 91910.

Ingles, J.E. (2004). *Landscaping principles and practices* (6th ed.). Clifton Park, NY: Thomson Delmar Learning.

Page, J.R. (2002). *Means site work & landscape cost data*. Kingston, Massachusetts: Author.

REFERENCE

Cohan, S.M. (2006). *Business principles of landscape contracting*. Upper Saddle River, NJ: Pearson Prentice Hall.

CASE STUDIES: LANDSCAPE DESIGN SCENARIOS

APPENDIX

These four case studies represent actual proposed landscape projects. Owner names have been omitted, and project information has been edited for consistent format, but each scenario is intended to provide students with an opportunity to consider and respond to a realistic project.

CASE STUDY #1
Old Neighborhood Residence

This is a small-scale home located in a relatively old, residential neighborhood near a university campus. The owner is a young, single college student who purchased the home with parental assistance. She currently rents rooms to other students to help pay the mortgage. She has a limited budget, but she is very interested in making landscape improvements to the property in order to expand her gardening activities, provide more space for socializing, and enhance the resale value since she plans to sell the home in a few years.

Base Map and Site Photos
The house is situated on a 138′ × 38′ lot (Figure A-1). Photo locations and site photos are shown in Figure A-2.

Owner Wish-List and Feedback

Desire a deck on the back of the house with enough room to entertain, grill, and study

There should be enough room for several chairs, the grill, and a small table. The yard should be accessible from the deck so that larger parties could overflow into the yard. Built-in seating would be a benefit since there is not a place to store outdoor furniture. Sun exposure would be nice for tanning, but shade would be nice for hot summer days. The air conditioner should be screened from deck views, and access will need to be considered or redesigned to the hose bib on the north side of the house.

Consider reorientation of side yard walk and house access

The back door is currently on the east side of the house. Future deck access should be considered from patio doors that would be installed in the dining room. In the meantime, the side yard steps and walk could use an improved layout, such as redesigned step locations or additional paving.

More privacy on front porch

The owner enjoys the friendly atmosphere of the neighborhood and spends a lot of time on her front porch. She would like some additional screening, however, so that the porch feels more private while still allowing views of neighbors and activities.

(continues)

CASE STUDY #1
(continued)

Old Neighborhood Residence

FIGURE A-1

Case Study #1–base map. *(Steven Rodie)*

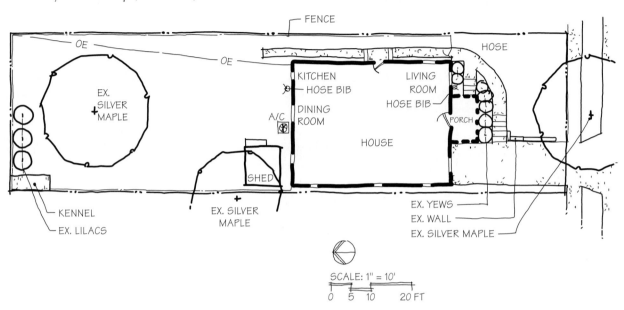

FIGURE A-2

Photo locations (a) for Case Study #1 and Photos #1 (b), #2 (c), #3 (d), #4 (e), and #5 (f). *(Steven Rodie)*

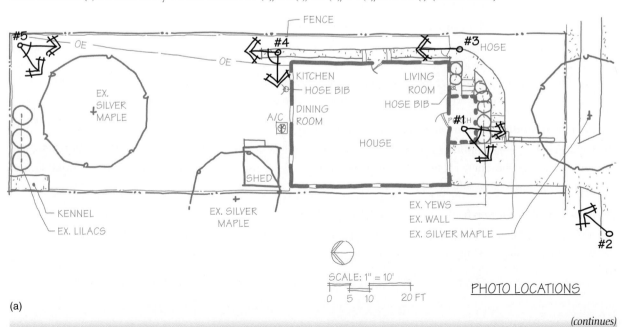

(a)

(continues)

Case Studies: Landscape Design Scenarios 391

CASE STUDY #1
(continued)

Old Neighborhood Residence

FIGURE A-2

Continued.

(b)

(c)

(d)

(e)

(continues)

CASE STUDY #1
(continued)

Old Neighborhood Residence

FIGURE A-2

Continued.

(f)

Ample open turf space

The fenced backyard needs to balance space between additional landscape plantings, a perennial flower garden area, and play space for two large, energetic dogs. The dogs tend to damage flowers; is there a way to give them the run of the backyard but keep them separated from the flowers?

Need backyard storage

The current shed is in need of replacement and could stay in the current location or be moved if a better place is found for storage. Proximity to the deck and house enhances accessibility.

Frontyard flower beds

The owner loves annual and perennial flowers and would like to dedicate a large, maintainable area to a flower bed in a visible, accessible frontyard location.

Inclusion of stone

The owner loves stone, whether it is paving, sculpture, or walls. She acknowledges that it can be expensive, but would like to incorporate some affordable stone somewhere in the yard.

Backyard privacy

Privacy is needed along the north property line in the backyard. Scattered lilacs currently provide some screening; additional lilacs could be added, or all new plantings can be proposed.

Consider change in fence material

The chain-link fence is functional but ugly. What possibilities exist to improve fence character or replace portions or the entire fence that might be affordable for a limited budget?

(continues)

CASE STUDY #1 (continued)

Old Neighborhood Residence

FIGURE A-3

Site issues for Case Study #1. *(Steven Rodie)*

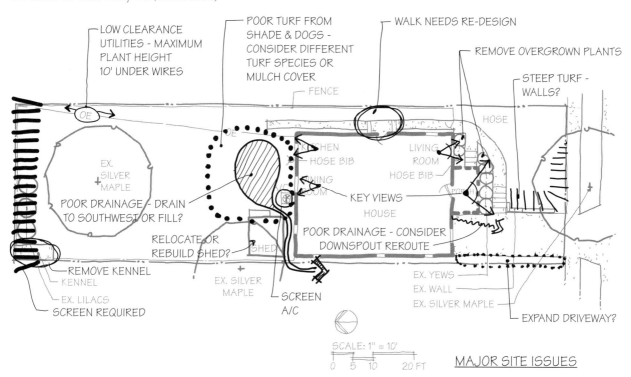

MAJOR SITE ISSUES

Important views

The two most important views are from the living room looking south and from the kitchen/dining room looking north. Landscape improvements would best be appreciated from these windows.

Site Inventory and Analysis

Important site issues are located and summarized in Figure A-3. Responses to selected site issues follow.

Sidewalk to backyard too narrow for mower, and mower must be lifted over stairs to get to backyard

Consider wider walk or stairs reconfiguration, ramp installation, or mower access from other side of house.

Frontyard turf is difficult to mow on slope next to existing retaining wall

Consider retaining walls to level slope area or creating planting area to eliminate sloping turf.

Drainage problem on north side of house

Regrade or add fill near house foundation to improve drainage to southwest corner of backyard; install French drain to route drainage to southwest corner if fill and regrading are not possible in deck and patio areas.

(continues)

CASE STUDY #1
(continued)
Old Neighborhood Residence

Existing kennel area in backyard is unnecessary
Eliminate the kennel, and use available space for additional property-line screening.

Turf hard to maintain on north side of house due to house and tree shade and to dog traffic
Portions will be replaced with deck and/or patio; replant required in dog play areas with shade-tolerant, traffic-tolerant turf species appropriate for climate (turf-type fescue).

Driveway is cracked and drains poorly
Consider rerouting downspout and replacing driveway with better slope for run-off.

Sidewalk leading to porch is partially covered by front shrubs
Shrubs are too large to trim effectively, so remove them and replace with other plantings.

Driveway is a little too narrow for adequate pedestrian circulation
Add paving along driveway edge (dry-laid pavers to complement existing brick color).

CASE STUDY #2
Acreage Residence

This is a new acreage residence built on a secluded rural property. The house was situated to take advantage of expansive views of a new pond as well as access to a dense stand of trees behind the house. The owners are a professional couple with no children; they work in a nearby small town. This is their dream home, so they are very interested in developing a landscape that takes advantage of their life-long experiences with gardens and plants. They are also willing to make a significant investment in their landscape, but they will likely do much of the work themselves over time.

Base Map and Site Photos
The area being considered for landscaping is approximately 100´ × 200´ (Figure A-4). The house is situated near the center of a 160-acre parcel (Figure A-5a). Photo locations and site photos are shown in Figure A-5b–i.

Owner Wish-List and Feedback
Enhance and protect views
Views from upper and lower decks and windows should be protected; any tree planting to frame views of surrounding landscapes should be carefully considered.

(continues)

CASE STUDY #2
(continued)
Acreage Residence

FIGURE A-4

Case Study #2—base map. *(Steven Rodie)*

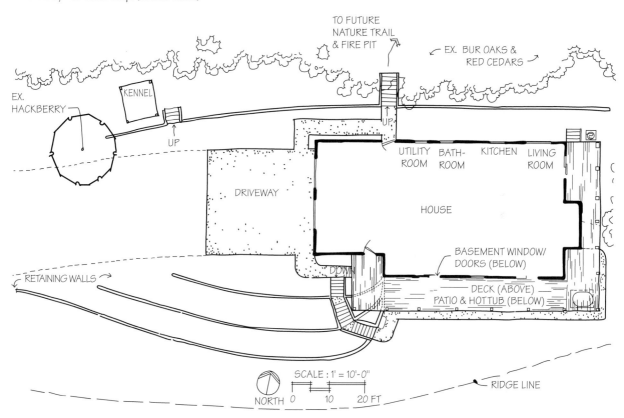

Use naturalistic plantings

Landscape plants should emulate the natural diversity and ecology of the rural setting. The owners are open to increasing plant diversity and using non-native plants, as long as they are proven adaptable to local growing conditions and are not invasive or highly favored by deer.

Develop a sense of entry into property

Naturalistic landscaping (minimally, some large native trees) should extend west of the house along the driveway to provide shade for parking.

Provide butterfly and bird (especially hummingbird) habitat

The owners enjoy watching butterflies and hummingbirds, and they would like to take advantage of outward views of the existing vegetation diversity on the property as well as enhance the variety of food and nesting availability for butterflies and birds.

(continues)

CASE STUDY #2 (continued)
Acreage Residence

FIGURE A-5

An aerial view of Case Study #2 (a); photo locations (b); and Photos #1 (c), #2 (d), #3 (e), #4 (f), #5 (g), #6 (h), and #7 (i). *(Jon Wilson)*

(a)

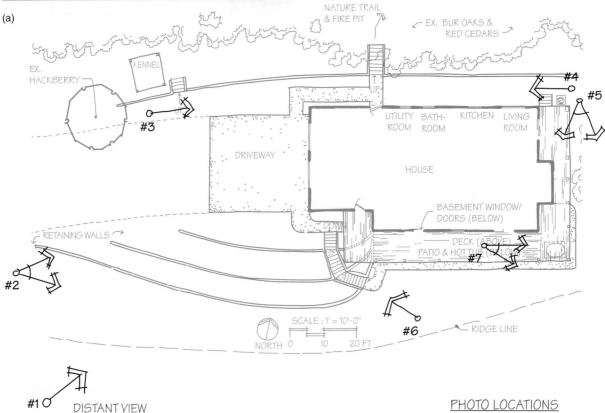
(b)

(continues)

Case Studies: Landscape Design Scenarios 397

CASE STUDY #2
(continued)
Acreage Residence

FIGURE A-5

Continued.

(c)

(d)

(e)

(f)

(continues)

CASE STUDY #2
(continued)
Acreage Residence

FIGURE A-5

Continued.

(g)

(h)

(i)

Minimize deer damage

Deer are very abundant, so deer-resistant landscape plants and deer deterrents will need to be included in design planting schemes.

Keep maintenance to a minimum

Plan for minimum maintenance once plants are established, and conserve water with drip irrigation, mulched planting beds, and use of drought-tolerant plants.

(continues)

CASE STUDY #2
(continued)

Acreage Residence

FIGURE A-6

Site issues for Case Study #2. *(Steven Rodie)*

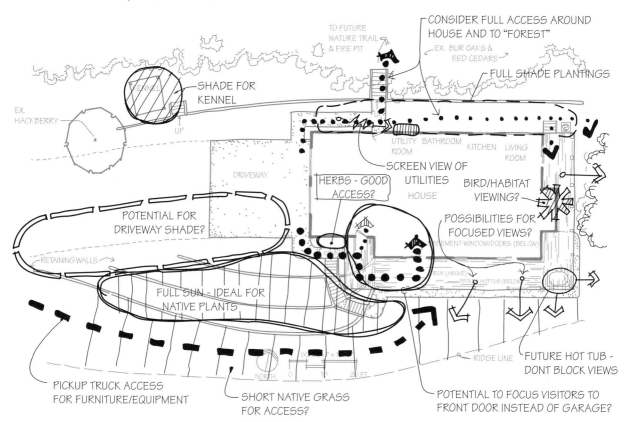

MAJOR SITE ISSUES

Plant requests

An herb garden should be considered in one tier of the walled planters; native perennials and grasses are a high priority; blues and purples are important colors, as well as lots of seasonal interest.

Site Inventory and Analysis

Important site issues are located and summarized in Figure A-6. Responses to selected site issues follow.

Provide shade for dog kennel

Consider using small to medium-sized trees to extend a portion of forest canopy behind house out to kennel area.

(continues)

CASE STUDY #2 (continued)
Acreage Residence

Screen view of utility equipment, vent pipe on north side of house
Consider using shade-tolerant plants to screen views while still providing access to equipment; pedestrian access is required along the entire north side of the house.

Keep access open to lower patio area and house walk-out basement
Equipment (pick-up truck) access is required for antique collectables delivery to lower basement patio doors; consider planting native turf band adjacent to house and walls.

Visually direct visitors to the front door of house on upper deck, rather than south garage door entrance
Provide an entry sequence along south side of house that will pull visitors to house front door; consider sculpture, container plants, paint door with low-contrast color, etc.

CASE STUDY #3
Suburban Neighborhood Residence

This residence is located in a suburban neighborhood with small to medium-sized homes and moderately sized lots. The owners have an energetic dog and a son in middle school. They picture themselves moving to a larger home in several years, so they are interested in landscaping that will promote curb appeal and salability and landscaping that will add to their quality of life while they remain in the house, but not require a significant investment. They both have very busy professional lives, so low maintenance is also a high priority for them.

Base Map and Site Photos
The house is situated on a 55´ × 120´ lot (Figure A-7). Photo locations and site photos are shown in Figure A-8.

Owner Wish-List and Feedback
Low-maintenance, sustainable landscaping
Overall, the owners would like to keep the landscaping somewhat low maintenance. Sustainable landscaping is important, particularly where water use is concerned. They do not have an irrigation system and sometimes have trouble keeping up with watering in extreme conditions. They prefer more of a natural-looking landscape (a very formal symmetrical look is not what they want).

Room for the dog
They have an Australian shepherd that has free run of the backyard. He needs at least one good, long open area to run in and a couple of places along the fence where he can get right up to the fence to say hi to his friends.

(continues)

CASE STUDY #3
(continued)

Suburban Neighborhood Residence

FIGURE A-7

Case Study #3—base map. Courtesy of Ernie Thackray. *(Steven Rodie)*

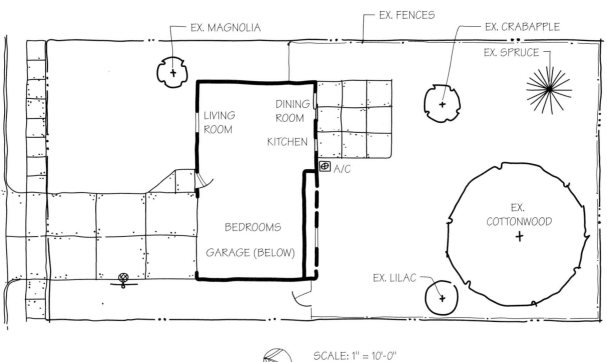

Frontyard plantings

One of the owners wants a tree; the other does not. Any tree in the front would have to be placed so that it does not block the view from the picture window on the main floor.

Maximize curb appeal

They plan on selling the home in 3 to 5 years. They want the frontyard landscape to really enhance the view from the street; however, it cannot be so overwhelming that it gives potential buyers the idea that it would be extremely high maintenance.

Backyard patio improvements

The number one priority for the family is some privacy from neighbors in the backyard. They want an area to entertain—something that is an extension of the dining area and that opens to the patio. Shade would be very important since the patio has a south exposure and direct sun exposure most of the day; extending the shade over to the southwest house corner

(continues)

CASE STUDY #3
(continued)
Suburban Neighborhood Residence

FIGURE A-8

Photo locations (a) for Case Study #3 and Photos #1 (b), #2 (c), #3 (d), #4 (e), #5 (f), and #6 (g). *(Steven Rodie)*

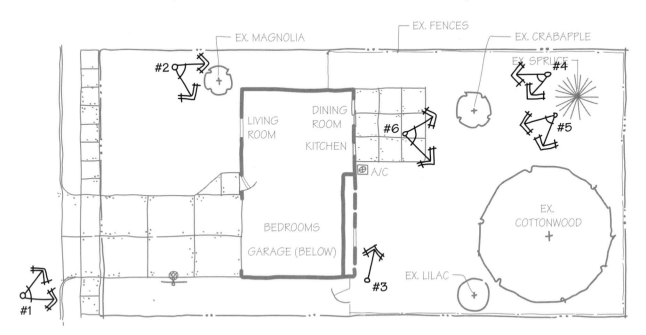

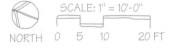

(a)

(b)

(c)

(continues)

Case Studies: Landscape Design Scenarios   403

CASE STUDY #3
(continued)

Suburban Neighborhood Residence

FIGURE A-8

Continued.

(d)   (e)

(f)   (g)

would be a good idea. They plan on getting a larger grill and smoker combo and would like it to be convenient to the back door but still partially screened from view.

Extend living space out into the yard

The owners would like a small, private sitting space somewhere in the yard. They like having birds and butterflies in the yard, but they want to keep bees away from the outdoor entertainment area.

(continues)

CASE STUDY #3
(continued)

Suburban Neighborhood Residence

Plant requests

They want all-season interest in the yard, and would also like a small herb garden conveniently located near the back door and patio. They would like to see a variety of foliage, using different shades of green and textures to create interest. Some plants with variegated leaves are fine, but not too many. Plants that have gray or silver foliage and purple flowering are also desired. The owners like a natural look with variety, but they are not looking for a hodge-podge English cottage garden look. Most-liked flower colors include purple/blue, whites, and yellows. Orange is okay, but they are not fond of pink and most reds. If red is used, it would have to be very dark or deep red. They are open to most ground covers, or vines, and small and medium shrubs, as long as they are not invasive. They like small trees, as long as they are not messy.

Site Inventory and Analysis

Important site issues are located and summarized in Figure A-9. Responses to selected site issues follow.

FIGURE A-9

Site issues for Case Study #3. *(Steven Rodie)*

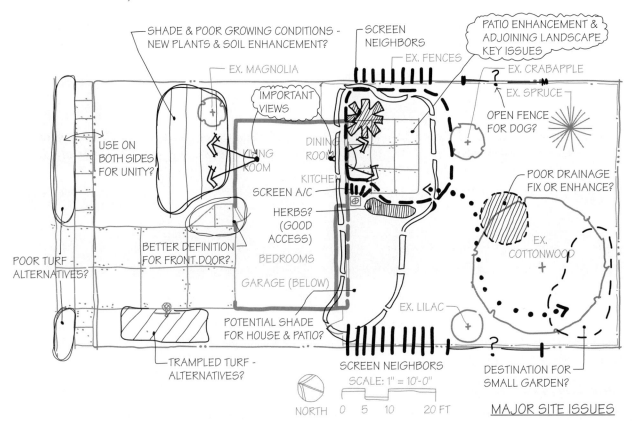

(continues)

CASE STUDY #3
(continued)

Suburban Neighborhood Residence

Basketball games in the driveway have trampled adjoining turf areas

Consider extending driveway paving, dry-laid pavers for a more decorative look, or replanting turf with structural foundation so that turf will hold up to occasional trampling.

Front of house is north facing and is shaded for most of the day

Consider shade-tolerant plants near house foundation that may also have to tolerate late-day hot summer sun.

There is very heavy clay soil in most of the frontyard; the previous owners tried planting two different varieties of maple and both died

Attempt to confirm specific circumstances for poor tree health so that similar conditions are not repeated (poor planting, poor stock, overwatering, etc.). Consider best locations for tree(s) in frontyard relative to framing a view and/or a front door accent; select appropriate tree species, and heighten success potential with proper planting and possible soil amendment in future rooting area.

Turf area between the sidewalk and the street is very hard to keep watered. The owners would consider turf replacement, but are concerned about how nonconformance with the neighborhood would effect selling the property.

Consider aeration or other soil improvements to existing turf, or replace with more drought-tolerant species. Consider using mulched groundcover plantings on both sides of walk to enhance unity of textures and colors.

The yard close to the south side of the house, particularly at the southwest corner of the house, is baked dry. It also slopes away from the house for drainage. The combination of the two conditions makes it very difficult to grow grass there. An alternative to turf is probably the way to go.

Consider replacing with landscape bed and mulched plantings as part of landscape bed development for entire backyard.

There is a little bit of drainage problem right in the middle of the backyard, but only when it rains a lot

Consider filling area with soil dug from landscape bedlines to eliminate ponding, or, depending on location, expand or deepen the area to enhance ponding for development of a rain garden.

The air conditioner next to the patio needs to be screened, as does the cable/phone pedestal in the southwest corner of the yard. In addition, they want to visually screen the neighbor's above-ground pool, pump, and deck on the west side of the property.

After proper clearances are considered for the utilities, plants, or structural screens (lattice, etc.) may provide the best screening, depending on available space. Privacy and screening of unsightly views from neighboring yards should be carefully assessed so that chosen locations and materials (hedge along the fence, or lattice with vine next to patio) strategically address specific views without limiting desired views for owners.

CASE STUDY #4
Large Lot Residence

This house is in an upscale neighborhood with moderate to large homes and large yards. The corner lot is open to the street on one side, and it adjoins a church property (parking lot and large mature trees) along the back property line. The couple who owns the property are university professors trained in wildlife and environmental biology, and they are very interested in creating a relatively natural landscape with high habitat value. They have one small child, who enjoys having lots of play space; they have no pets.

Base Map and Site Photos

The house is situated on a 75′ × 210′ lot (Figure A-10). Photo locations and site photos are shown in Figure A-11.

Owner Wish-List and Feedback

House configuration

The house was built by a contractor who apparently oriented the front of the house to the backyard. As a result, the front door and entrance adjacent to the driveway do not seem visually well defined. The owners are open to landscaping ideas that would better balance the frontyard and enhance the front door.

FIGURE A-10

Case Study #4–base map. *(Steven Rodie)*

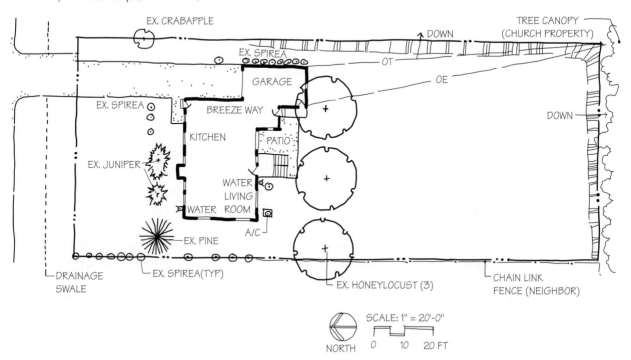

(continues)

CASE STUDY #4
(continued)

Large Lot Residence

FIGURE A-11

Photo locations (a) for Case Study #4 and Photos #1 (b), #2 (c), #3 (d), #4 (e), #5 (f), and #6 (g). *(Steven Rodie)*

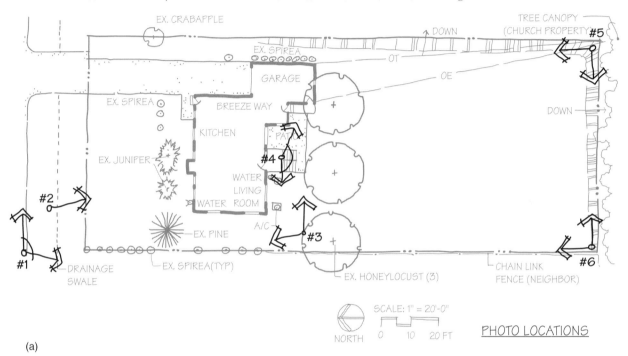

(a)

(b)

(continues)

CASE STUDY #4 (continued)
Large Lot Residence

FIGURE A-11

Continued.

(c)

(d)

(e)

(continues)

CASE STUDY #4 (continued)
Large Lot Residence

FIGURE A-11

Continued.

(f)

(g)

Play space for son
Provide space initially, but this will become a less important priority as he grows older; transition of space into different uses should be considered.

Enhance backyard living space
The owners are struggling with how to add trees and privacy while maintaining enough sun for prairie plants and some vegetables. The vegetable garden should not be visible by the neighbors. They would like to consider adding a deck or screened porch off the back of the house in the future. It is also important to have high habitat value in the landscape.

Environmental soundness
They would like to see drought-tolerant native and adapted plants used as much as possible. They dislike roses of any kind and do not want invasive species or plants that require lots of fertilizer or pesticides.

Plant requests
They would like a big deciduous tree (preferably native) in the frontyard. In addition, they would like small, flowering trees planted near the south property line in the backyard (at least one of the large shade trees located there has suffered recent severe squirrel damage). They also like fruiting trees and shrubs, and they would like plants that provide winter color and interest.

Site Inventory and Analysis
Important site issues are located and summarized in Figure A-12. Responses to selected site issues follow.

(continues)

CASE STUDY #4
(continued)

Large Lot Residence

FIGURE A-12

Site issues for Case Study #4. *(Steven Rodie)*

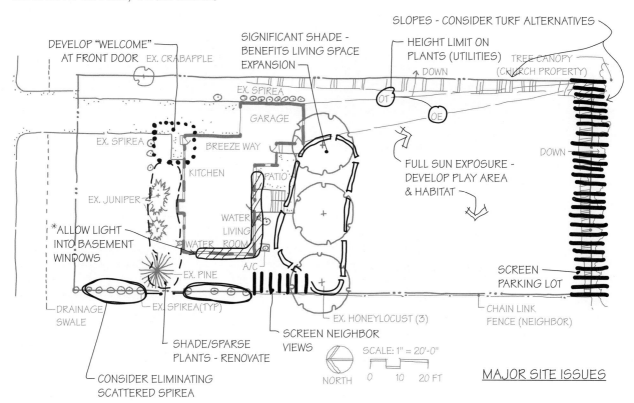

Lack of privacy due to the openness of the yard and adjoining church parking lot
Consider specific views from house windows or outdoor locations that require privacy, and strategically assess best locations (near viewpoint or along property boundaries) for plants or structural (lattice with vines, fence, etc.) solutions.

Foundation plantings and turf adjacent to house in frontyard are growing poorly, due to shaded conditions
Landscape beds should be expanded to remove turf; and shade-tolerant plants should be selected.

Backyard terracing has created relatively steep slopes on east and south property lines
Consider planting native or adapted woody groundcover plants to help define yard edges and eliminate turf slopes.

Lack of privacy between property and neighbor to the west
Consider plantings on house corner that help frame views of house while providing a neighborly visual screen for patio privacy.

Existing row of three trees provides significant shade
Consider value of shade versus need for full-sun conditions for prairie-type plants; is it possible to leave trees for living space shading while allowing sufficient sunny areas for plantings in remainder of yard?

GLOSSARY

A

analogous colors—Those colors adjacent to each other on the color wheel; for example, yellow-orange is analogous to yellow and to orange.

arbor—An open, airy, architectural structure generally constructed from wood or metal. It can have posts with cross beams and sometimes a second layer of cross beams to create a crisscross pattern. Arbors are generally smaller than pergolas.

architect's scale—Contains 11 different scales, all of which represent 1 foot; for example, the $\frac{1}{4}$ scale means that a $\frac{1}{4}$ inch represents 1 foot on the drawing.

B

balance—This design principle addresses the concept of visual equilibrium; types of balance include symmetrical or asymmetrical.

baseline measurement—A measurement method that documents the linear dimensions of features, such as window and door locations, along a set distance (baseline). It is a more accurate and efficient measurement method than taking individual measurements of each feature.

bid—An offer to provide services, such as the design and installation of the project, for a set price.

break even—The point where all costs associated with the project have been recovered, but no profit has been gained.

C

CAD—Computer-aided design.

change order—Describes in detail the amendment or change to the landscape contract, including the change in price.

chroma—A measure of the actual hue content of a color; sometimes called intensity, saturation, or purity.

color-rendering—Application of color to a black-and-white drawing.

complementary colors—Those colors located opposite one another on a color wheel; for example, purple and yellow.

composite wood—A product made from a combination of recycled plastic, wood products, and glue (resin) that is moisture- and insect-resistant and does not require sealing, painting, or staining.

concrete paver—A uniform-shaped piece of concrete commonly used for driveways, patios, and other outdoor ground level surfaces; available in various colors and shapes.

contingency—An allowance that serves as extra insurance for a landscape contractor; it helps to account for unknowns associated with a specific project and the unpredictable nature of landscape construction in general.

cost—For a landscape contractor, the cost of a project includes labor, materials, equipment, overhead, and sometimes subcontractor costs needed to complete the project.

covenant—A legally binding property agreement implemented by a land developer that limits individual landowner practices for the ultimate aesthetic or functional benefit of an entire development.

cut stone—Stone cut to size at the quarry, split faced if required, and polished or finished to different specifications; also called dimension stone or wall stone.

D

deadman—A building element used to further the structural integrity of retaining walls; commonly used with timber retaining walls and involves anchoring a T-shaped structural member deep into the soil behind the wall.

deciduous—This type of plant drops its leaves each fall in response to the shortened day length and colder temperatures that accompany this seasonal change.

design concept—The glue that unifies the ideas inherent in a design; it represents the essence of what is expected from the design and what the design should or should not do.

design contract—A legal document signed by the client and designer to complete a landscape design.

Design Planning Questionnaire—A form used by a designer to summarize client personalities, desires, and knowledge of site conditions. It usually consists of a series of questions (short answer, multiple choice, etc.) pertaining to site microclimate, soils, and drainage issues, etc. along with personal and family needs, particular uses for outdoor spaces, required sizes of spaces, maintenance requirements, and the ultimate goals and objectives that are to be addressed through the landscape design.

design process—A creative, problem-solving decision framework that includes an initial identification of the problems and issues to be addressed (accept situation); a thorough inventory of existing information and the potential opportunities and constraints associated with the information (analyze); definition of the design goals and objectives to be met with the design (desired outcomes); creative thought-processing of a wide range of solutions (generate ideas); selection of the best idea or combination of ideas (select); implementation of the idea (implement); and follow-up to determine what worked or did not work (evaluation).

design program—The documented wishes and desires expressed by the client, including specific landscape problem-solution, quantification of space and human comfort requirements, and aesthetic and character requirements for landscape plants, features, and spaces.

design theme—A design theme reinforces the landscape framework and helps unify adjoining areas. A theme is established through the consistent use of ground-plane patterns and shapes that are developed through visual and physical connections with existing landscape features.

direct cost—Costs that can be directly attributed to a specific landscape project.

direct measurement—The most common technique for measuring that documents the distance between two points (e.g., width of a sidewalk) or two objects (between the fence and side of the house).

direct job overhead—Overhead costs that can be directly attributed to a specific landscape project, such as a building permit fee for that site or dumpster rental fee; these costs are not actually part of the installation of the landscape, but they are still required for the project.

dressed dimension—The dimension of wood after it has dried and is planed.

E

easement—A specified area inside a property boundary (usually linear in shape) where access is allowed by companies or persons who do not own the property; examples include utility company access for utility repairs, areas left open for drainage, or a land owner requiring access to an adjacent property.

emphasis—This design principle creates focal points that draw the eye to specific landscape locations when the landscape is viewed as a whole.

engineer's scale—Contains six different scales divided into decimal parts or units of 10. The scale marked 10 means that the inch is divided into 10 parts.

estimate—An approximation of what price a client will pay for the landscape project (i.e., design and installation). It is important to clarify with the client that the actual price may be higher or lower, depending on a number of factors that could affect the project.

evergreen—This type of plant keeps its leaves or needles all year long; evergreens are divided still further into broadleaf evergreens, such as rhododendrons and hollies, or needled evergreens, such as pines and spruces.

F

fieldstone—Stone that has not been milled and has retained its natural, uneven shape and edges.

flagstone—A generic term used for flat stone slabs that range in thickness from $\frac{1}{2}$ inch to 3 inches.

form—Associated with three-dimensional objects; made from connected lines and the way these lines are arranged.

freestanding wall—A decorative wall similar to fences. It can define a boundary, direct traffic, or obscure a view.

French curve—A rigid template used to draw curves.

functional diagram—Also referred to as a concept plan or concept diagram, a functional diagram uses informal graphics in bubble shapes to designate the approximate sizes and functional relationships between landscape areas and elements. It provides an effective tool for initial arrangement, layout, and scaling of design ideas.

G

general overhead—The expenses associated with operating a business and that cannot be directly attributed to a specific landscape project such as insurance, rent, and utilities; also referred to as indirect costs.

graphic symbol—A drawn symbol that incorporates a consistently applied combination of lines, dots, and value contrasts to represent specific landscape information or elements.

ground plane—Also called the floor of an outdoor room; it includes patios, walkways, decks, and steps.

H

hardiness zone—A classification system in the United States developed by the United States Department of Agriculture to depict average minimum winter temperatures based on the lowest temperatures recorded for a 20-plus year average.

heat zone—A classification system developed by the American Horticultural Society to depict the average number of days above 86° F (30° C).

herbaceous plant—This type of plant produces soft, succulent stems that die back to the ground each year in the winter. Each spring, a herbaceous plant starts growing from the ground up and reaches its full size by the end of the growing season.

hue—The name of a color, such as red or blue.

hyper-tufa—A strong, lightweight material used to make trough gardens.

I

indirect cost—The expenses associated with operating a business and that cannot be directly attributed to a specific landscape project such as insurance, rent, and utilities.

L

labor cost—The wages, benefits, and labor burden of employees.

landscape contractor—A business person who provides landscape installation services.

landscape preference factors—Visual and spatial factors that have been shown through research to best predict human preference for specific landscape characteristics and settings. They include coherence (a sense of structure and order), complexity (enough to generate interest without visual chaos), legibility (a sense that one could find the way into and out of a landscape and would be welcome in the landscape), and mystery (a sense that landscape exploration will yield more information and interest).

landscape use area—An area of a landscape typically identified by how it will be used or viewed. Public, private, and utility areas are the three most typically defined in residential landscapes.

line—Lines can be horizontal, vertical, diagonal, or curved. It is how they are used individually or in combination that gives an object, including a landscape, dimension.

line and dot quality—The basic character of lines and dots defined by the clarity and consistency of the applied pen or pencil; a high-quality line should have definite start and end points and a consistent width; a quality dot should have a clean point and reflect consistency in a rounded shape and size.

line weight hierarchy—The application of different line weights in a drawing to create visual contrasts between drawing elements and represent the character and relative importance of elements in a drawing (wall height, importance of paving versus planting bed, etc.)

lump sum bid—A bid that includes one all-inclusive figure for the landscape project.

M

materials cost—The cost of materials for the landscape project.

microclimate—Landscape areas that exhibit variable growing conditions, including temperature, light, wind, humidity, and precipitation relative to a regional or local climate.

N

negative space—Space within a landscape that is unoccupied or empty.

nominal dimension—The dimension of wood when it is milled and cut into certain sizes (e.g., 1″ × 4″, 2″ × 4″, 2″ × 10″, 6″ × 8″, etc.).

O

order—The design principle that addresses the overall framework for a design.

overhead plane—Also called the ceiling of an outdoor room; includes arbors and pergolas.

P

paraline drawing—A two-dimensional drawing that has been given height and a three-dimensional representation through the use of projected parallel lines; also known as an axonometric drawing, the drawing gives the impression that the viewer is looking at the landscape from above.

parallel ruling straightedge—Similar to a T-square, except that it is fixed to the drawing surface and moves over the area via small cables or guides. The bar remains locked at a consistent angle across the drawing as it is moved up and down on the drafting table.

pergola—An open, airy, architectural structure generally constructed from wood or metal. It can have posts with cross beams and sometimes a second layer of cross beams to create a criss-cross pattern. Pergolas often cover patios, decks, or long walkways, and they help define a garden space.

perspective drawing—A drawing that reflects in a single view the realistic three-dimensional qualities of a landscape, including depth, spatial configurations, vanishing points, and a horizon line.

plot plan—A scaled plan that summarizes all of the survey information documented for a lot. In addition to reflecting property boundary lengths and the orientation of the boundaries and property to true north, it also includes rights-of-way, setbacks, and easements.

pondless water feature—A water feature that does not have a collection reservoir above ground.

positive space—Positive space is occupied or filled space in a landscape design.

preliminary design—The process or finished drawing that converts a loose, freehand, functional diagram into a more refined, but still preliminary, design drawing. The drawing reflects clear spatial organization, plant masses and locations, and the framework for design theme development.

price—What a contractor will charge a client to complete a landscape project; it is cost plus profit.

production rate—The time required to complete a specific task.

profit—Revenue generated after the project costs are deducted from the price.

proportion and scale—This design principle refers to the size relationship between different elements within the landscape.

proposal—Similar to a bid, it is an offer to complete landscape services for a set price.

prospect-refuge—A universally preferred landscape setting characteristic where a person is able to see what is going on in the landscape (prospect) from a location that is relatively hidden from view and provides a sense of protection (refuge).

R

repetition—This design principle is created in a landscape when anything is repeated, including color, form, texture, a particular plant, a mass of similar plants, or a plant composition.

restorative characteristics—Characteristics reflected within landscape settings that enhance human mental and emotional restoration. They include features that engage energized focus (fascination), provide a place that is physically or mentally removed from a location of stress (being away or escape), provide an escape location with a sense of unlimited boundaries (extent), and provide for human comforts (compatibility).

retainer fee—A fee, often 10–20 percent of the total design fee, paid by the client and required before work on the project starts.

retaining wall—A wall designed and constructed to withstand the weight of the retained soil, as well as the pressure of the water contained in that soil.

rhythm—This design principle relates to the organized movement within a landscape and addresses the factors of time and movement within the space.

right-of-way—Public property where streets, roads, and sometimes alleys are located. Right-of-way is normally wider than the street surface and typically includes the sidewalk, the parking strip or drainage swale along the street, and the trees planted near the street.

S

section-elevation graphic—A drawing illustrating landscape information that provides a side view of a landscape at a specified vantage point; it typically reflects the landform, plants, and other features along and just in front of/behind the vantage point location.

segmental wall unit—A precast concrete unit that can be used for both freestanding and retaining walls.

setback—The minimum required distance between a specified property line and any portion of a structure or site improvement. Setbacks are normally required from the front, back, and sides of a property, and they determine the locations of walls and fences as well as buildings.

sheet layout—The overall configuration of information on a landscape drawing, typically including the title block, scale, north arrow, plant list, notes, the landscape plan, and supplemental/detail drawings.

site analysis—The process of identifying all of the key physical, climatic, and environmental factors on a given site, and determining the design opportunities and constraints that are derived from the factors.

site survey—(also referred to as plat) A drawing that combines the plot plan information (property lines, easements, setbacks, etc.) with accurately located site features, including the house footprint, driveway, sidewalks, etc.

slope—A change in elevation along a specified distance in the landscape.

soil pH—A measure of the acidity or alkalinity of a soil; measured on a scale of 1.0 to 14.0, where 7.0 is considered neutral.

soil texture—Determined by the relative proportion of sand, silt, and clay particles.

sustainable design—A design process or product that defines landscapes with the following characteristics: reduced and or efficient consumption of energy, water, and other resources; maximized recycling of resources to minimize waste; maintained or enhanced local ecological structure, function, and biodiversity; and an overall perspective of design from the viewpoint of nature as the designer.

T

template—A thin, flat, plastic tool with openings of different shapes.

texture—The perceived appearance or feel of a landscape or plant surface. In a landscape drawing, it is represented by an organized pattern of lines and dots that enhance value contrasts and define relative smooth or rough surfaces in adjacent drawing areas.

trace—Thin, translucent paper.

triangulation—A measurement method using two known points (such as two house corners) to locate a third point (such as a tree trunk); variations of triangulation can be used to locate the ends of an element (such as a hedge) or a series of points, such as those that form a curving bedline.

T-square—A technical drawing instrument used primarily as a guide for drawing horizontal lines, or used in combination with a triangle to draw vertical lines; its name comes from the general shape of the instrument.

U

unity—The design principle that creates a link between the plants, hardscapes, and house. It creates a sense of interconnectedness within the design composition.

V

value—Describes how light or dark a color is; sometimes called brightness, lightness, or luminosity.

value contrast—The relative amount of light and dark areas in a drawing. White adjacent to black creates the highest contrast, clarity, and potential interest. Differences in value identify changes in materials, define edges, and enhance the three-dimensional quality of drawings.

vellum—A high-quality paper made from 100 percent pure white rag stock.

vertical plane—Also called the walls of an outdoor room; includes fences and walls.

W

woody plant—This type of plant produces woody stems that continue to increase in diameter and length each year. A woody plant continues to grow where it left off at the end of the previous growing season.

INDEX

A

Absolute scale, 157
Access, 166–167
Acid-etching concrete, 329
Acute angles, 238
Advertisements, 102
Aesthetic principles
    applied, 159–163
    balance, 154–156
    emphasis, 159
    framework, 146–147
    in lighting, 347–348
    order, 146–147
    plant selection
        color considerations, 280–284
        flowering considerations, 284–285
        form considerations, 271–276
        four-season plants, 271
        growth rate, 287
        lifecycle considerations, 269–271
        mature size, 286–287
        texture considerations, 276–277
        type considerations, 268–269
    preliminary design, 225
    proportion, 156–158
    repetition, 150–151
    rhythm, 151–153
    scale, 156–158
    size, 156–158
    unity, 154
Aesthetics, 107
Aluminum edging, 317
American Horticultural Society Plant-Heat Zone Map, 263
Analogous color, 281
Analysis. *See also* Site inventory and analysis
    aspects of, 116, 118
    concept and, 73
    site, 200–201
Angles, tools for, 17–19
Angular gravel, 317
Annuals, 260, 289
Architect's scales, 13–15
Architecture, 144
Art elements
    color as, 135–139
    combining, 134
    form, 133–134
    line, 133
    texture as, 139–140

Artemesia *(Artemesia schmidtiana)*, 284
Assessments
    behavioral, 121–123
    humanistic, 124–127
Asterisks, 220
Asymmetrical balance, 154–155
AutoCAD, 23–24
Automobile circulation, 212
AutoSketch, 24
Awareness, 132

B

Backflow preventer, 188
Backlighting, 349–350
Baldcypress *(Taxodium distichum)*, 268–269
Balusters, 332
Basalt, 319
Baseline measurement, 191–193
Base maps
    case study, 196–198
    defined, 2
    documentation, case study, 196–198
    drafting, 94
    existing information, 183–187
    information missing from, 195–198
    lot drawing, 184
    measured information, 187–196
    plot plans, 184–186
Behavioral assessment, 121–123
Berms, 165
Beveled edge, 326
Bids
    client's view, 373–374
    defined, 372
    design/installation projects, 380–381
    estimates *versus*, 372–373
    lump sum, 374
    sample, 373
Biennials, 270
Big-leaf hydrangea *(Hydrangea macrophylla)*, 276
Black-and-white graphics, 39
Black plastic, 317
Blender markers, 9
Blotter, 9
Bluestone, 319
Boundary bearings, 184
Boxwood *(Buxus sempervirens)*, 276
Brain hemispheres, 83, 112
Branch density, 275

Branching habit, 274–275
Break even, defined, 370
Brick
    defined, 319
    standards, 322–323
    use, 322
    versatility of, 340
Broom finish, 327
Bubble shapes
    defined, 219
    use areas, 221–222
Budgets, 180–182
Business
    computer software, 384–385
    Country Landscapes model
        application of, 383–384
        company profile, 381
        costs, 382–383
        process, 382
        project components, 382
        terms and definitions, 381
    design service only
        contracts, 367–368, 371–374
        payment, 369–370
        pricing, 366
        time schedule, 369
    design/build services, 375
    design/installation services
        bid development, 380–381
        contingency, 378–379
        labor costs, 376–377
        materials costs, 376
        overhead, 378
        pricing approaches, 375–376
        production rate example, 377
        profits, 379–380
Business model, 366

C

CAD. *See* Computer-aided design
Cameras, 23
Canopy, 52, 225
Chain-link fencing, 338
Change order, 380
Chroma, 137–138
Circle templates, 21
Circulation
    access and, 166–167
    automobile, 212
    patterns, 295
    pedestrian, 212

Clients
  as resource, 199
  bids and, 373–374
  communications, 174–182
  communications with
    budget discussions, 180–182
    design planning questionnaire, 174, 176
    design program, 217
    effective, 172
    final meeting, 370
    first meeting, 176–182, 366
    initial contact, 174
    new landscapes, 178
    site analysis and, 199
  contracting with, 367
  feedback, 225
  site features related to, 212–214
Cockspur hawthorn *(Crataegus crusgalli)*, 274
Cognitive preferences, 164
Coherence, 124–125
Cohesive plant masses, 143, 146–147
Cohesive whole, 140
Color
  analogous, 281
  applications, 81
  basics, 73, 81, 139
  chroma, 137–138
  complementary, 137, 281–282
  concrete, 320–321, 328
  dimensions, 136–138
  house, 212–213
  hues, 136–137
  laying, 84
  monochromatic planting schemes, 282–283
  Munsell system, 135–136
  plant selection and, 280–284
  triadic planting schemes, 283–284
  value, 137
Color Drawing, 137
Colored pencils, 7
Color graphics, 39
Color wheels, 280
Columnar skyrocket juniper *(Juniperus scopulorum)*, 272
Communications, designer and client
  budget discussions, 180–182
  design program, 217
  effective, 172
  final meeting, 370
  first meeting, 176–182, 366
  initial contact, 174
  new landscapes, 178
  questionnaires, 174, 176
  site analysis and, 199
Compact yews *(Taxus)*, 295
Compass, 18
Compatibility, 128
Complementary color, 137, 281–282

Complexity, 124–125
Composite wood, 331–332
Computer-aided design (CAD)
  AutoCAD based, 23–24
  benefits of, 100
  estimating, 25–27
  estimator function, 386
  graphics enhancing, 26–27
  key features, 25
  linked, 25–27
  overview, 23–24
  purchasing, 27–28
  QuickCAD, 24
  summary, 102–103
  tradeoffs, 101–102
  types of, 23–27
Conceptual plant communities, 298–300
Concrete
  poured-in-place
    binding agents, 320
    coloring agents, 320–321
    enhancing techniques, 328–330
    pouring skills, 328
    site preparation, 327
  precast, 342
Concrete pavers
  beveled edge, 326
  characterization, 324
  popularity of, 322
Constructed features, 212–214
Consultation, 365
Containers
  materials, 357–358
  placement, 358
  plants in, 359–360
  types, 357
  uniqueness, 358–359
Contingency allowance, 378–379
Contour information, 187
Contour line, 58
Contracts
  amendments to, 380
  defined, 374
  proposal *versus*, 374
  terminology
    bids, 372–374
    costs, 371
    estimates, 372–373
    price, 371
    proposals, 374
Cool colors, 136
Copyright infringement, 100
Cosmos, 268–269
Costs
  Country Landscapes model, 382–383
  defined, 370
  materials, 376
  price *versus*, 371

Country Landscapes
  business model
    application of, 383–384
    costs, 382–383
    process, 382
    project components, 382
    terms and definitions used by, 381
  company profile, 381
Curves, creating, 19–20
Cut stone, 319

D

Dashed arrows, 219
Deadheading, 359
Deadmen, 335
Deciduous plants
  defined, 53
  leaves, 268–269
  lifecycle, 268
  symbols, 54
Decks, 307, 309
Desiccation, 269
Design concepts
  arbor and pergola, 344–346
  art elements
    color, 135–139
    combining forms, 134
    line, 133
    texture, 139–140
  basics, 216
  case study, 218
  common misconceptions, 108–113
  components of, 107–108
  defined, 107, 216
  in master plan, 246
  learning to, 132–133
  walls, 336–337
Design planning questionnaire
  benefits of, 174, 176
  example, 257–258
Design principles
  aesthetic, 146–147, 150–159
  functional, 163–170
  overarching, 141–146
  summary, 148
Design process
  analysis, 116–118
  case study
    application, 119–121
    base map, 198
    client feedback, 182
    concept and program, 218
    introductions, 175
    site inventory, 215–216
  client-designer communications
    budget questions, 180–182
    importance of, 173
    initial contact, 174
    meeting outcomes, 179–180

INDEX 419

meeting strategies, 176–179
questionnaires, 174–176
Country Landscapes examples, 250, 253–354
defined, 113
final package, 241, 245–248
functional diagrams, 217–224
initial contact, 174
inventory and analysis, 198–216
Kinghorn Gardens examples, 251–252, 254–255
market segment, 176
Mulhall's Nursery examples, 250–251, 254
Perennial Gardens examples, 249–250, 252–253
preliminary phase
  client feedback, 225
  function, 224
  graphics, 241
  ground-plane framework, 229–230
  refinement, 225–229
  theme development, 234–241
  theme selection, 231–234
renovation example, 252–253
shortcuts, 173
site documentation
  common site measurements, 191–195
  existing information, 183–187
  measured information, 187–189
  measurements tools, 189–191
  photos/videos, 182–183
  removed, 195–198
site inventory and analysis
  constructed features, 212–214
  distinct steps of, 199–201
  homeowner as resource, 199
  location factors, 201–203
  natural and physical features, 203–212
  process considerations, 198–199
  sustainable design with, 214
stages, 113–121
Design services
  business terms, 370
  computer software, 384–386
  contracts
    client meetings, 367
    content details, 367
    sample, 368
    terminology, 371–374
  estimates
    bids *versus*, 372–373
    computer software for, 384–385
    defined, 372
    developing, 374
    manual measuring, 386–388

installation services with
  bid development, 380–381
  contingency, 378–379
  labor costs, 376–377
  materials costs, 376
  overhead, 378
  pricing approaches, 375–376
  production rate example, 377
  profits, 379–380
payment, 369–370
pricing, 366
professional networks, 370
referrals, 370
time schedule, 369
Design themes
  case study, 242–243
  development process, 234–240
  function of, 231
  selecting, 231–234
Design tools
  compass, 18
  computer-aided. *See* Computer-aided design
  dusting brush, 11–12
  erasers, 10–11
  lettering guide, 20
  markers, 9–10
  papers, 2–3
  pencils, 4–7
  pens, 8–9
  protractors, 17
  rolling ruler, 16
  scales, 12–17
  T-square, 2
  templates, 20–22
  triangles, 17
Design/build services, 375
Designers
  approaches of, 119
  client communications
    budget questions, 180–182
    first meeting location, 179
    first meeting strategy, 176–179
    initial contact, 174
    new landscapes, 178
    outcomes, 179–180
    planning questionnaire, 174, 176
    renovated landscapes, 178
    deadline considerations, 109, 111
    technical experts and, 165
Diagrams. *See* Functional diagrams
Diazo process, 3
Digger's Hotline, 189
Digital cameras, 23
Digital capturing tools, 23
Digitizers, 386
Dioecious plants, 285
Direct costs, defined, 370
Direct measurement, 191
Dot quality, 38–40

Downlighting, 350
Downspout locations, 205–206
Drafted graphics, 33–35
Drafting, 85
Drafting tape, 3–4
Drainage, 204–206
Drawing leads, 5–6
Drawing scale, 89
Drawing sets, 97–99
Drawing surface, 2
Drawings. *See* Freehand drawings; Sketching
Dressed dimensions, 330
Driveways, 212, 310–312
Dry-laid pavers, 219
Dry-stack walls, 341
Dusting brush, 11–12
Dusty miller *(Senecio cineraria)*, 284
DynaSCAPE, 26

E

Easements, 185
East side, 209
EasyEst software, 385
Emphasis in design, 159
Engineer's scale, 15–16
Environment
  plant selection and, 260–267
  urban, 300–301
Erasers, 10–11
Eraser shield, 11
Erasing technique, 11
Erosion, 295–296
Escape, 127
Estimates
  bids *versus*, 372–373
  computer software for, 384–385
  defined, 372
  developing, 374
  manual measuring, 386–388
  software for, 25–26
European hornbeam *(Carpinus betulus)*, 274
Evaluation process, 118
Expression, 133
Extent, sense of, 127–128

F

Fascination, 127
Fasteners, 331
Fences
  function, 333
  materials, 337–338
  privacy, 333–334
  sun exposures, 209
Final design. *See* Master plans
Fire pits, 356
Flagstone, 319, 321

INDEX

Flexible curve, 19
Flexible liners, 352
Floating walls, 336
Flowering habit, 284–285
Flowing bedlines, 235
Focal points
   location, 160
   multiple, 159
Footing, 341
Form
   and function blend, 144
   function of, 133–134
   plant, 271–276
Found objects, 361
Fountains, 354
Four-season plants, 271
Framework in design, 146–147
Frameworks, 229
Freehand drawings. *See also* Sketching
   hard-line drawing *versus*, 33–35
   lines, 38–42
   in master plan, 246, 249
   mechanics, 36
   paper choice, 2–3
   section-elevation
      defined, 62
      guidelines, 62–63
      styles, 68–72
   sketching and, 82–85
   symbols
      defined, 52
      patterns, 55
      plant, 52–59
      size, 54
      spacing, 55
      standard, 52
      style, 53
   texture, 46–51
   value contrast, 43–46
Freestanding walls, 334
French curves, 20
Fresco, 361
Frost heave, 317
Frost occurrence, 263
Fruiting habit, 284–285
Function, 107, 144
Functional diagrams
   case study, 227–228
   circulation documentation, 222–223
   defined, 217
   developing, 220–223
   drawing, 94
   elements, 219–220
   function, 219
   information in, 219–220
   overhead plane, 223
   overlay, 96
   summary, 223
   trees, 223
   views documentation, 222–223

Functional principles
   in lighting, 347
   in preliminary design, 225–226
   irrigation needs, 167–169
   maintenance needs, 167
   outdoor living space needs, 166–167
   people's needs, examples, 164
   plant selection
      defining spaces, 291, 293
      views, 293–295
   sustainability, 169
   topographical, 165–166
   wildlife habitat value, 169
Furniture, 360–361
Furrow edging, 319

G

Garden embellishments
   categorization, 357
   containers, 357–360
   found objects, 361
   function, 350–351
   furniture, 360–361
   special effects, 361–362
   statuary, 357
   style of, 356–357
Gardens for People, 132
Gazebos, 346–347
General overhead, 378
Geogrids, 335
Ginkgo *(Ginkgo biloba)*, 285
Grading, 165
Granite, 319
Graphic enhancing software, 26–27
Graphic markers, 9
Graphics
   analysis and concept, 73
   base maps, 94
   black-and-white, 39
   CAD. *See* Computer-aided design
   color, 39
   computer, 34
   consistent inconsistency, 40
   copyright infringement, 100
   dot quality, 38–40
   drafting, 85
   drawing type applications, 93–94
   final designs, 97–100
   freehand, 33–35
   functional diagrams, 94, 96–97
   hard-line, 33–35
   importance, 31–33
   in master plan, 246–248
   landscape photography, 103
   layout, 89–93
   lettering, 85–88
   line quality, 38–40
   line weight hierarchy, 40–43
   paraline drawings, 72
   perspective drawings, 72–73

   preliminary designs, 94–97, 241
   section-elevation, 62–72
   sheet mechanics, 89–93
   site inventory and analysis, 94–95
   sketching, 82–83
   standards, 36
   texture, 46–52
   value contrast, 43, 45–46
Graphic symbols
   bubbles, 57
   defined, 52
   functional diagrams, 219–220
   guidelines, 60–62
   hardscape, 59
   overlapping, 56
   pattern, 55–57
   size, 54
   spacing, 55–57
   standard, 52
   style, 53
   textures, 56–57
   water textures, 59
Grasses, ornamental, 289
Gravel
   characterization, 316–317
   edging materials, 317–319
   types, 317
Gridded vellum, 2–3
Groundcovers, 57–58, 293
Ground-plane framework, 229
Ground plans, 307
Grubbing, 377

H

Habitat
   theory, 121–122
   wildlife, 169
Hand trowel, 327
Hard-line graphics, 33–35
Hardiness zone, 261–262
Hardscapes
   decks, 307, 309
   defined, 107
   driveways, 310–312
   garden features
      embellishments, 356–360
      fire pits, 356
      found objects, 361
      furniture, 360–361
      location, 350–351
      special effects, 361–362
      water, 352–356
   ground plan, 307
   materials
      brick, 322–323
      composite products, 331–332
      concrete pavers, 323–326
      edging, 317–319
      gravel, 316–317
      natural stone, 319–321

poured concrete, 326–330
summary, 339
wood, 330–331
overhead plane, 343–347
patios, 307
steps, 310–312
symbols, 59
types of, 306
vertical plane, 332–343
walkways, 309
Hatching, 58
Heat zones, 262–263
Height measurement, 194
Herbaceous plants
lifecycle, 268, 288–290
maintenance, 288–290
perennials, 268, 290
High-traffic steps, 312–313
History, in landscape design, 144
Holly *(Ilex)*, 285
Honeysuckle *(Lonicera tatarica)*, 275
House color, 212–213
Hues, 136–137
Humanistic assessments, 124–127
Humans. *See* People
Hyper-tufa containers, 359

I

Imagination, 133
Implementation in design, 118
Income, generating, 365
Indirect costs, defined, 370
Ink pens, 8–9
Inking templates, 21
Installation graphic, 34
Installation services
bid development, 380–381
contingency, 378–379
labor costs, 376–377
materials costs, 376
overhead, 378
pricing approaches, 375–376
production rate example, 377
profits, 379–380
Integrated pest management (IPM), 365
Intermediate lines, 234
Inventory and analysis. *See* Site inventory and analysis
IPM. *See* Integrated pest management
Irrigation
function of, 167
maintenance, 169
requirements, 168

J

Japanese Pagodatree *(Styphnolobium japonicum)*, 261–262
Junipers, 295
Junipers *(Juniperus)*, 280

K

Kentucky coffee tree *(Gymnocladus dioica)*, 277

L

LANDCADD, 24
Landscape drawings, 34
Landscape timbers, 343
Landscapes. *See also* Design concepts
conceptual plant communities, 298–300
functional uses in, 290–298
preferences
behavioral assessments, 121–123
cognitive, 164
humanistic assessments, 124–127
physical, 164
restorative, 164
replicating nature in, 300
restorative characteristics, 127–129
structures connection to, 160
templates, 21
urban environment, 300–301
utility purposes, 295–297
views and, 293–295
Lawn seeding, 377
Layout, 89–93
Leads
hardness, 5–6
holder, 6
Leaf characteristics, 277–280
Left-brain approach, 112
Legibility, 125
Lettering
alternatives, 85–86
characteristics, 85
style criteria, 86–88
Lettering guide, 20
Light deflection, 296–297
Lighting
advances in, 347
aesthetic techniques, 347–350
functional, 347
Light patterns, 301
Limestone, 319
Limitations, in landscape design, 145–146
Lines
contour, 58
different, combining, 297
function of, 133
grid framework, 234
quality, 38–40
symbols, 219
weight hierarchy, 40–43
weights, 4–6
Linked design software, 25–27
Littleleaf linden *(Tilia cordata)*, 291
Lot drawing, 184

Lots
measurements, 185
variations, 184
Lump sum bid, 374

M

Maintenance
consultation, 365
container plants, 359–360
irrigation, 169
needs, 167
plant selection and, 287–290
wall, 337
water features, 355
Markers, 9–10
Marketing
creative strategies, 365
selling *versus*, 174
Market segment, 176
Mass, 160
Master plans
case study, 247–248
checklist, 245–246
contents, 246
graphics, 97–100
Materials. *See also* specific products
costs, 376
decks, 307, 309
driveway, 311–312
fence, 337–338
hardscapes, 316–318
thickness chart, 323
wall, 340–343
water features, 352
Measurements
baseline, 191–193
common site, 191–195
direct, 191
height, 194
lot, 185
site
additional, 188–189
recording, 187
techniques, 191–195
tools, 189–191
tools, 12–17
Measuring wheel, 386
Mechanical pencils, 6
Metal fencing, 338
Microclimates
creating, 267
defined, 108
impact of, 208
plant selection and, 267
residential, 209
Microstation, 24
Millimeter pencils, 6
Monochromatic color schemes, 282–283
Mosaics, 361–362

Moss rose *(Portulaca grandiflora)*, 269
Movement, directing, 161–162
Multiple focal points, 159
Munsell Color-Order System, 135–136
Mycorrhizae, 301
Mylar
  base maps on, 94
  characterization, 3
  reproduction and, 3

N

Native plants
  care of, 303
  communities, 298
  nonnative plants *versus*, 304
  urban soils and, 303–304
  value of, 302–303
Natural features, 203
Natural stone
  base preparation, 321
  characteristics, 320
  defined, 319
  history of use, 319
  thickness range, 321
  types, 319, 321
  walls built with, 340–341
Nature
  in landscape design, 145
  replicating, 300
Negative space, 140
New residential landscapes, 178
Nominal dimensions, 330
Nonnative plants, 302–304
Nonprimary steps, 313
North arrow, 89, 93
North side, 209

O

Obelisks, 357
Objectives, developing, 118
Off-site views, 213–214
On-site views, 213–214
One Call, 189
Order in design, 146–147
Ornamental grasses, 289
Ornamental trees, 298
Outdoor fireplaces, 356
Outdoor space
  focus of, 107
  functional principles, 166–167
  meeting people needs with, 162–163
  people's relationship to, 121–129
  scale, 158
Overarching principles
  blending form and function, 144
  defined, 141
  local elements, inclusion of, 144–146
  simplicity, 142–144

Overhead planes
  categories, 378
  hardscapes
    arbors, 344–346
    defined, 343
    design considerations, 344
    gazebos, 346–347
    pergolas, 344–346
  trees and, 223
Oversized walls, 336

P

Pagoda dogwood *(Cornus alternifolia)*, 275
Pallet fee, 376
Paper types, 2–3
Paraline drawings, 72
Parallel ruling straightedge, 2
Parking, 212
Patina, 317
Patios, 307
Patterns, 295
Paving materials, 323
Pea gravel, 317
Pedestrian circulation, 212
Pencils
  initial design and, 90
  smudging by, 11–12
  types, 4–9
Pens. *See* Ink pens
People
  functional design and, 164
  movement/views, directing, 161–162
  outdoor space relationship to
    humanistic assessment, 124–127
    landscape preference, 121–123
    needs considerations, 162–163
    restorative values, 127–129
  plant care by, 301
Perception, 132
Perennials
  lifecycle, 269–270
  maintenance, 290
  symbols for, 57–58
Pergolas, 344–345
pH, 265–266
Photocopying, 3
Photography, 182–183
Photo-realistic simulations, 100
Physical preferences, 164
Pin oak *(Quercus palustris)*, 275
Planimeter, 386
Plants
  annuals, 269
  biennials, 270
  branching habit, 274–275
  cohesive masses, 143, 146–147
  conceptual communities, 298–300
  containers, 359–360
  dioecious, 285
  four-season, 271

  functional diagrams, 223
  growth rate, 287
  herbaceous, 288–290
  leaf characteristics, 277–280
  native, 302–304
  nonnative, 302–304
  perennials, 269–270
  proper care, 301
  schedule, in master plan, 246
  size relationship and, 157–158
  space definition by, 291–293
  texture, 276–277
  types, 268–269
  views and, 293–294
  woody, 268, 288
Plant selection
  aesthetic purposes, 297–298
  aesthetic qualities
    branching habit, 274–275
    color, 280–284
    flowering/fruiting habit, 284–285
    form, 271–276
    four-season plants, 271
    impact of, 267–268
    mature size, 285–287
    plant types, 268–269
    texture, 276–277, 280
  environmental considerations, 260–267
  form considerations, 271–276
  functional aspects, 290–298
  lifecycle considerations, 269–271
  maintenance requirements, 287–290
  microclimate, 267
  site analysis and, 260
  size considerations, 285–287
  soil considerations, 264–266
  sun exposure, 266–267
  temperature considerations, 261–264
Plant symbols
  bubbles, 57–58
  drawing to scale, 21–22
  styles, 53–54
Plat. *See* Site survey
Plot plans
  defined, 184
  easements, 185
  right-of-way, 184–185
  setbacks, 186
  site survey, 186–187
  variances, 186
Pondless water feature, 355
Positive space, 140
Poured-in-place concrete
  aggregate materials, 320
  characterization, 320
  coloring agents, 320–321
  defined, 319
  site preparations, 327–328
  techniques to enhance, 328–330

Precast blocks, 319
Precast concrete units, 332
Preliminary designs
    aesthetic refinement, 225
    case study, 244–245
    client feedback, 225
    defined, 224
    functional refinement, 225–226
    graphic style, 241
    ground-plane framework, 229
    spatial refinement, 226, 228–229
    theme establishment, 231–240
Prevailing winds, 184
Price
    contract development, 366–370
    costs *versus*, 371
    design/installation services, 375–380
    steps determining, 366
Primary lines, 234
Privacy fences, 333–334
PRO Landscape, 25
Problem solving, 116
Production rate, 376, 377
Professional networks, 370
Profit, defined, 370
Profit margins, 379–380
Program
    case study, 218
    defined, 216
    function, 217
Project scale, 109
Property lines, 184
Prospect-refuge theory, 122, 213
Protractors, 17
Public areas, 220–221
Pull-lists, 365

Q

QuickCAD, 24

R

Radial balance, 156
Rag stock, 2
Rainfall, 263–264
Rebar, 327
Referrals, 370
Relative scale, 156–157
Renovation
    challenges, 178
    example, 252–253
    irrigation, 168
Repetition in design, 150–151
Restorative characteristics, 127–129
Restorative effect, 164
Retaining walls
    deadmen in, 335
    defined, 334
    structure, 334–335
Rhododendrons, 275

Rhythm
    categories of, 151
    function of, 151
    human movement and, 152–153
Right-brain approach, 83, 112
Right-of-ways, 184–185
Rigid curve, 19
Riser height, 314–315
Rolling ruler, 16–17
Rosettes, 270

S

Sandstone, 319
Sandwich soils, 300
Scale
    defined, 156
    outdoor settings, 158
    plant materials and, 157–158
    plant symbol to, 21–22
    project, 109
    types of, 156–157
Scales
    architect's, 13–15
    drawing, 89
    engineer's, 15–16
    importance of, 16
    selecting, 12–13
    using, 14
Scarify, 377
Seasonal rainfall, 263–264
Secondary lines, 234
Section-elevation graphics
    abstract style, 70–72
    definitions, 62–63
    guidelines, 62–67
    style elements, 68–70
Segmental walls. *See* Precast
    concrete units
Sense of place, 145
Serviceberry *(Amelanchier arborea)*, 271
Services. *See* Design services
Setbacks, 186
Shade exposures, 209–212
Shadows, 45
Sheet layout, 89–90
Sheet mechanics, 89–93
Simplicity, 142–144
Site documentation
    base maps
        existing sources, 183–187
        measured information, 187–189
        techniques, 189–191
        tools, 189–191
    features
        client-related, 212–214
        constructed, 212–214
        drainage, 204–206
        microclimates, 208–212
        slopes, 206–207
        soils, 203

        topography, 204–206
        wind, 203–204
    functional diagrams, 222–223
    location, 201–203
    photos, 182–183
    videos, 182–183
Site inventory and analysis
    case study, 216
    documentation
        constructed features, 212–214
        location, 201–203
        natural features, 203–212
    drawing, 94–95
    homeowner's input, 199
    opportunities, 198–199
    plant selection and, 260
    steps, 199–201
    sustainable design and, 214
Site measurement
    additional, 188–189
    recording, 187
    techniques, 191–195
    tools, 189–191
Site restrictions, 212
Site survey, 186–187
Size
    in design, 156–158
    use areas, 221–222
Sketching. *See also* Freehand drawings
    ability, developing, 83
    advantages, 82
    as right-brain function, 83–85
    color and, 81
    landscape budgets, 181
Sketchup software, 24
Slopes
    calculations, 206–207
    estimating, 194–195
    function, 165
    guidelines, 205
    light exposure and, 267
    site survey exclusion of, 187
    terminology, 206–207
Smudging, limiting, 11–12
Software
    business, 384–385
    design. *See* Computer-aided design
Soils
    basics, 264–266
    classification, 203
    ideal, 264
    pH, 265–266
    structure, 265
    texture, 265
    urban, 300–301
Solid-line arrows, 219
South side, 209
Space. *See also* Outdoor space
    defining, 291–293
    as design element, 140
    establishing, 160

Spatial refinement, 226, 228–229
Spirea *(Spiraea spp.)*, 275
Split-faced stone, 319
Spotlighting, 349
Staghorn sumac *(Rhus typhina)*, 275
Stamped concrete, 328–329
Star-shaped symbols, 220
Statement of design intent. *See* Design concepts
Statuary, 357
Steel edging, 317
Stenciling concrete, 329
Steps
  as design feature, 316
  function, 312
  high-traffic, 312–313
  nonprimary, 313
  riser height, 314
Stippling, 58
Stone. *See* Natural stone
Structures
  base map documentation, 212–213
  landscape connection to, 160
Sun exposures
  influences, 209–211
  plant selection and, 266–267
  variations in, 211
Sustainability
  design solutions, 214
  incorporating, 169–170
Sweet William *(Dianthus barbatus)*, 270
Symbols. *See* Graphic symbols
Symmetrical balance, 154

T

Technical experts, 165
Technical pens, 8
Templates, 20–21
Tender perennials, 270
Text, 89
Textures
  categories, 139
  defined, 46, 139
  effective use of, 46–50
  guidelines, 60–62
  patterns, 51–52
  plants, 276–277
  soil, 265
  symbols and, 56–57
  water, 59
Themes. *See* Design themes
Thornless honeylocust *(Gleditsia triacanthos)*, 291
3-D perspective drawing, 35
Timbers. *See* Landscape timbers
Time schedule, 369

Title blocks, 89–92
Tools. *See* Design tools
Topography
  defined, 62
  drainage and, 204–206
  existing, working with, 165–166
Topsoils, 300
Trace, 2
Tracing paper, 90
Trees
  canopy, 209, 225
  ornamental, 298
  uplighting, 349
Triadic planting schemes, 283–284
Triangles, 17
Triangulation, 193
T-squares, 2
Turf, 296
Turf area symbols, 57–58

U

Unity in design, 154
Uplighting, 349
Urban environment, 300–301
Urns, 358–359
Use areas
  identifying, 220–221
  sizing, 221–222
Utility lines, 189, 213

V

Value contrast
  creation, 46
  highest, 43, 45
  section-elevation graphics, 68
  shadows, 45–46
Variances, 186
Vellum
  base maps on, 94
  characterization, 2–3
  reproduction and, 3
Veneer surface, 341
Vertical measurement, 194
Vertical plane hardscapes
  characterization, 332
  fences, 333–334
  materials, 337–343
  walls, 334–343
Videos, 23, 182–183
Views
  directing, 161–162
  framing, 294
  functional diagrams of, 222–223
  landscape, 160

off-site, 213–214
on-site, 213–214
screening, 214, 293–294
Vinyl fencing, 338
Visual simplicity, 143
Void, 160

W

Walkways, 309
Walls
  design considerations, 336–337
  dry-stack, 341
  footing, 341
  freestanding, 334
  hidden costs, 341
  maintenance, 337
  materials, 340–343
  multiple uses, 337
  retaining, 334–335
  segmental, 342
  veneer surface, 341
Warm colors, 136
Water
  preference for, 122
  textures, 59
Waterfalls, 354
Water features
  categories, 353
  defined, 352
  functions, 355–356
  history of use, 352
  lighting, 350
  location, 351
  maintenance, 355
  technology, 352
  types of, 353–356
Weeping crabapple, 274
West side, 209
Wet areas, 117
White space, 84
Wildlife habitat, 169
Wind, 184, 203–204
Wood
  characterization, 318
  composite products, 331–332
  dimensions, 330
  edging, 318
  function, 330
  selecting, 331
Wood, Grant, 111
Woody plants, 268, 288
Wright, Frank Lloyd, 144

Y

Yellow corydalis *(Corydalis lutea)*, 263